AF411328

# Recent Innovations in Post-harvest Preservation and Protection of Agricultural Products

# Recent Innovations in Post-harvest Preservation and Protection of Agricultural Products

Editor

**Dirk E. Maier**

MDPI • Basel • Beijing • Wuhan • Barcelona • Belgrade • Manchester • Tokyo • Cluj • Tianjin

*Editor*
Dirk E. Maier
Iowa State University
USA

*Editorial Office*
MDPI
St. Alban-Anlage 66
4052 Basel, Switzerland

This is a reprint of articles from the Special Issue published online in the open access journal *Agriculture* (ISSN 2077-0472) (available at: https://www.mdpi.com/journal/agriculture/special_issues/agricultural_products_preservation_protection).

For citation purposes, cite each article independently as indicated on the article page online and as indicated below:

LastName, A.A.; LastName, B.B.; LastName, C.C. Article Title. *Journal Name* **Year**, *Volume Number*, Page Range.

ISBN 978-3-0365-3291-2 (Hbk)
ISBN 978-3-0365-3292-9 (PDF)

# Contents

# About the Editor

**Dirk E. Maier** is a Professor and Postharvest Engineer in the departments of Agricultural & Biosystems Engineering as well as Food Science & Human Nutrition at Iowa State University. He is responsible for leading an internationally recognized research and outreach program in postharvest engineering and technology applied to global food security and nutrition. He serves as the Director of the Consortium for Innovation in Post-Harvest Loss & Food Waste Reduction, which aims to achieve the sustained, scalable implementation of appropriate methods to preserve, process, package, and transport nutritious foods. His current graduate students are from the U.S., Ivory Coast, DR Congo, Rwanda, Uganda, Ghana, and the Philippines.

 *agriculture*

*Editorial*

# Recent Innovations in Post-Harvest Preservation and Protection of Agricultural Products

**Dirk E. Maier * and Hory Chikez**

Department of Agricultural and Biosystems Engineering, Iowa State University, 3325 Elings Hall, 605 Bissell Road, Ames, IA 50011-3270, USA; horych@iastate.edu
* Correspondence: dmaier@iastate.edu

**Citation:** Maier, D.E.; Chikez, H. Recent Innovations in Post-Harvest Preservation and Protection of Agricultural Products. *Agriculture* 2021, 11, 1275. https://doi.org/10.3390/agriculture11121275

Received: 3 December 2021
Accepted: 8 December 2021
Published: 15 December 2021



## 1. Consortium for Innovation in Post-Harvest Loss and Food Waste Reduction

Food loss and waste is a global problem that negatively impacts the bottom lines of producers and agri-businesses, wastes limited resources, and contributes to climate change. The Foundation for Food and Agriculture Research (FFAR), The Rockefeller Foundation, Iowa State University, University of Maryland, Wageningen University and Research, Volcani Center, Zamorano University, Stellenbosch University, University of São Paulo, University of Nairobi, and Kwame Nkrumah University of Science and Technology partnered to establish the Consortium for Innovation in Post-Harvest Loss and Food Waste Reduction.

The Consortium is committed to training the next generation of food system leaders, researchers, and entrepreneurs. Undergraduate and graduate students from these institutions are conducting innovative research that improves drying, handling, storage, and distribution, develops monitoring and tracking technology, extends shelf-life and minimizes spoilage, and changes behavior and practices to reduce post-harvest loss and food waste from field to fork. Innovative entrepreneurs trained by these institutions are commercializing technology, adding value to agricultural crops, and developing nutritious food products.

## 2. Review Process

All articles published in this Special Issue "Recent Innovations in Post-Harvest Preservation and Protection of Agricultural Products" underwent peer review by independent subject matter experts in the field of post-harvest science, technology, engineering and management.

## 3. Recent Innovations in Post-Harvest Preservation and Protection of Agricultural Products: Summarized Articles by Area

a. Stored Product Protection

(1) Determine grain quality and pesticide residue concentrations of maize stored in porous versus hermetic storage bags. Maize stored in air-tight (hermetic) bags were shown to have higher grain quality and lower aflatoxin and pesticide residue concentrations than maize stored in porous woven polypropylene bags. Educating smallholder farmers on the benefits of hermetic storage bags, and promoting adoption of this innovative chemical-free protection technology, should continue to be a priority among supply chain actors to ensure food-safe maize from producers to consumers [1]–Consortium;

(2) Apply dynamic controlled atmosphere technologies to reduce incidence of physiological disorders and maintain quality of apples. 'Granny Smith' apples stored under repeated low oxygen stress (RLOS) in combination with ultra-low oxygen (ULO) or controlled atmosphere (CA) conditions, and under dynamic controlled atmosphere (DCA) conditions in combination with chlorophyll fluorescence (CF) treatment had significantly ($p < 0.05$) higher flesh firmness and total soluble

solids. The post-harvest treatments and storage conditions reduced superficial scald by possibly suppressing the oxidation of volatiles implicated in its development [2]–Consortium;

(3) Investigate effects of hot-air and freeze drying on the physicochemical, phytochemical, and antioxidant capacity of dried pomegranate arils during long-term cold storage of whole fruit. Results from this one-time experiment showed that quality attributes such as color, total phenolic content (TPC), total anthocyanin content (TAC), and radical scavenging activity (RSA) improved distinctly due to freeze-drying and subsequent storage at $7 \pm 0.3\ °C$ and $92 \pm 3\%$ relative humidity. Freeze-drying was therefore recommended over hot-air drying as the preferred preservation treatment [3]–Consortium;

(4) Analyze different storage conditions in terms of profitability based on market prices for pears during three storage seasons. Storage conditions had a strong influence on perishable fruit quality parameters. They were found to affect most visibly mass loss and incidence of postharvest diseases and disorders. The storage of 'Conference' cultivar pears for 180 days in normal atmosphere was not economically viable, even when the fruit was subjected to treatment with 1-methylcyclopropene (1-MCP), a synthetic plant growth regulator used commercially to slow down fruit ripening. However, it was profitable to store 'Conference' pears under controlled atmosphere conditions each season, no matter whether 1-MCP was applied or not [4].

b.    Post-Harvest Handling and Drying

(5) Evaluate a 500 kg portable column dryer with a biomass burner heat source for maize drying. Indicators such as drying rate, drying efficiency, and moisture extraction rate were used to assess technical operations performance. Results showed that maize moisture content was reduced from 22.3% to $13.4\% \pm 2.6\%$ in 5 h at an average drying rate of 1.81 percentage points per hour with a drying efficiency of 64.7%. Utilization of such low-capacity mobile dryers to provide drying services was found to be economically viable based on net present value analysis resulting in internal rates of return (IRR) above 70%, pay-back periods (PBP) of less than two years, and positive benefit-cost ratios (BCR) greater than 2.5. Affordable access to drying services in maize-growing communities has potential to improve the socio-economic status of smallholder maize farmers in sub-Saharan Africa [5]–Consortium;

(6) Analyze the effect of vibration on grape berry drop during vertical transportation and of different packaging materials on grape clusters during robotic placement. Dropping and shattering of grape berries reduces quality during harvest and post-harvest handling. This study developed an objective method to observe and analyze damage and detachment force for cluster fruits during robotic post-harvest handling. Higher speeds and acceleration excitations during vertical transportation tests increased hanging force positively ($R^2 = 0.92$) while the force after striking the grape cluster with packaging materials decreased negatively ($R^2 = 0.97$) and the corresponding index of berry deflection increased. High-speed camera images revealed that rigid plastic boxes caused maximum deflection of grape berries, with the highest change in force of 8.6 N after impact. Experimental results showed a negative correlation between hanging force signals and the force after impact of the cluster, with a goodness of fit of $R^2 = 0.95$ at different speeds [6].

c.    Crop End-Use Quality Sensors

(7) Effect of numbers and placement of temperature sensors on aeration cooling of a stored grain mass. Results predicted by a 3D finite element computational model demonstrated that temperature cables in the center or near the edges of the silos were not representative of average temperatures in the grain mass,

resulting in too infrequent or excessive aeration, respectively. Placement of "wireless" sensors at fixed grain depths but randomized horizontally along the diameter resulted in similar average temperatures, while an increase in randomized sensor numbers reduced variability among years of weather data simulated [7]–Consortium;

(8) Use of near infrared hyperspectral imaging to evaluate color, firmness, and soluble solid content (SSC) of Korla fragrant pears. This study acquired hyperspectral imaging data for 200 samples to construct statistical evaluation models for predicting these quality parameters using iteratively retaining informative variables (IRIV) and least square support vector machine (LS-SVM) analysis. Results demonstrated that the combination of IRIV and LS-SVM can be used to predict values for color parameter, a *, firmness, and SSC to define grade of Korla fragrant pears with correlation coefficients of the validation set measuring 0.927, 0.948, and 0.953, respectively [8].

d.  Post-harvest loss reduction

(9) Evaluate the effects of five harvest and post-harvest technologies (harvesting tools, cold stores, plastic crates, fruit fly traps, ground tarps) promoted by the Rockefeller Foundation Yieldwise Initiative (YWI) on post-harvest loss (PHL) incurred at three stages of the mango value chain (harvest, transportation, point of sale) in Kenya. Results indicated that plastic crates used to transport or store mangos and fruit fly traps used to attract and kill fruit flies were statistically significant ($p < 0.05$) in reducing PHL at the point of sale. Interestingly, no statistical evidence of PHL reduction was observed from smallholder farmers using harvesting tools, cold stores, and ground tarps [9]–Consortium;

(10) Assess four on-farm maize storage technologies with and without chemical protectant in two locations of the Republic of Benin. The analysis showed that in central and northern Benin hermetic bags and polypropylene bags recorded less storage losses and were more profitable than improved and closed clay earth granaries and unsealed metal silos. Gastight (hermetic) bag storage technology recorded the lowest post-harvest loss in the two locations when grain was initially treated with the chemical protectant 2% pirimiphos-methyl (central 9.42 ± 4.64%, northern 2.69 ± 0.77%) versus without (central 11.71 ± 2.78%, northern 7.71% ± 1.74%). Maize stored in woven polypropylene bags recorded losses due to insect pests with chemical protectant (northern 4.02 ± 1.23%) versus without (northern 9.64 ± 2.73%). Financial analysis indicated that the most profitable storage technologies were hermetic bags without an initial chemical treatment in central Benin, a more humid region, and woven polypropylene bag with an initial chemical protectant treatment in northern Benin, a more arid region [10];

(11) Review of mango fruit processing options for small-scale processors in low-income countries. Processing mango fruit into a number of shelf-stable food products makes the seasonal fruit more broadly available to consumers year-round. Research and food product development have resulted in several unique processed mango products with specific qualities and nutritional attributes in demand by consumers. These include pulp (puree), juice concentrate, ready-to-drink juice, nectar, wine, jams, jellies, pickles, smoothies, chutney, canned slices, chips, leathers, and powder. Minimum processing of mango fruit as a fresh-cut product is popular among health-conscious consumers. Mango pulp and powder can be used to enrich or flavor secondary products such as yoghurt, ice cream, beverages, and soft drinks. Byproducts of mango processing, such as peel and kernels, are rich in bioactive compounds including carotenoids, polyphenols, and dietary fibers, can be used in food fortification and manufacture of animal feeds. This adds value to the fruit while reducing food loss and waste [11]–Consortium.

## 4. Outlook with Regard to Continued Research in Post-Harvest Preservation and Protection

Despite continued progress, several challenges pertaining to reducing post-harvest loss and food waste reduction remain unresolved and need further basic and applied research including:

(1)     Electricity and financing to reliably and affordably power the refrigerated and controlled atmosphere storage chains to ensure perishable agricultural crops can be preserved with net-zero carbon dioxide equivalent emissions by 2030;

(2)     Non-chemical technologies and practices to mitigate spoilage agents and protect stored products from post-harvest quality degradation and food safety pathogens;

(3)     Alternative energy drying (dehydration) technologies and practices to reduce moisture content (water activity) of agricultural crops to safe storage levels as close to the producer as possible and preserve them for handling, storage, processing, packaging, transportation and marketing throughout the supply chain.

**Author Contributions:** Conceptualization, writing—original draft preparation, writing—review and editing, were contributed equally by the authors. Funding acquisition, D.E.M. All authors have read and agreed to the published version of the manuscript.

**Funding:** Funding to support publication of research results by members of the Consortium for Innovation in Post-Harvest Loss and Food Waste Reduction was provided by The Rockefeller Foundation (2018 FOD 004), the Foundation for Food and Agriculture Research (DFs-18-0000000008), and the Iowa Agriculture and Home Economics Experiment Station.

**Institutional Review Board Statement:** Not applicable.

**Informed Consent Statement:** Not applicable.

**Data Availability Statement:** Not applicable.

**Acknowledgments:** The authors would like to thank the funders and collaborators of the Consortium and all manuscript contributors and peer reviewers of this special issue of Agriculture.

**Conflicts of Interest:** The authors declare no conflict of interest.

## References

1.     Nyarko, S.; Akyereko, Y.; Akowuah, J.; Wireko-Manu, F. Comparative Studies on Grain Quality and Pesticide Residues in Maize Stored in Hermetic and Polypropylene Storage Bags. *Agriculture* **2021**, *11*, 772. [CrossRef]

2.     Kawhena, T.; Fawole, O.; Opara, U. Application of Dynamic Controlled Atmosphere Technologies to Reduce Incidence of Physiological Disorders and Maintain Quality of 'Granny Smith' Apples. *Agriculture* **2021**, *11*, 491. [CrossRef]

3.     Adetoro, A.; Opara, U.; Fawole, O. Effect of Hot-Air and Freeze-Drying on the Quality Attributes of Dried Pomegranate (*Punica granatum* L.) Arils During Long-Term Cold Storage of Whole Fruit. *Agriculture* **2020**, *10*, 493. [CrossRef]

4.     Łysiak, G.; Rutkowski, K.; Walkowiak-Tomczak, D. Effect of Storage Conditions on Storability and Antioxidant Potential of Pears cv. 'Conference'. *Agriculture* **2021**, *11*, 545. [CrossRef]

5.     Obeng-Akrofi, G.; Akowuah, J.; Maier, D.; Addo, A. Techno-Economic Analysis of a Crossflow Column Dryer for Maize Drying in Ghana. *Agriculture* **2021**, *11*, 568. [CrossRef]

6.     Faheem, M.; Liu, J.; Chang, G.; Abbas, I.; Xie, B.; Shan, Z.; Yang, K. Experimental Research on Grape Cluster Vibration Signals during Transportation and Placing for Harvest and Post-Harvest Handling. *Agriculture* **2021**, *11*, 902. [CrossRef]

7.     Plumier, B.; Maier, D. Effect of Temperature Sensor Numbers and Placement on Aeration Cooling of a Stored Grain Mass Using a 3D Finite Element Model. *Agriculture* **2021**, *11*, 231. [CrossRef]

8.     Liu, Y.; Wang, T.; Su, R.; Hu, C.; Chen, F.; Cheng, J. Quantitative Evaluation of Color, Firmness, and Soluble Solid Content of Korla Fragrant Pears via IRIV and LS-SVM. *Agriculture* **2021**, *11*, 731. [CrossRef]

9.     Chikez, H.; Maier, D.; Sonka, S. Mango Postharvest Technologies: An Observational Study of the Yieldwise Initiative in Kenya. *Agriculture* **2021**, *11*, 623. [CrossRef]

10.     Sissinto Gbenou, E.; Patrice Adégbola, Y.; Manhoussi Hessavi, P.; Zossou, S.; Biaou, G. On-Farm Assessment of Maize Storage and Conservation Technologies in the Central and Northern Republic of Benin. *Agriculture* **2021**, *11*, 32. [CrossRef]

11.     Owino, W.; Ambuko, J. Mango Fruit Processing: Options for Small-Scale Processors in Developing Countries. *Agriculture* **2021**, *11*, 1105. [CrossRef]

 *agriculture*

*Article*

# Experimental Research on Grape Cluster Vibration Signals during Transportation and Placing for Harvest and Post-Harvest Handling

Muhammad Faheem [1,2], Jizhan Liu [1,*], Guozheng Chang [1], Irfan Abbas [1], Binbin Xie [1], Zhu Shan [1] and Kaiyu Yang [1]

[1]  Key Laboratory of Modern Agricultural Equipment and Technology, Ministry of Education, Jiangsu University, Zhenjiang 212013, China; engr.faheem@uaf.edu.pk (M.F.); 2111916017@stmail.ujs.edu.cn (G.C.); 5103180336@stmail.ujs.edu.cn (I.A.); 2111816004@stmail.ujs.edu.cn (B.X.); 2112016015@stmail.ujs.edu.cn (Z.S.); 2221916041@stmail.ujs.edu.cn (K.Y.)

[2]  Department of Farm Machinery and Power, University of Agriculture, Faisalabad 38000, Pakistan

*  Correspondence: liujizhan@163.com; Tel./Fax: +86-511-88797338

Citation: Faheem, M.; Liu, J.; Chang, G.; Abbas, I.; Xie, B.; Shan, Z.; Yang, K. Experimental Research on Grape Cluster Vibration Signals during Transportation and Placing for Harvest and Post-Harvest Handling. *Agriculture* **2021**, *11*, 902. https://doi.org/10.3390/agriculture11090902

Academic Editor: Francesco Marinello

Received: 28 July 2021
Accepted: 14 September 2021
Published: 19 September 2021

**Publisher's Note:** MDPI stays neutral with regard to jurisdictional claims in published maps and institutional affiliations.

**Abstract:** Berry dropping or shattering is an important factor during the harvest and post-harvest handling of fresh eating grapes until they reach the supermarkets. There are a lot of methods to measure post-harvest placing damage and the detachment force for single fruits. However, until now, there has been no objective method to observe and analyze the berry dropping mechanism of cluster fruits during robotic post-harvest handling. Therefore, in this paper, the effect of a cluster's vibration on berry drop during vertical transportation and the impact of different packaging materials on fresh grape clusters during robotic placing were analyzed. For this purpose, a lead screw lathe, along with an attached actuator, three grape cluster samples (0.48, 0.50, 0.53 kg), three packaging materials (rigid plastic box, corrugated fiberboard box, expandable polystyrene box), four transportation speeds (0.4, 0.6, 0.8, 1.0 m/s), and four acceleration excitations (6, 8, 10, 12 m/s$^2$) that were given in a mechanical system (actuator) were studied. In order to analyze the berry drop mechanism of grape clusters before and after the impact with packaging material, a force sensor and high-speed video camera were used. It was concluded from the vertical transportation test that with the increase in speed and acceleration excitations, the change in hanging force increased positively ($R^2 = 0.92$). Additionally, the force after the striking of the grape cluster with packaging materials decreased negatively ($R^2 = 0.97$), and the corresponding index of berry deflection increased. It was also observed from the high-speed camera images that rigid plastic boxes caused the maximum deflection of the grape berries, with the highest change in force of 8.6 N after the impact. Experimental results showed a negative correlation between the hanging force signals and the force after impact of the cluster, with a goodness of fit of $R^2 = 0.95$ at different speeds. Overall, the proposed findings can be used as a reference study for improving robotic post-harvest handling, providing a useful visual and technical understanding of the berry fall susceptibility of cluster fruits, and can be used to develop a post-harvest robotic placing tool for avoiding berry drop damage on both industrial and farm levels.

**Keywords:** grapes; cluster fruits; packaging materials; transportation and placing; excitation; vibration; signals

## 1. Introduction

Grapes occupy an important place in the world; their annual output has reached about 79 million tons [1]. Grape cultivars can be categorized into four main groups for food usage: table grapes, wine grapes, sweet juice grapes, and raisin grapes [2]. Among these, table grapes occupy an important place in global fresh cluster fruit production. Because of their high quality, attractiveness, and numerous nutritional facts [3], more than 65% of grapes produced are consumed as a fresh eating fruit [4]. The yield of fresh eating table grapes is enhanced by the development of the grape industry [5].

Table grapes are not a climacteric fruit, and in the process of harvesting and post-harvest operations such as storage, packaging, transportation, and logistics, table grapes undergo serious mechanical loads due to impact, collision, and long-term vibration. These mechanical loads cause the berry falling of clusters. Therefore, it is of great significance to analyze the berry drop mechanism of grape clusters during post-harvest operations to improve the quality and shelf life of grape clusters and reduce the economic loss of the producer. The harvest and post-harvest handling of fresh eating table grapes are in the form of clusters or bunches until they reach the supermarkets. As to manual or robotic handling during the harvesting and post-harvest of fresh grape clusters, the mechanism and phenomenon of damage are totally different from single stem fruits [6]. Usually, grape clusters are clamped and cut from the main rachis. The hanging cluster, after cutting, needs to be transported to a basket or box, then unloading and placed into a basket or bulk bin to complete the on-site transportation. Additionally, for long-distance transportation or post-harvest operations, grape clusters need to be handled several times [7–15]. So the probability of berry drop greatly increases. The loss caused by the berry drop and decay of fruit grain is up to 20% to 30%. The integrity of fresh cluster fruit and non-destructive evaluation are two major quality criteria of grape clusters [16]. The problem of berry fall seriously affects their shelf life and marketability [10,17,18], which has become a serious problem that has plagued the table grape industry chain for a long time. Additionally, it has become a key obstacle to the development and control of machinery and robotic equipment for grape cluster post-harvest handling. Therefore, basic research on grape cluster vibration signals can help to predict and grasp the berry dropping mechanism during robotic transportation and placing. This study will help suggest an effective means of control that has important scientific significance and commercial value.

Mechanical loads have been known for many years as a major factor causing post-harvest losses and damage to many single stem fruits. The dynamic impact of collision is the main cause of single stem fruit damage. Many studies have been carried out on the impact damage of various kinds of single stem fruits all over the world. The most common methods to determine impact loading damage are as follows: (1) drop tests [19–23]; (2) pendulum action, either by attaching a fruit to the pendulum [24,25] or by hitting the fruit with a pendulum tipped with a specific shape impactor [26,27]; (3) electronic fruit or impact recording devices [28–30]. However, the main problem with these methods is that vibrations make it difficult to accurately record force and deformation during impact due to shorter time periods to observe the mechanism. Therefore, high-speed cameras are more frequently used to observe quality and damage for impact and vibratory research into different fruits [8,21,30–34]. Impact damage is mainly caused by the factors such as the type of packaging surface onto which the single stem fruit drops, drop height, and the velocity at the moment of collision [35–40]. However, impact damage for single fruits is totally different from cluster fruits because cluster fruits are gripped and cut from the main rachis. Therefore, vibration transmissions during transportation and excitation transmissions due to the impact of packaging surfaces on the cluster fruits are totally different from single stem fruits.

The post-harvest operations affect the quality of table grapes through direct contact with packaging materials and machine components. A large quantity of table grapes is wasted just because of damage such as berry fall and fruit decay. Berry fall or decay is mostly caused by impact loads during the mechanical handling, packaging, storage, and transport of table grapes [41]. During fresh fruit transport and handling, dynamical loads cause, by far, the most fruit decay and shatter damage because these loads are higher in incidence and magnitude than static loads [42,43]. The berry drop (shatter) of the grape cluster during and after harvest is related to its physiological process and physical function. Vibration and impact in each operation can lead to fruit stalk detachment. There are three categories of grape berry drop: (1) berry shatter, which consists of a detachment of berries from the main rachis due to the fragile tissue structure of the stalk; (2) wet drop, that is, berries are sloughed from the stems and attached to the pedicel because of the short and thin berry brush [44,45]; (3) dry drop or abscission, which is caused by the formation of an abscission zone (AZ) in the grape, which develops at the junction between the pedicel and berry [4].

At present, chemical methods are widely used in the grape industry to solve the problem of grape berry drop and berry decay [46–49]. However, in recent years, more and more attention has been paid to non-chemical methods. Researchers have been trying to explore the relationship between mechanical harvesting and berry damage to optimize mechanical handling methods for the reduction of berry fall and berry damage. For instance, Pezzi et al. used an electronic fruit to investigate the collision of fresh grapes during mechanical harvesting and transportation [50,51], and Yue et al. found that the drop impact of grape berries has significant effects on physiological quality during storage and transportation [52]. Bian et al. studied the influence of drop height on the dielectric properties of red globe grapes [53], and Vinokur et al. found that the berry fall rate is directly proportional to the free-fall height [41]. Jung et al. evaluated the effect of vibration stress on the quality of packaged grapes by simulated transportation [2], and Vallone et al. measured the effect of mechanical harvesting of grapes using an instrumented sphere [54]. Deng et al. developed a mathematical model that predicts grape berry drop during storage [4], and Fischer et al. determined the critical frequencies for grape and strawberry fruit shattering during transportation for distribution [55,56]. Lu designed tests, such as an emergency stop test in vertical fall, to observe the fruit collision of grape clusters with a piezoelectric film [57], and Hao et al. found that the greater the vibration acceleration, the greater the damage to Kyoho grapes during storage and road transportation [58]. Demir et al. calculated the natural frequency of grape and berry drop during simulated transportation [59]. The above studies deal with the collision between single berries and placing surface and the effect of drop height on berry damage. However, the impact of mechanical load on harvest and post-harvest quality and berry drop of fresh grape clusters in term of vibration has still many research gaps.

Grape cluster vibration plays a vital role in the process of mechanical harvesting and post-harvest handling of table grapes because it will cause berry fall and berry damage. To explore the influence of excitation on vibration, a simulation modeling method is very important, in addition to experimental methods. Simulation can help us to study the transmission route of the excitation and vibration of the cluster. Additionally, simulation can help us to realize the effect of mechanical handling on the grape cluster. To find out the vibration range under different conditions, Kondo et al. designed a low-speed tomato vibrating test system and modeled the panicle tomato [60], and Liu et al. found that acceleration and deceleration are the reasons for vibration in grape fruit clusters; they also found the relationship between the angle deviation of the grape cluster and the excitation transmission through a high-speed camera [61]. Liu et al. designed a compound mechanical model of grape clusters and carried out simulation and experimental analyses under different excitations in horizontal transportation to observed the swing angle of each berry [8]. Faheem et al. found the relationship of the swing angle of clusters with hanging force during linear robotic transportation of the whole grape cluster at different excitations [62]. No specific studies have been done on the impact of packaging materials and the effect of the cluster's vibration on the berry drop mechanism during robotic vertical transportation and placing of the whole grape cluster.

The damage due to the impact and vibration of the grape cluster has a significant effect on control losses during robotic post-harvest handling on the industrial as well as farm level. In this context, the main purpose of the study is to analyze the berry dropping mechanism during vertical transportation and the placing of the whole grape cluster. The behavior of the grape cluster's berries and the hanging force signals (the force that bears the weight of the gripped grape cluster against gravity during robotic transportation) under different speed and acceleration excitations were observed and analyzed. The effect of different packaging materials on the berries from the top and bottom sides of the cluster was analyzed using a high-speed video camera. Additionally, the relationship of the cluster's mass with forces before and after the impact was analyzed to understand the berry drop mechanism. Overall, this study provides theoretical support to the industries by optimizing the berry falling loss of different cluster fruits during robotic post-harvest handling and suggests a safe packaging material and excitation at which the cluster vibrates with less magnitude. Hence, berry drop will be reduced.

## 2. Materials and Methods

### 2.1. Structure of Fresh Grape Clusters

Cluster fruits have some special features compared to single-stem fruits. Fresh table grapes develop as clusters (bunches), with each berry attached to the pedicel through rachis and sub rachis, which contain vascular bundles (also known as the cap stem). The stem unites the berry with the rachis, as shown in Figure 1. This union is very important to avoid loss of berries (dropping or shattering) [62].

**Figure 1.** Structure of table grape cluster with and without berries. (**a**) Table grape cluster; (**b**) stalk fruit structure.

### 2.2. Different Excitations and Behaviour of Grape Cluster Fruit during Robotic Transportation and Placing

When a grape cluster is transported towards a box after harvesting, the excitation is mainly caused by the start–stop of the mechanical system that transfers the gripper to the main rachis of the grape cluster, when the position of the main rachis of the cluster deviates from the point parallel to the gripper, as shown in Figure 2. Thus, the grape cluster starts to vibrate, and bending of the main rachis happens. In the operation of speedy robotic transportation, this bending of rachis would cause a severe load on the pedicel. When the load exceeds the connection strength, then the berry fall starts. During the speedy robotic placing of the grape cluster into the box, excitations come from the packaging materials, which is transmitted into the whole cluster, as shown in Figure 2. These excitations apply a load on the connection point between the pedicel and the berry so that berry shattering happens. Thus, the berry shatter of grape clusters is a highly complex problem of multiple loads during the whole robotic transportation placing cycle. Therefore, in this paper, the main focus is to analyze the vibration behavior of hanging grape clusters during vertical robotic transportation and placing cycles. The effect of different speed and acceleration excitations on the cluster during vertical transportation and the impact of different packaging materials on the cluster were observed and analyzed.

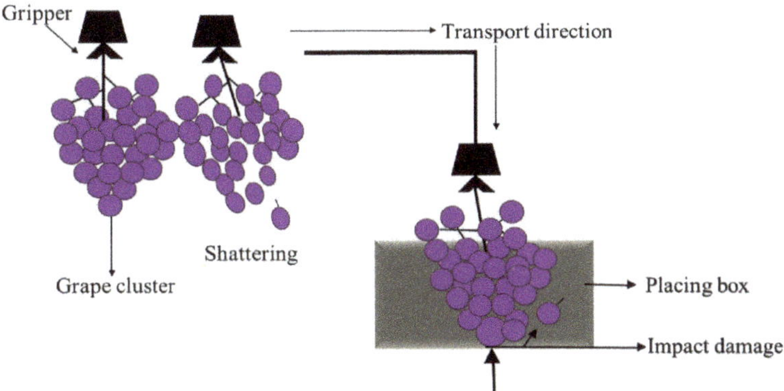


**Figure 2.** Different excitation transmissions and damage of grape cluster fruit during robotic transportation and placing cycles.

### 2.3. Experimental Materials

The experiment was carried out in the Key Laboratory of Modern Agricultural Equipment Engineering, designated by the Ministry of Education, Jiangsu University, Zhenjiang, China. Three artificial grape clusters with different shapes and masses (0.48, 0.50, 0.53 kg), measured with a digital weight balance (BP Professional Electronic Balance BP-6228, accuracy 0.01 g), were used as experimental materials. There were about 70 to 80 berries in each cluster that were filled manually with soil to maintain a similar mass to that of real berries, and each berry mass was 0.005 to 0.007 kg [63,64]. Since the real cluster will be damaged in experiments under different speeds and accelerations, which will lead to a change in conditions, the results of different excitation treatments cannot be put together to compare and analyze [62]. For 40 years, artificial fruits have been built similar to real agricultural produce in order to measure the mechanical load caused by harvest and post-harvest handling systems [30]. Three different packaging materials—(a) a rigid plastic box, (b) a corrugated fiberboard box, and (c) an expandable polystyrene box—were used for the experimental placing test [35]. The experimental setup and materials are shown in Figure 3. A lead screw lathe with a fabricated 1 DOF (degree of freedom) actuator, a gripper, and a single-axis force sensor (model: MIK LCS1; weight range 0–5 kg with 0.03% FS) that was fixed in between the gripper and the hanging grape cluster were used for the measurement of magnitude of the cluster's vibration in terms of forces acting on the grape cluster before and after the impact in real-time. The behavior of grape cluster movement during and after the impact with the packaging materials was recorded with a high-speed camera system (Olympus, i-speed L1), as shown in Figure 3a [65]. The camera system was adjusted to 1000 fps (frames per second), with a black and white screen and a light source for high visibility. It was placed at a distance of 3.5 m from the hanging grape cluster.

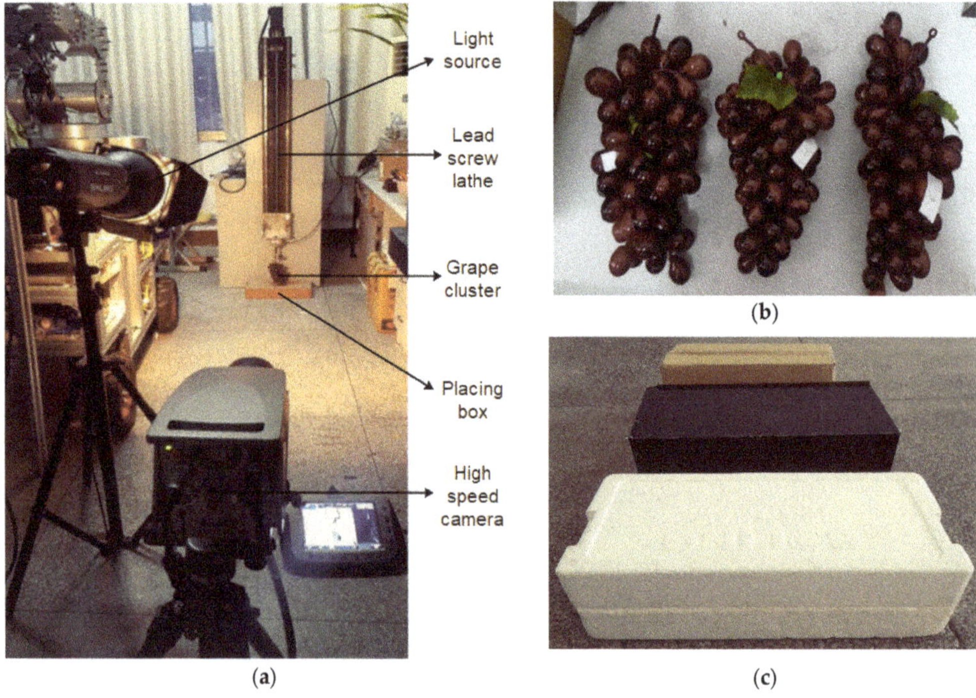


**Figure 3.** Experimental materials: (**a**) experimental setup; (**b**) 3-grape clusters; (**c**) 3-packaging materials.

### 2.4. Experimental Methods

### 2.4.1. Vertical Transportation of Grape Cluster

Three different grape clusters were used as an experimental material because the study's aim was to analyze the vibration mechanism of the hanging grape cluster in different zones of vertical transportation, which was difficult to do using real grape clusters. For understanding the berry drop mechanism of grape clusters due to vibration, an actuator is moved linearly on the rails of the lead screw lathe according to the input speed and acceleration excitations from the PLC (programmable logic control), as shown in Table 1. A vertical start and stop transportation test setup of grape clusters was constructed, as shown below in Figure 4a. Due to the actuator movement, the excitations are transferred from the gripper towards the berries through the main rachis, and the hanging grape cluster starts to vibrate. The magnitude of the cluster's vibration during vertical transportation is determined from the force sensor signals. The excitation displacement or stroke length was adjusted to 0.8 m from the start to the stop point [62]. Additionally, the vibration characteristics of the grape clusters during the different phases of vertical transportation were analyzed.

**Table 1.** Different excitation treatments applied to the actuator for the transportation of grape clusters.

| Speed (m/s) | Accelerated Speed (m/s$^2$) | Accelerated Time (ms) | Duty Cycle (s) |
|---|---|---|---|
| 0.4 | 6 | 66.66 | 2 |
| | 8 | 50 | 2 |
| | 10 | 40 | 2 |
| | 12 | 33 | 2 |
| 0.6 | 6 | 100 | 1.33 |
| | 8 | 75 | 1.33 |
| | 10 | 60 | 1.33 |
| | 12 | 50 | 1.33 |
| 0.8 | 6 | 133 | 1 |
| | 8 | 100 | 1 |
| | 10 | 80 | 1 |
| | 12 | 66 | 1 |
| 1 | 6 | 166 | 0.8 |
| | 8 | 125 | 0.8 |
| | 10 | 100 | 0.8 |
| | 12 | 83 | 0.8 |

**Note:** A duty cycle is the time taken by the actuator to complete one stroke (stroke/time) from the start to stop positions.

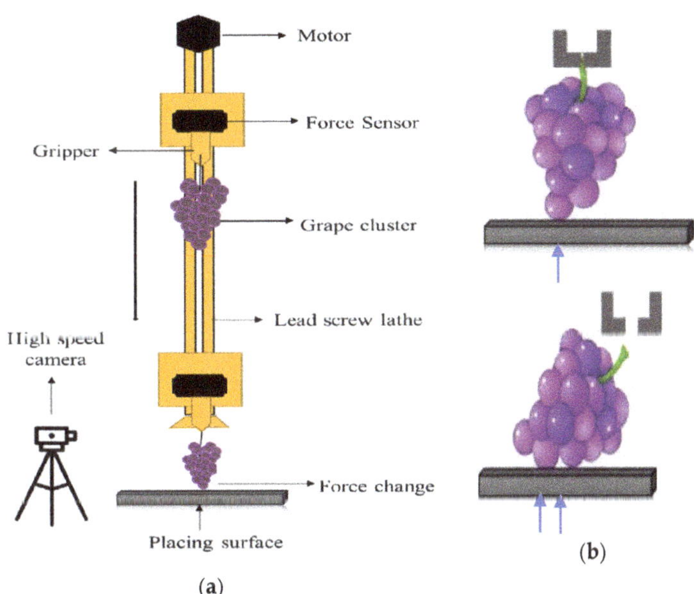


**Figure 4.** Vertical start and stop transportation placing cycle: (**a**) vertical transportation; (**b**) placing view.

## 2.4.2. Placing of Grape Cluster

Three different packaging materials (rigid plastic box, corrugated fiberboard, expandable polystyrene) were used to measure the change in force signals during the robotic placing of the grape cluster. The deflection of the berries during placing and the state of the whole grape cluster were observed from the videos and images of a high-speed camera, as shown in Figure 4a. The mechanism of the grape cluster berry drop was observed and explained from these videos and images. The main rachis of the grape cluster was fixed in a gripper, different speed and acceleration excitations were transmitted to the actuator, and it moved from the start to the stop position. In the robotic placing phase of the grape cluster, the excitation comes from the placing surface, which is measured by the single-axis force sensor, as shown in Figure 4b. During these treatments, the change in force signals during the placing of the three grape cluster samples with different masses (0.48, 0.50, 0.53 kg) was also determined. The behavior of the top berry and the bottom berry was also observed accordingly; the safest packaging material and excitation treatment, at which there are low chances of berry deflection or berry fall, were realized.

## 3. Results

The behavior of the grape cluster during vertical transportation and placing was observed. The details of the results obtained from the experiments are explained in the following sections.

### 3.1. Vibration Characteristics during Vertical Transportation and Placing of Grape Cluster

Based on the force signals, the vibration characteristics of the grape cluster during vertical transportation and placing under different excitations are observed in this section. The details of the vibration characteristics of the grape cluster and its berries are explained in the subsequent sections.

### 3.1.1. Different Stages of Vertical Transportation

Due to the actuator movement from the start position to the stop position, as shown in Figure 5a, the hanging grape cluster started to vibrate, and the magnitude of the cluster's vibration in terms of hanging force was measured. Hanging force was measured in the form of analog signals. These force signals of the hanging cluster during vertical transportation and placing were divided into 5 phases, as shown in Figure 5b, for a better understanding of the dropping mechanism or berry deflection of the hanging grape cluster at different speed and acceleration excitations.

1.  Phase 1 (stationary phase of the grape cluster): In this phase, the force signals showed that the hanging grape cluster is in the stationary position with its static weight (calibrated value of the hanging grape cluster).
2.  Phase 2 (accelerating phase): In this phase, when excitations were applied to the actuator (IDOF manipulator), it started to move, and the hanging grape cluster suddenly vibrated due to the movement of the actuator. Hence, the magnitude of the force signals is observed as high in this phase due to the high vibration of the cluster.
3.  Phase 3 (constant speed phase): In between Phase 2 and Phase 4, the force signal curve depicts that the cluster vibrates with constant amplitude due to the short interval of time.
4.  Phase 4 (deaccelerating phase): In the next phase, when the actuator was going to stop, a dramatic decrement in the magnitude of the force signals was observed, which shows the hanging grape cluster has reached a minimum position (lowest position) during the stop phase of the actuator's motion. This is due to the excitation coming from the packaging surface to the whole cluster, which causes an impact on the whole cluster; the berries started to deflect in this phase.
5.  Phase 5 (placing phase): After the stop of the actuator in Zone 5, all the excitation energy comes from the packaging material storing the grape cluster; the impact with the grape cluster becomes the reason for the berries falling and the bending of the main rachis.

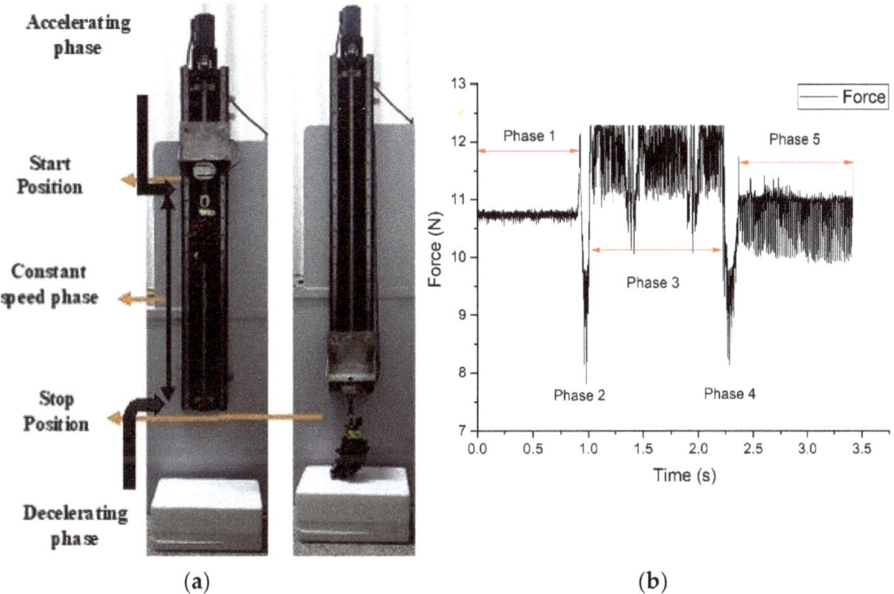


(**a**)  (**b**)

**Figure 5.** Vibration characteristics of the grape cluster during vertical transportation: (**a**) phases of the grape cluster; (**b**) force signals.

### 3.1.2. Different Excitation Transmissions during Placing

It is difficult to observe the deflection mechanism of berries during vertical transportation and placing with the naked eye due to the short interval of time. Therefore, a high-speed camera was used to analyze the deflection of berries at different excitations during the transportation placing cycle, as shown in Figure 6. When the grape cluster collides with the packaging surface, the excitation moves from the packaging material to the whole grape cluster, which causes damage in the form of berry fall and decay. There are two types of berry dropping mechanisms observed from the high-speed photography images during the placing of the grape cluster at different speed and acceleration excitations [8] applied to the hanging grape cluster, as defined below. Excessive bending of the main rachis was observed during the robotic placing of the grape cluster. The impact of the packaging surface on the whole grape cluster after contact was divided into two categories, as shown in Figure 6c.

1. Bottom Berries

The impact excitation comes from the packaging materials after the collision, which caused the deflection of the bottom barriers due to the torsional load in between the berry and the pedicel.

2. Top Berries

Excitations are transmitted to the whole cluster after striking the packaging materials, and upper barrier deflection was observed due to the vibration or bending of the main rachis.

(a)    (b)    (c)

**Figure 6.** Load on the berries of the grape cluster during the placing phase: (**a**) grape cluster; (**b**) transportation load; (**c**) placing load.

### 3.2. Effect of Different Factors on Vibration Signals

In this section, the effect of different factors such as speed, acceleration of the actuator on the vibration signals of the grape cluster during vertical transportation is studied. The experimental effect of different packaging materials on the grape cluster at different speed and acceleration excitations is explained. Additionally, the effect of the cluster's mass on berry deflection during vertical transportation and placing is analyzed. The details about the effect of these factors on the vibration signals are given in the subsequent sections.

### 3.2.1. Effect of Different Speeds during Vertical Transportation

Figure 7 shows the force signals of the hanging grape cluster during different phases of vertical transportation under different speed excitations (0.4, 0.6, 0.8, 1.0 m/s). Based on the results, the magnitude of force for the grape cluster was observed at maximum at the start position (acceleration of the actuator) for four speeds, i.e., 12.2, 12.25, 12.28, 12.4 N. This was due to the sudden excitation coming from the actuator and transmitted towards the hanging grape cluster at the start; the cluster vibrated with high magnitude. Additionally, the magnitude of force increases with an increase in the speed of the actuator. It was observed from Figure 7 that the force signals suddenly decreased because gravity helped to reduce the weight of the cluster. After that, the signals showed that the cluster moved downward with constant speed and collided with the packaging material. After colliding with the packaging surface, the force signals suddenly became low because the excitation that comes from the packaging surface was transmitted to the whole cluster. Then, force signals showed some fluctuations after the impact due to the deflection of the cluster, and, at last, the force signals showed that the cluster reached the stop position, which showed the calibrated value of the grape cluster. These results suggest that the accelerating phase of the actuator caused the vibration of the cluster and more berry drop at this phase of robotic vertical transportation of the grape cluster.

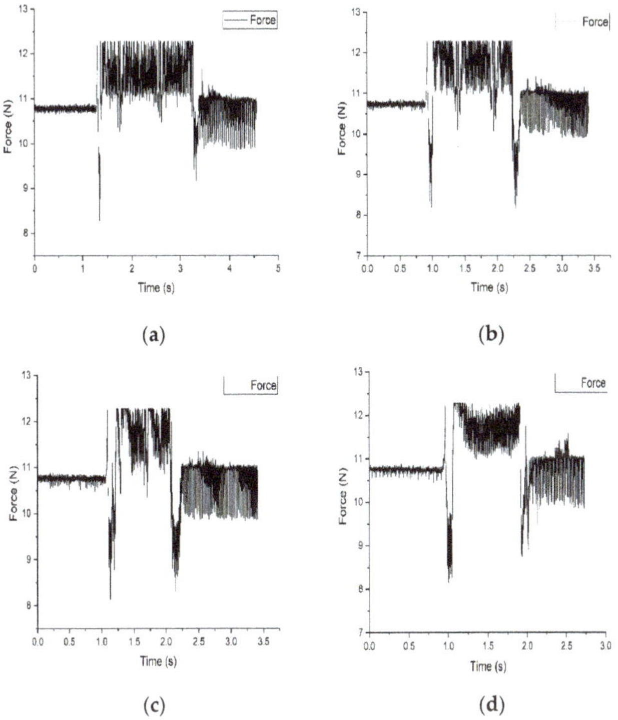

**Figure 7.** Effect of different speeds on the force of the grape cluster during vertical transportation test: (**a**) 0.4 m/s; (**b**) 0.6 m/s; (**c**) 0.8 m/s; (**d**) 1.0 m/s.

### 3.2.2. Effect of Acceleration Excitations during Vertical Transportation

It was observed from the experiments that at input acceleration excitations (6, 8, 10, 12 m/s$^2$), the amplitude of the cluster's vibration was maximum at the start of the actuator movement. It was due to the sudden motion of the actuator and the friction of the guide rails [66] on which the actuator moves. It was observed, secondly, from the experimental analysis, that the magnitude of the cluster's hanging force in the starting phase of the actuator was positively correlated with the input acceleration and mass of the cluster. The magnitude of hanging force was observed to be high at the acceleration excitation of 12 m/s$^2$ at the following speeds of the actuator, i.e., (0.4, 1.0 m/s), as shown in Figure 8. It was also observed from the experimental analysis that the optimized acceleration excitation was 10 m/s$^2$, at which the deflection of the berries was minimum. This is due to the cluster swinging in one direction, with minimum twisting of the main rachis. Therefore, the chances of berry deflection will be minimized. These results suggest that the acceleration phase causes serious vibrations of the grape cluster during robotic vertical transportation, and more berry deflection occurs in this phase. In robotic vertical transportation, the acceleration excitations did not affect the vibration of the grape cluster too much, so the cluster's vibration is always low in vertical downward transportation compared to horizontal transportation of the hanging grape cluster, as measured in our previous research [62].

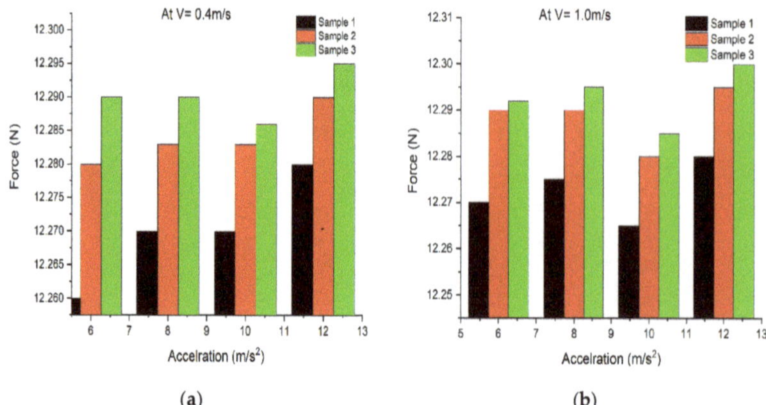



(**a**)  (**b**)

**Figure 8.** Effect of acceleration excitations on the peak force of the cluster during vertical transportation at different speeds, such as (**a**) 0.4 m/s; (**b**) 1.0 m/s.

### 3.2.3. Effect of Packaging Materials on the Berry Deflection of Cluster

In the placing phase of the grape cluster, the time interval of the stop or deaccelerating phase of the actuator was too short, so the berry dropping mechanism of the whole cluster could not be observed properly. Therefore, a high-speed photography camera was used for the analysis. It was observed from the high-speed photography images that the deflection of the cluster's berries in terms of torsion from the bottom and bending from the upper side increases with an increase in speed and acceleration excitations of the actuator. It was also observed from Figure 9 that the effect of the rigid plastic box on the deflection of the upper and lower berries was at a maximum, and it showed small force signals, i.e., 8.8 N, compared to expandable polystyrene and corrugated fiberboard boxes (9.6 and 9.35 N). This is due to the maximum excitations that transmit from the rigid box surface towards the whole grape cluster. It can also be seen from Figure 9 that the fluctuations in the force signals are more after striking with the rigid plastic box due to the large deflection of the whole cluster. These force signal results indicate that the choice of packaging materials can significantly reduce the possibility of berry drop damage during the robotic placing of cluster fruits.

(**a**)

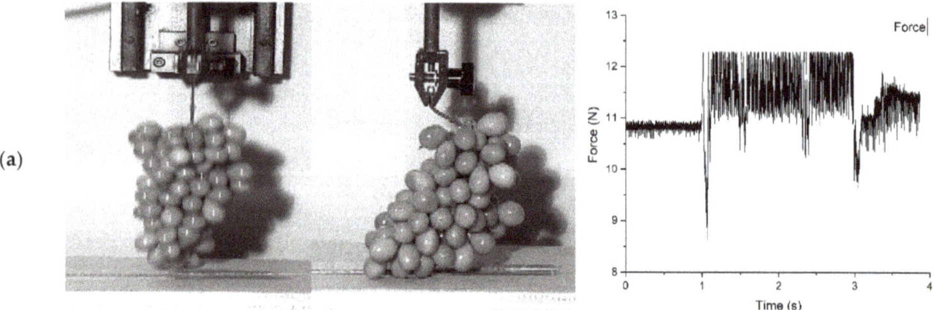



**Figure 9.** *Cont.*

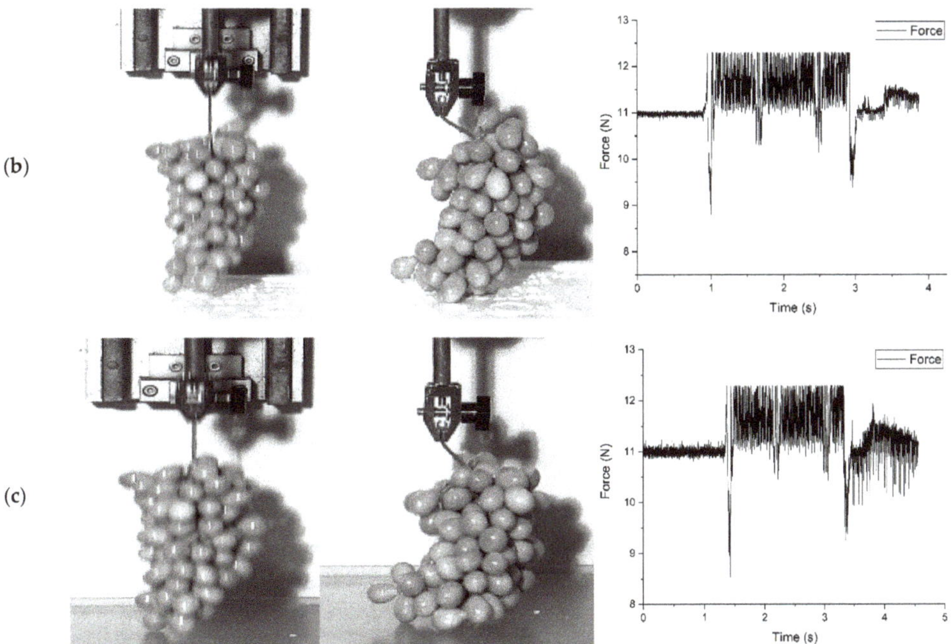

**Figure 9.** High-speed camera-based determination of cluster deflection and force signals after striking with different placing materials: (**a**)force change signals with corrugated fiberboard; (**b**) force change signals with expandable polystyrene; (**c**) force change signals with a rigid plastic box.

### 3.2.4. Effect of the Cluster's Mass on Berry Deflection during Placing

Figure 10 shows the high-speed camera images of three different grape clusters (0.48, 0.50, 0.53 kg) during placing on a rigid plastic box, along with the corresponding results of the force sensor. According to the results, with the increment in the mass of the grape cluster, the deflection of the berries was observed to be less. This was due to the excitations absorbed by the grape cluster after colliding with the packaging material; it decreased with the increase in the mass of the cluster. The grape cluster with a mass of 0.48 kg showed the maximum change of force or load, i.e., 8.20 N from the calibrated force value of that sample, 10.8 N after colliding with the rigid plastic box, as compared to two other grape cluster samples with different masses, i.e., 0.50 and 0.53 kg, with changes in force signals of 8.75 and 9.28 N. The calibrated force of these two cluster samples was 11 and 11.15 N respectively. These results suggest that fruit mass is an important component of the momentum that affects the detachment forces of the berries; the higher the fruit mass, the higher the momentum and hang force, and the lower the berry shattering during placing due to the compactness of the berries.

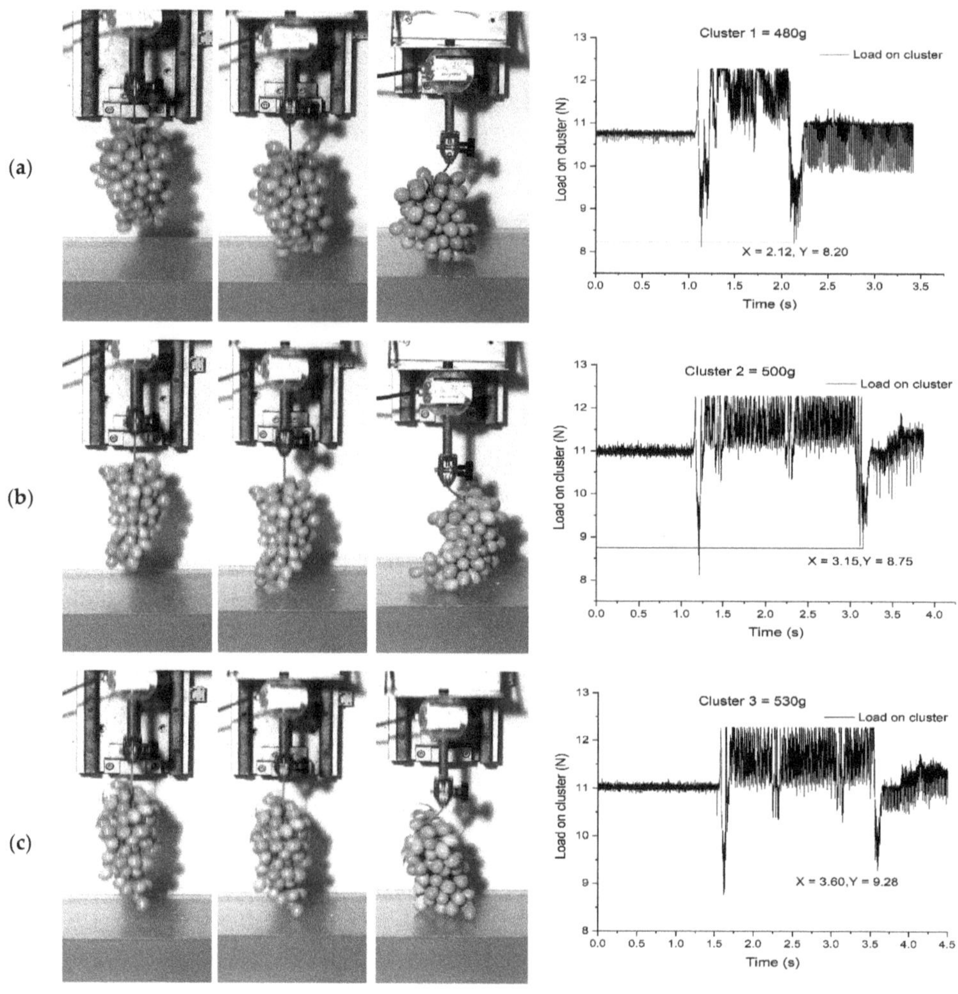



**Figure 10.** Effect of the cluster's mass on force change during placing of grape clusters: (**a**) 480 g; (**b**) 500 g; (**c**) 530 g.

### 3.2.5. Behavior of Top and Bottom Berries during Placing

Figure 11 shows the behavior of the top berry and the bottom berry during the placing of the grape cluster at an excitation speed of 1.0 m/s. The selected top berry is shown in red color; the bottom berry is in yellow. It can be seen from Figure 11 that at the start of the placing, there is no bending of the main rachis; hence, a small deflection of the berries at the top and bottom sides of the cluster was observed. When the cluster collided with the packaging material, the top berry deflected from the initial position due to the load coming from the bending of the main rachis; the bottom berry was also displaced from the initial position due to torsion between the pedicel and the berry. At the end of the placing phase, the top berry continued to deflect in some other direction. These continuous changes in the position of the berries decrease the connection strength between the berry and the pedicle and become the reason for berry drop in speedy robotic placings of grape clusters on both industrial and farm levels. These dropping of berries due to deflection during placing can be controlled by a force feedback mechanism.

**Figure 11.** Behavior of berry deflection during robotic placing. Top berry in (red color) and bottom berry (yellow color).

### 3.2.6. Relationship between the Cluster's Force before and after Impact

It was observed from the experimental analysis that with the increase of speed, i.e., (0.4, 0.6, 0.8, and 1.0 m/s) of the actuator, the corresponding average magnitude of hanging force for all three grape cluster samples increased linearly (with $R^2$ = 0.92, 0.97, 0.98) at the start of the actuator's motion. Additionally, the magnitude of the force after impact with all three packaging surfaces decreased linearly (with $R^2$ = 0.99, 0.97, 1), as shown in Table 2 and Figure 12. This was due to the reason that the packaging surface bears the weight of the grape cluster during the stop of the actuator, which caused the berries' deflection from both the top and bottom sides of the cluster. It is easy to conclude that the higher the hanging force of the grape cluster, the greater the deflection of the cluster after striking the packaging surface, and more berry deflection occurs. There is a negative correlation between hang force and force after the impact of the cluster with a goodness of fit of $R^2 = 0.95$ at different speeds, as shown below in Figure 13. These results show that the cluster with high mass showed a high magnitude of hang force, but it strikes the packaging material with low deflection of the berries due to the compactness of the berries in the cluster with high mass.

**Table 2.** Hang and impact force under different speed intensities.

| Speed (m/s) | Hang Force (Sample 1) (N) | Hang Force (Sample 2) (N) | Hang Force (Sample 3) (N) | Impact with Corrugated Box (N) | Impact with Expandable Polystyrene Box (N) | Impact with Rigid Plastic Box (N) |
|---|---|---|---|---|---|---|
| 0.4 | 12.2 | 12.3 | 12.41 | 9.6 | 9.3 | 8.5 |
| 0.6 | 12.25 | 12.35 | 12.49 | 9.5 | 9.25 | 8.4 |
| 0.8 | 12.28 | 12.4 | 12.56 | 9.41 | 9.12 | 8.3 |
| 1.0 | 12.4 | 12.42 | 12.6 | 9.28 | 9.05 | 8.2 |
| $R^2$ | 0.92 | 0.97 | 0.98 | 0.99 | 0.97 | 1 |

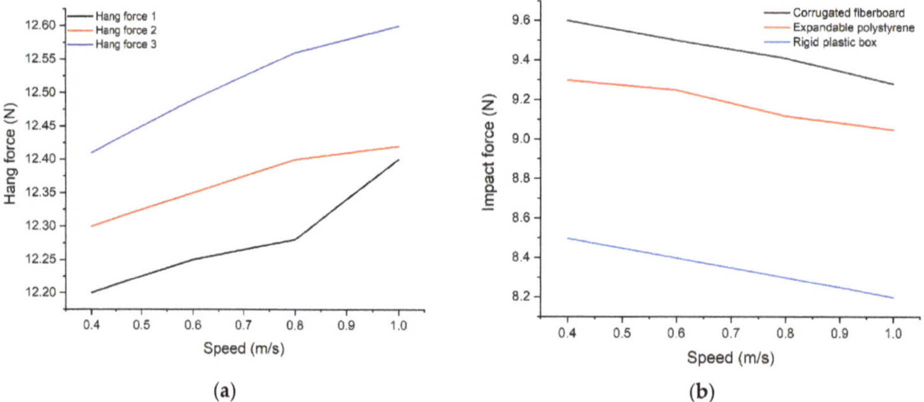


**Figure 12.** Effect of different speeds on the changes of forces during transportation and impact: (**a**) hanging force; (**b**) impact force.

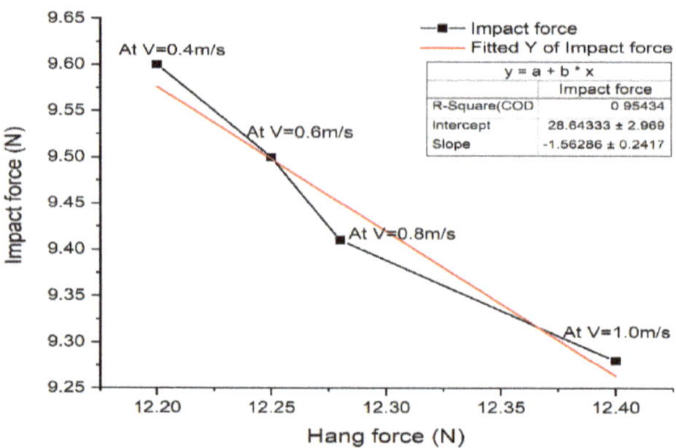


**Figure 13.** Relationship between hang force and impact force.

## 4. Conclusions

In this paper, our aim was to observe the effect of different speed and acceleration excitations on the berry drop mechanism during the vertical transportation and placing of grape clusters. Force sensor signals and high-speed photography images were used for the verification of the deflection mechanism of the berries under different packaging materials. The following conclusions were drawn from this study:

1. It is concluded that the accelerating phase of the actuator causes high vibrations of the cluster during vertical transportation, and its magnitude increases with an increase in speed, so the hanging force signals also increase. The results of the peak force signals of the grape cluster during vertical transportation at different speeds (0.4, 1.0 m/s) suggest that the optimum acceleration excitation is 10 m/s$^2$, at which berry deflection is observed to be at a minimum.
2. It is concluded that with an increase in speed and acceleration excitation, the magnitude of force signals after colliding decreases due to the excitations coming from the packaging surface, which causes more berry deflection to occur.
3. It is concluded from the force sensor signals that rigid plastic boxes deflect the whole cluster most compared to expandable polystyrene and corrugated fiberboard boxes.
4. The behavior of the upper and lower berries was observed from the high-speed photography images during the placing of the whole cluster, and it is concluded that the deflection of the upper berries is due to the excitations coming from bending of the main rachis, and the deflection of lower berries is due to the torsional load on the junction between the pedicel and the berry.
5. It is observed from the vibration signals that with the increment in the mass of the cluster, the deflection of the berries decreases during placing.

Thus, research in this paper can act as a guide for the harvest and post-harvest robotic handling of cluster fruits such as grapes, cherries, blueberries, and litchi. The proposed study can be used to realize optimal control of the transportation and placing of the robot. Overall, this study provides technical and theoretical support for the industrial needs of low-loss robotic handling of different fresh-eating cluster fruits that are consumed abundantly. In the future, dynamic impact or collision between grape cluster berries during robotic post-harvest handling will be discussed.

**Author Contributions:** Conceptualization, J.L. and M.F.; methodology, M.F. and J.L.; software, M.F. and G.C.; validation, M.F., J.L., and G.C.; formal analysis, M.F. and J.L.; investigation, M.F., J.L., and I.A.; resources, M.F., J.L., and I.A.; data curation, J.L., M.F., and B.X.; writing—original draft preparation, M.F.; writing—review and editing, M.F., J.L., Z.S., and K.Y.; visualization, Z.S., and I.A.; supervision, J.L.; project administration, J.L.; funding acquisition, J.L. All authors have read and agreed to the published version of the manuscript.

**Funding:** The research was supported by grants from the National Science Foundation of China (Grant No. 31971795) and a project funded by the Priority Academic Program Development of Jiangsu Higher Education Institutions (No. PAPD-2018-87).

**Institutional Review Board Statement:** Not applicable.

**Informed Consent Statement:** Not applicable.

**Data Availability Statement:** Not applicable.

**Acknowledgments:** The authors are grateful to the National Science Foundation of China. The first author gives many thanks to the China Scholarship Council (2017GXZ026592) for providing 36 months of scholarship for studying in the People's Republic of China.

**Conflicts of Interest:** The authors declare no conflict of interest.

## References


1. FAO. FAOSTAT—Food and Agriculture Organization of the United Nations (FAO). Selection Criteria Grapes, All Countries, Prod. Quant. 2018. Available online: http//www.fao.org/faostat/en/#data/QC2020 (accessed on 5 June 2021).
2. Jung, H.M.; Lee, S.; Lee, W.-H.; Cho, B.-K.; Lee, S.H. Effect of vibration stress on quality of packaged grapes during transportation. *Eng. Agric. Environ. Food* **2018**, *11*, 79–83. [CrossRef]
3. Rizzuti, A.; Aguilera-Sáez, L.M.; Gallo, V.; Cafagna, I.; Mastrorilli, P.; Latronico, M.; Pacifico, A.; Matarrese, A.M.S.; Ferrara, G. On the use of ethephon as abscising agent in cv. Crimson seedless table grape production: Combination of fruit detachment force, fruit drop and metabolomics. *Food Chem.* **2015**, *171*, 341–350. [CrossRef]
4. Deng, Y.; Wu, Y.; Li, Y.; Zhang, P.; Yang, M.; Shi, C.; Zheng, C.; Yu, S. A mathematical model for predicting grape berry drop during storage. *J. Food Eng.* **2007**, *78*, 500–511. [CrossRef]
5. Luo, L.; Tang, Y.; Zou, X.; Wang, C.; Zhang, P.; Feng, W. Robust grape cluster detection in a vineyard by combining the AdaBoost framework and multiple color components. *Sensors* **2016**, *16*, 2098. [CrossRef] [PubMed]
6. Liu, J.; Peng, Y.; Faheem, M. Experimental and theoretical analysis of fruit plucking patterns for robotic tomato harvesting. *Comput. Electron. Agric.* **2020**, *173*, 105330. [CrossRef]
7. Hussein, Z.; Fawole, O.; Opara, U.L. Harvest and postharvest factors affecting bruise damage of fresh fruits. *Hortic. Plant J.* **2020**, *6*, 1–13. [CrossRef]
8. Liu, J.; Yuan, Y.; Gao, Y.; Tang, S.; Li, Z. Virtual model of grip-and-cut picking for simulation of vibration and falling of grape clusters. *Trans. ASABE* **2019**, *62*, 603–614. [CrossRef]
9. Lichter, A.; Gabler, F.M.; Smilanick, J.L. Control of spoilage in table grapes. *Stewart Postharvest Rev.* **2006**, *6*, 1–10.
10. Chen, R.; Wu, P.; Cao, D.; Tian, H.; Chen, C.; Zhu, B. Edible coatings inhibit the postharvest berry abscission of table grapes caused by sulfur dioxide during storage. *Postharvest Biol. Technol.* **2019**, *152*, 1–8. [CrossRef]
11. Hussein, Z.; Fawole, O.; Opara, U.L. Preharvest factors influencing bruise damage of fresh fruits—A review. *Sci. Hortic.* **2018**, *229*, 45–58. [CrossRef]
12. Hu, X.; Xu, Y.; Liu, J. Design of vibration monitoring system and its application in grape fatigue damage research. In Proceedings of the 2nd International Conference on Control and Computer Vision, Jeju Island, Korea, 15–18 June 2019; pp. 141–149.
13. Jobbagy, J.; Kristof, K.; Schmidt, A.; Krizan, M.; Urbanovicova, O. Evaluation of the mechanized harvest of grapes with regards to harvest losses and economical aspects. *Agron. Res.* **2018**, *16*, 426–442.
14. Costa, W.V.D.; Elorza, P.B.; Garrido-Izard, M. Impact of local conditions and machine management on grape harvest quality. *Sci. Agric.* **2019**, *76*, 353–361. [CrossRef]
15. Fernando, I.; Fei, J.; Stanley, R.; Enshaei, H. Measurement and evaluation of the effect of vibration on fruits in transit-review. *Packag. Technol. Sci.* **2018**, *31*, 723–738. [CrossRef]
16. Gross, K.C.; Wang, C.Y.; Saltveit, M.E. *Agriculture Handbook Number 66: The Commercial Storage of Fruits, Vegetables, and Florist and Nursery Stocks*, 5th ed.; United States Department of Agriculture, Agriculture Research Service: Washington, DC, USA, 2016.
17. Mingjuan, L.; Xiangrong, Y.; Rende, W.; Yayuan, Z.; Jian, S.; Zhichun, L.; Changbao, L. Study on fruit quality and physiology and biochemistry of grapes during cold storage. *South J. Agric. Sci.* **2013**, *44*, 1883–1889.
18. Nicolosi, E.; Ferlito, F.; Amenta, M.; Russo, T.; Rapisarda, P. Changes in the quality and antioxidant components of minimally processed table grapes during storage. *Sci. Hortic.* **2018**, *232*, 175–183. [CrossRef]
19. Komarnicki, P.; Stopa, R.; Kuta, Ł.; Szyjewicz, D. Determination of apple bruise resistance based on the surface pressure and contact area measurements under impact loads. *Comput. Electron. Agric.* **2017**, *142*, 155–164. [CrossRef]
20. Stropek, Z.; Gołacki, K. Bruise susceptibility and energy dissipation analysis in pears under impact loading conditions. *Postharvest Biol. Technol.* **2020**, *163*, 111120. [CrossRef]
21. Hu, G.; Chen, J. Transverse anisotropy mechanical properties and drop test of apple. In Proceedings of the 2020 ASABE Annual International Virtual Meeting, Omaha, NE, USA, 13–15 July 2020; p. 1.
22. Yousefi, S.; Farsi, H.; Kheiralipour, K. Drop test of pear fruit: Experimental measurement and finite element modelling. *Biosyst. Eng.* **2016**, *147*, 17–25. [CrossRef]
23. Shafie, M.M.; Rajabipour, A.; Mobli, H. Determination of bruise incidence of pomegranate fruit under drop case. *Int. J. Fruit Sci.* **2017**, *17*, 296–309. [CrossRef]
24. Opara, L.U.; Al-Ghafri, A.; Agzoun, H.; Al-Issai, J.; Al-Jabri, F.; Opara, U.L. Design and development of a new device for measuring susceptibility to impact damage of fresh produce. *N. Z. J. Crop. Hortic. Sci.* **2007**, *35*, 245–251. [CrossRef]
25. Wang, W.; Zhang, S.; Fu, H.; Lu, H.; Yang, Z. Evaluation of litchi impact damage degree and damage susceptibility. *Comput. Electron. Agric.* **2020**, *173*, 105409. [CrossRef]
26. Van Zeebroeck, M.; Tijskens, E.; Liedekerke, P.; Deli, V.; Baerdemaeker, J.; Ramon, H. Determination of the dynamical behaviour of biological materials during impact using a pendulum device. *J. Sound Vib.* **2003**, *266*, 465–480. [CrossRef]
27. Stropek, Z.; Gołacki, K. Impact characteristics of pears. *Postharvest Biol. Technol.* **2019**, *147*, 100–106. [CrossRef]
28. Öztekin, Y.B.; Güngör, B. Determining impact bruising thresholds of peaches using electronic fruit. *Sci. Hortic.* **2019**, *262*, 109046. [CrossRef]
29. Praeger, U.; Surdilovic, J.; Truppel, I.; Herold, B.; Geyer, M. Comparison of electronic fruits for impact detection on a laboratory scale. *Sensors* **2013**, *13*, 7140–7155. [CrossRef] [PubMed]

30. Surdilovic, J.; Praeger, U.; Herold, B.; Truppel, I.; Geyer, M. Impact characterization of agricultural products by fall trajectory simulation and measurement. *Comput. Electron. Agric.* **2018**, *151*, 460–468. [CrossRef]
31. Stropek, Z.; Gołacki, K. Quantity assessment of plastic deformation energy under impact loading conditions of selected apple cultivars. *Postharvest Biol. Technol.* **2016**, *115*, 9–17. [CrossRef]
32. Horabik, J.; Beczek, M.; Mazur, R.; Parafiniuk, P.; Ryżak, M.; Molenda, M. Determination of the restitution coefficient of seeds and coefficients of visco-elastic Hertz contact models for DEM simulations. *Biosyst. Eng.* **2017**, *161*, 106–119. [CrossRef]
33. Liang, N.; Ni, F.; Zhang, K.; Tang, Y.; Hu, Y. Optimized installation angle and distance of a grading channel for dried jujube fruit with a push-pull actuating mechanism. *Comput. Electron. Agric.* **2018**, *150*, 134–142. [CrossRef]
34. Jinwu, W.; Han, T.; Jinfeng, W.; Dongxuan, J.; Xin, L. Measurement and analysis of restitution coefficient between maize seed and soil based on high-speed photography. *Int. J. Agric. Biol. Eng.* **2017**, *10*, 102–114.
35. Xia, M.; Zhao, X.; Wei, X.; Guan, W.; Wei, X.; Xu, C.; Mao, L. Impact of packaging materials on bruise damage in kiwifruit during free drop test. *Acta Physiol. Plant.* **2020**, *42*, 1–11. [CrossRef]
36. Zhou, J.; He, L.; Karkee, M.; Zhang, Q. Effect of catching surface and tilt angle on bruise damage of sweet cherry due to mechanical impact. *Comput. Electron. Agric.* **2016**, *121*, 282–289. [CrossRef]
37. Du, D.; Wang, B.; Wang, J.; Yao, F.; Hong, X. Prediction of bruise susceptibility of harvested kiwifruit (*Actinidia chinensis*) using finite element method. *Postharvest Biol. Technol.* **2019**, *152*, 36–44. [CrossRef]
38. Fu, H.; He, L.; Ma, S.; Karkee, M.; Chen, D.; Zhang, Q.; Wang, S. 'Jazz' apple impact bruise responses to different cushioning materials. *Trans. ASABE* **2017**, *60*, 327–336. [CrossRef]
39. Deng, W.; Wang, C.; Xie, S. Impact peak force measurement of potato. *Int. J. Food Prop.* **2020**, *23*, 616–626. [CrossRef]
40. Hussein, Z.; Fawole, O.A.; Opara, U.L. Bruise damage susceptibility of pomegranates (*Punica granatum*, L.) and impact on fruit physiological response during short term storage. *Sci. Hortic.* **2019**, *246*, 664–674. [CrossRef]
41. Vinokur, Y.; Rodov, V.; Levi, A.; Kaplunov, T.; Zutahy, Y.; Lichter, A. A method for evaluating fruit abscission potential of grapes and cherry tomato clusters. *Postharvest Biol. Technol.* **2013**, *79*, 20–23. [CrossRef]
42. Mohsenin, N.N. *Physical Properties of Plant and Animal Materials*; Routledge: Abingdon, UK, 1986.
43. Kupferman, E. *Minimizing Bruising in Apples*; Postharvest Information Network, Washington State University, Tree Fruit Research and Extension Center: Wenatchee, WA, USA, 2006.
44. Fahe, C.; Xin, Y.; Weiyi, Z. Study on relationship between pedicel structure and berry abscission of 'Xinjiang wuhebai' grape cultivars. *J. Xinjiang Agric. Univ.* **2000**, *1*, 44–48.
45. Youmei, W.; Xuezeng, H.; Yu, L.; Jianchuan, R. Postharvest berry abscission and storage of grape fruit. *Acta Phytophysiol. Sin.* **1992**, *18*, 267–272.
46. Zahedipour, P.; Asghari, M.; Abdollahi, B.; Alizadeh, M.; Danesh, Y.R. A comparative study on quality attributes and physiological responses of organic and conventionally grown table grapes during cold storage. *Sci. Hortic.* **2019**, *247*, 86–95. [CrossRef]
47. Zhang, Z.; Xu, J.; Chen, Y.; Wei, J.; Wu, B. Nitric oxide treatment maintains postharvest quality of table grapes by mitigation of oxidative damage. *Postharvest Biol. Technol.* **2019**, *152*, 9–18. [CrossRef]
48. Vázquez-Hernández, M.; Navarro, S.; Sanchez-Ballesta, M.T.; Merodio, C.; Escribano, M.I. Short-term high $CO_2$ treatment reduces water loss and decay by modulating defense proteins and organic osmolytes in Cardinal table grape after cold storage and shelf life. *Sci. Hortic.* **2018**, *234*, 27–35. [CrossRef]
49. Liguori, G.; Sortino, G.; Gullo, G.; Inglese, P. Effects of modified atmosphere packaging and chitosan treatment on quality and sensorial parameters of minimally processed cv. 'Italia' table grapes. *Agronomy* **2021**, *11*, 328. [CrossRef]
50. Pezzi, F.; Caprara, C.; Bordini, F. Transmission of impacts during mechanical grape harvesting and transportation. *J. Agric. Eng.* **2008**, *39*, 43–48. [CrossRef]
51. Caprara, C.; Pezzi, F. Measuring the stresses transmitted during mechanical grape harvesting. *Biosyst. Eng.* **2011**, *110*, 97–105. [CrossRef]
52. Yue, X.; Wu, P.; Wang, S.; Liu, Y.; Su, H. Experimental analysis of drop and vibration damage during grape storage and transportation. *Packag. Eng.* **2019**, *40*, 9–18.
53. Bian, H.; Tu, P. The influence of drop height on the dielectric properties of Red Globe grapes. *Food Ferment. Ind.* **2013**, *39*, 154–157.
54. Vallone, M.; Alleri, M.; Bono, F.; Catania, P. Acceleration assessment during mechanical harvest of grapes using a non commercial instrumented sphere. *Chem. Eng. Trans.* **2017**, *58*, 277–282.
55. Fischer, D.F.; Craig, W.L.; Ashby, B.H. Reducing transportation damage to grapes and strawberries. *J. Food Distrib. Res.* **1990**, *21*, 193–202.
56. Fischer, D.; Craig, W.L.; Watada, A.E.; Douglas, W.; Ashby, B.H. Simulated in-transit vibration damage to packaged fresh market grapes and strawberries. *Appl. Eng. Agric.* **1992**, *8*, 363–366. [CrossRef]
57. Lixin, L. Dynamic mechanical model of fruit under drop impact. *Eng. Mech.* **2009**, *4*, 228–233.
58. Hao, D. Research on the Key Factors Affecting the Quality of Kyoho Grapes during Storage and Transportation. Master's Thesis, Tianjin University of Commerce, Tianjin, China, 2016.
59. Demir, F.; Kara, Z.; Carman, K. Table grapes transport simulation study by Bardas (*Vitis vinifera* L.) cultivar grown in Karaman Turkey. In Proceedings of the 2nd International Symposium on Sustainable Development, Sarajevo, Bosnia and Herzegovina, 8 June 2010; pp. 456–463.

60. Kondo, N.; Tanihara, K.; Shiigi, T.; Shimizu, H.; Kurita, M.; Tsutsumi, M.; Chong, V.K.; Taniwaki, S. Path planning of tomato cluster harvesting robot for realizing low vibration and speedy transportation. *Eng. Agric. Environ. Food* **2009**, *2*, 108–115. [CrossRef]

61. Liu, J.; Tang, S.; Shan, S.; Ju, J. Simulation and test of grape fruit cluster vibration for robotic harvesting. *Trans. Chin. Soc. Agric. Mach.* **2016**, *47*, 1–8.

62. Faheem, M.; Liu, J.; Chang, G.; Ahmad, I.; Peng, Y. Hanging force analysis for realizing low vibration of grape clusters during speedy robotic post-harvest handling. *Int. J. Agric. Biol. Eng.* **2021**, *14*, 62–71. [CrossRef]

63. Peacock, B.; Simpson, T. *The Relationship between Berry Weight, Length and Width for Five Table Grape Varieties*; University of California Cooperative Extension (UCCE): Tulare County, CA, USA, 2017; Publ. TB1-95.

64. Mack, J.; Rist, F.; Herzog, K.; Töpfer, R.; Steinhage, V. Constraint-based automated reconstruction of grape bunches from 3D range data for high-throughput phenotyping. *Biosyst. Eng.* **2020**, *197*, 285–305. [CrossRef]

65. Wang, Z.; Zhang, Y.; Wang, Q.; Dong, K.; Yang, S.; Jiang, Y.; Zheng, J.; Li, B.; Huo, Y.; Wang, X.; et al. Dynamics of droplet formation with oscillation of meniscus in electric periodic dripping regime. *Exp. Therm. Fluid Sci.* **2021**, *120*, 110250. [CrossRef]

66. Rekhviashvili, S.; Pskhu, A.; Agarwal, P.; Jain, S. Application of the fractional oscillator model to describe damped vibrations. *Turk. J. Phys.* **2019**, *43*, 236–242. [CrossRef]

*Article*

# Comparative Studies on Grain Quality and Pesticide Residues in Maize Stored in Hermetic and Polypropylene Storage Bags

**Samuel Kofi Nyarko [1], Yaw Gyau Akyereko [1,2], Joseph Oppong Akowuah [3,*] and Faustina Dufie Wireko-Manu [1]**

[1] Department of Food Science and Technology, Faculty of Biosciences, Kwame Nkrumah University of Science and Technology, Kumasi 03220, Ghana; kofinyarkosamu-el65@gmail.com (S.K.N.); akyereko.edu@gmail.com (Y.G.A.); fdbaah@yahoo.com (F.D.W.-M.)

[2] Department of Food and Post-Harvest Technology, Faculty of Applied Science and Technology, Koforidua Technical University, P.O. Box KF-981, Koforidua 03420, Ghana

[3] Department of Agricultural and Biosystems Engineering, College of Engineering, Kwame Nkrumah University of Science and Technology, Kumasi 03220, Ghana

* Correspondence: akowuahjoe@yahoo.co.uk

**Abstract:** The conventional method of grain storage involving the use of polypropylene bags in conjunction with pesticides and hermetic bags are paramount in developing countries. However, there is limited information on grain quality and pesticide residue concentration of maize stored in such bags. This work determined grain quality and pesticide residue concentrations of maize stored in polypropylene and hermetic storage bags. Maize samples stored for a period of one year in polypropylene and hermetic bags were obtained from three major maize growing communities in the Ashanti region of Ghana and were analyzed for grain quality, aflatoxin content and pesticide residue concentration using standard methods. The amount of diseased, discolored, broken, insect-damaged, stained, germinated, shriveled, total defective, inorganic and organic matter of maize stored in hermetic bags was significantly lower than that of polypropylene. Levels of aflatoxin in maize stored in the polypropylene bags were significantly higher (13.9 ppb–20 ppb) than in maize stored in the hermetic bags (0.90 ppb–2.6 ppb). Out of 35 pesticides screened, only lambda-cyhalothrin was detected in polypropylene bags and deltamethrin in hermetic bags. The presence of these pesticide residues may be due to their long-lasting abilities. Levels of lambda-cyhalothrin residues were above the maximum residue limit (MRL) of 0.02 mg/kg, but have no significant effect on health. Deltamethrin residue concentrations in hermetically stored maize samples were below the MRL. In conclusion, maize grains stored in hermetic bags have higher grain quality and lower aflatoxin and pesticide residue concentrations than polypropylene bags. Education and promotion on the utilization of hermetic bags should be a priority in storing and supplying safe maize grains to consumers.

**Keywords:** maize grain storage; hermetic storage bags; polypropylene storage bags; quality attributes; pesticide residues

**Citation:** Nyarko, S.K.; Akyereko, Y.G.; Akowuah, J.O.; Wireko-Manu, F.D. Comparative Studies on Grain Quality and Pesticide Residues in Maize Stored in Hermetic and Polypropylene Storage Bags. *Agriculture* **2021**, *11*, 772. https://doi.org/10.3390/agriculture11080772

Academic Editor: John M. Fielke

Received: 25 June 2021
Accepted: 5 August 2021
Published: 13 August 2021

**Publisher's Note:** MDPI stays neutral with regard to jurisdictional claims in published maps and institutional affiliations.

## 1. Introduction

Maize (*Zea mays* L.) is the most extensively produced and consumed cereal, accounting for over half of Ghana's entire grain harvest [1]. It is principally cultivated by small-scale farmers in most of the agro-ecological zones of Ghana, who depend predominantly on rainfall. Obaatanpa, Mamaba, Dadaba and Aburohoma are the common maize varieties grown by most farmers in Ghana [2]. Ghana's maize production capacity currently stands at 2.76 million MT with an annual growth rate of 8.06% [3]. Maize is used in many Ghanaian staples: poultry feed formulation, maize-grit production, alcohol brewing, baby food and breakfast cereal production [1,3]. There are basically two maize growing seasons (major and minor) along the transitional areas of Ghana, and usually one harvest season coincides with the rainy period, which threatens grain quality, particularly with respect to mold growth and insect pest infestation since most farmers rely on the sun for drying [4].

Maize storage is a major issue in maize production and a key contributing factor to post-harvest losses of maize across the globe. According to a report by FAOSTAT (Rome, Italy) [5], post-harvest losses of maize stand at 30%, with major causes being drying inefficiencies, poor post-harvest management, overdue harvesting and poor storage systems. Studies by Opit et al. [2] and Likhayo et al. [4] discovered that storage of maize in warehouses in Ghana is being affected by insects such as weevils (*Sitophilus zeamais* Motschulsky) and larger grain borer (*Prostephanus truncatus* Horn) owing to inappropriate moisture content, temperature and storage material and high gaseous exchange. Efforts by researchers in reducing post-harvest losses of maize have resulted in formulation of pesticides and invention of storage bags, silos, warehouses and others. However, studies have revealed that due to the high-cost nature and inefficiencies of some of these methods, farmers still use pesticides which are less expensive but have other deleterious effects on consumers' health. In late 2010, 15 farmers living in the Upper East region of Ghana died from consuming cereals suspected to have been treated with pesticides [6]. As a result, some of these pesticides have been banned in Ghana, yet some farmers use them secretly.

Efforts to mitigate the use of pesticides in cereal or maize storage led to discovery of hermetic storage bags, which provide a pesticide-free and cost-effective storage system [7,8]. Hermetic storage technology operates on the principle of depleting $O_2$ and accumulating $CO_2$ concentrations in the interior of the bags by virtue of grain and pest metabolism [4]. Walker et al. [9] emphasized that the hermetic phenomenon thwarts evaporation and gaseous exchange, thereby adjusting the interior composition ($O_2$ and $CO_2$) of the container to eliminate insect pests. Nevertheless, the concept of hermetic storage has not been fully accepted by many stakeholders along the maize value chain because little information has been published on the relative advantages of hermetic bags over traditional (polypropylene) bags with respect to grain quality and pesticide residues.

A national approach in Ghana to attain food security via the introduction of the "Planting for Food and Jobs" initiative may increase maize production to meet domestic consumption as well as international market demands. Nonetheless, the full prospects of the initiative may be unattainable unless stakeholders along the maize supply chain are provided with the capacity to preserve maize and market excesses for profit [8]. The current overreliance of smallholder maize farmers on the storage of maize in polypropylene bags with pesticide application is overwhelming and has the tendency to increase the effect of pesticides on the health of the Ghanaian consumers and the international market. It is therefore vital to provide information on the concentrations of pesticide residues and the quality of maize stored in hermetic and traditional (polypropylene) storage bags in Ghana to offer farmers and other stakeholders the opportunity to make informed storage choices for maize and promulgate laws with respect to eliminating pesticide use for maize preservation, for the safety of consumers.

This research work seeks to determine and compare the quality characteristics and the pesticide residue concentrations of maize stored in hermetic and traditional (polypropylene) bags in the leading maize producing areas in the Ashanti region of Ghana.

## 2. Materials and Methods

### 2.1. Study Area

This research work was conducted in three leading localities (Ejura, Abofour and Asante-Akyem Agogo) in the Ashanti region of Ghana notable for their high volumes of maize production. The three municipalities were chosen as study areas because of their significant production, active trade and involvement in the maize supply chain.

### 2.2. Sampling

Maize samples stored for a period of one year were collected from traditional polypropylene bags and hermetic bags in two different warehouses in each locality. A random sampling was done by taking a minimum of the square root of the number of bags in the warehouse, to get a fair representative sample of the total consignment. Samples were collected from bags by inserting a sampler from the top to the bottom of the bag at multiple random points. They were then sealed separately in zip-lock bags and labelled.

*2.3. Maize Grain Quality Determination*

Moisture content, physical quality, pesticide residues and aflatoxin content of maize samples were determined in the Laboratory of the Pesticide Division of the Ghana Standards Authority, Accra, as follows:

2.3.1. Determination of Moisture Content

Sampled maize grains were combined and mixed thoroughly, and about 500 g was milled using a standard laboratory mill (SUS304, China) to obtain uniform particle sizes. About 50 g of the milled sample was analyzed for moisture content using a Dicky-John Instalab® 700 proximate analyzer (IL7101FG, USA). The procedure was triplicated and their mean values recorded for each sample.

2.3.2. Determination of Maize Grain Physical Quality

Cone and quartering method were used to sub-divide the mixed sample several times to get a representative amount of 100 g. Grains were sorted according to stated parameters or defects (diseased, discolored, broken/chipped, insect damaged, stained, germinated, shriveled, other grains, total defective, inorganic matter, organic matter). After thorough sorting, the percentage defects were then calculated using Equation (1):

$$\text{Percentage defective grains} \; = \; \frac{\text{Weight of defective grains (g)}}{\text{Weight of sample (g)}} \times 100 \qquad (1)$$

The analysis was conducted in triplicate and the resulting average percentages for the various defects were then recorded and compared with the Ghana Standards Authority's grading specification.

2.3.3. Aflatoxin Analysis

- Source of Reagents
  ENVIROLOGIX QUIKSCAN® DB5 Buffer solution and Sodium Lauryl Sulphate were obtained from Portland-USA, and 50% *v/v* ethanol solution was prepared using absolute ethanol and distilled water.
- Cleaning of Glassware
  Preceding the analysis, glassware was cleaned using Ecolab® Food grade detergent and washed with deionized water. They were further cleaned with acetone, dehydrated and stored in dust free cabinets until required.
- Extraction and Purification
  The sample was mixed thoroughly to achieve a homogenous mixture. About 500 g of the sample was milled (SUS304, China) to attain a granulation of 841 microns. Then, 25 g of the milled sample was weighed into a beaker. One packet of extraction powder was added to the flour as well as 50 mL of 50% ethanol. It was then shaken vigorously for 2 min by hand and allowed to settle for 2 min for a clear separation into lipid and aqueous phases. Finally, 100 µL was pipetted from different portions of the lipid phase into a centrifuge tube and centrifuged for 1 min in a LabniqueTM centrifuge (Spinplus-6, China).
- Analysis
  First, 200 µL of buffer solution was pipetted into a reaction vial, and 100 µL of the clarified extract containing the analyte was pipetted, added and mixed thoroughly. A test strip was then added to the vial and allowed to run for 5 min. Test strips were immediately cut at the top of the arrow tape and inserted into a QuickScan® reader

with barcode facing down. A corn high sensitivity Matrix Group was selected and the result was read.

### 2.3.4. Pesticide Residue Analysis

Pesticide residue analysis was done using the QuCHERS method of analysis, which involved extraction, purification and quantification of the extract via the chromatographic method as described below.

1. Extraction

   Five grams of comminuted sample was weighed into a 50 mL centrifuge tube and 1 mL of deionized water was added, and then it was vortexed for 30 s. About 10 mL of acetonitrile was added, and it was vortexed again for 60 s. A mixture containing 4 g of 0.2 g magnesium sulphate anhydrous, 1 g of 0.05 g sodium chloride, 1 g of 0.05 g trisodium citrate dihydrate and 0.5 g of 0.03 g disodium hydrogen citrate sesquihydrate was added before being immediately vortexed for 1 min and centrifuged (Spinplus-6, China) for 5 min at 3000 U/min.

2. Dispersive Solid Phase Extraction

   A 6 mL aliquot of the extract was pipetted into a polypropylene centrifugation tube containing 150 mg primary and secondary amine and 900 mg magnesium sulphate. The tube was closed and shaken strongly for 30 s and centrifuged for 300 s at 3000 U/min. For matrixes containing low amounts of fat, freezing out and addition of 150 mg of carbon-18 was done.

About 4 mL of the cleaned extract was pipetted into a round bottom flask and the pH adjusted quickly to 5 by adding 40 μL of 5% formic acid solution in acetonitrile (*v/v*), and the filtrate concentrated below 40 °C on a rotary evaporator and 1 mL of ethyl acetate was added for re-dissolution. About 20 μL of 1% polyethylene glycol solution in ethyl acetate (*v/v*) was added and the extract transferred into a 2 mL standard opening vial for quantitation via GC-ECD and GC-PFPD. Qualitative confirmation for positive detection was done via GC/MS.

### 2.4. Data Analysis

Data from the study was analyzed for mean values, standard deviation and significant difference at 95% confidence level, using a one-way ANOVA from Statistical Package for Social Science (SPSS) software version 20.0.

## 3. Results

### 3.1. Grain Quality of Maize Samples

Grain quality traits employed in domestic transaction tend to imitate the grading categories stipulated by Ghana Standards Authority (GSA). Prominent quality parameters such as visual appearance, moisture content, grain color, dryness and cleanness are mostly paid attention to by stakeholders along the maize supply chain. The fraction of insect-damaged grains and organic and inorganic material are also considered by producers, aggregators, retailers and processors as the main maize grain quality parameters with reference to the Ghana Standards Authority criteria. Grain quality is also a function of the wholesomeness of grains as verified by several examinations from visual appearance to complex laboratory analysis. The quality standards for grain are country-specific, as different countries have different grades and standards to facilitate marketing and commercial values of produce [10].

Moisture content of maize is normally beyond 18% at harvest, and further reduction to 13% (ideal for storage) is achieved through drying [11]. Maize samples stored in hermetic and polypropylene bags for a period of one year from the three study locations recorded moisture content in the range of 10.9–12.1% and 13.8–14.9%, respectively. The moisture contents of maize stored in polypropylene bags were above the GSA specification ($\leq$13%), whereas those of hermetic storage were within the acceptable range.

All the maize grains sampled from the polypropylene and hermetic bags recorded low concentrations of organic and inorganic matter, stained, germinated and other grains (Figure 1). Maize grains sampled from the hermetic bags had lower defects in the grain quality attributes. The presence of other grains and organic and inorganic matter in a bag of maize informs the quality (purity) and level of adulteration. For this study both polypropylene and hermetic bags recorded low amounts of other grains (0.78 and 0.75%), inorganic matter (2.17 and 1.31%), organic matter (5.27 and 3.19%) and shriveled grains (3.52% and 1.45%), respectively. The differences were not significantly different ($p < 0.05$), implying equivalent levels of defects. However, the percentages of diseased grains in the polypropylene bags (1.54%) were significantly ($p < 0.05$) higher than those in the hermetic bags (0.04%). Similar observations of significant difference ($p < 0.05$) were recorded for polypropylene and hermetic bags for discolored maize grains (3.70% and 0.79%), broken/chipped maize (7.85% and 6.04%), insect-damaged grains (9.75% and 1.96%) and total defective grains (30.49% and 9.58%), respectively.

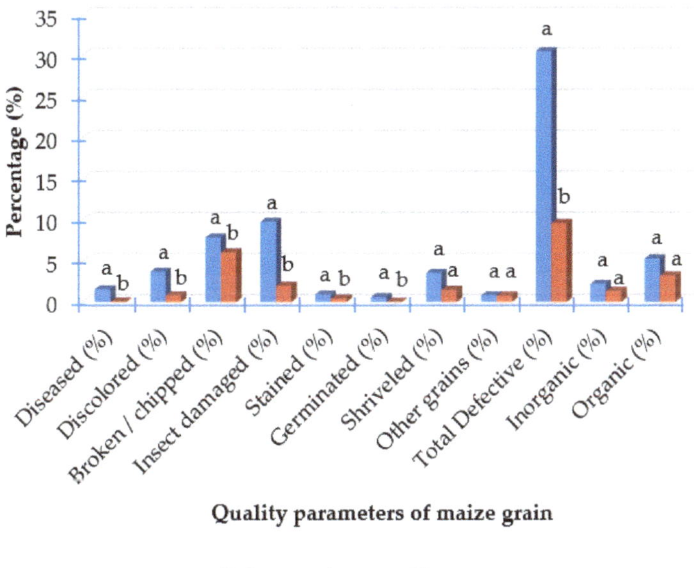


**Figure 1.** Percentage defects in grains stored in hermetic and PP bags. Means within the graph followed by the same letter are not significantly different at $p < 0.05$ (ANOVA: *t*-test).

### 3.2. Aflatoxin Content of Maize Samples

Levels of aflatoxin in maize samples stored in polypropylene and hermetic bags ranged between 0.9 and 20 ppb, as shown in Figure 2. Results from the aflatoxin analysis revealed that maize sampled from polypropylene bags had relatively higher aflatoxin levels (13.9–20 ppb) than those from hermetic bags (0.90–2.60 ppb). These values were significantly different ($p < 0.05$). Similar observations were made for aflatoxin levels in grains sampled at Ejura, PP bag (18 ppb), hermetic bag (2.6 ppb), and Asante Akyem Agogo, PP bag (13.9 ppb), hermetic bag (1.6 ppb), respectively.

All maize samples stored in the traditional (polypropylene) bags had aflatoxin concentrations above the recommended limit (15 ppb) for human consumption as reported by Omari et al. [12] whereas those of hermetic storage bags were below the limit.

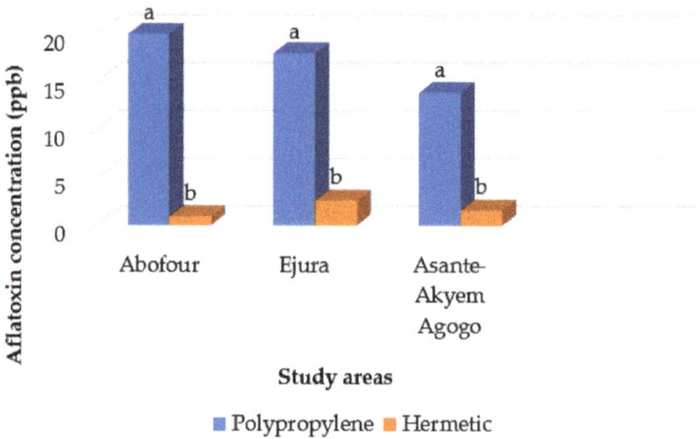



**Figure 2.** Aflatoxin concentration of maize stored in hermetic and polypropylene storage bags. Means within the graph followed by a different letter are significantly different at $p < 0.05$ (ANOVA: *t*-test).

### 3.3. Pesticide Residue Concentration in Maize Samples

Detected concentrations of the various pesticide residues in each maize sample from the three research areas (Abofour, Ejura and Asante-Akyem Agogo) are presented in Figures 3 and 4. Thirty-five pesticide residues were analyzed for two storage bags from each of the study areas. A total of 33 residues representing about 94.29% of the residues that were screened were found absent. These included bifenthrin, chlorpyrifos, dimethoate, permethrin, fenvalerate, profenofos, delta-HCH, fenpropathrin, p,p'-DDT, cyfluthrin, fonofos, ethoprophos, malathion, methoxychlor, chlorfenvinphos, heptachlor, lindane, p,p'-DDD, fenitrothion, dieldrin, endosulfan sulfate, alpha-endosulfan, p,p'-DDE, endrin, aldrin, beta-endosulfan, beta-HCH, diazinon, methamidophos, pirimiphos-methyl, gamma chlordane and parathion. The absence of organochlorine pesticide residues in the maize samples could be attributed to farmers' adherence to the ban on the application of organochlorine pesticides [13].

Lambda-cyhalothrin was detected in all maize samples stored in polypropylene and all were above the EU maximum residue limit of 0.02 mg/kg. Deltamethrin residues were detected in hermetic bags and were below the EU maximum residue limit of 2.0 mg/kg as reported by Milne [14].

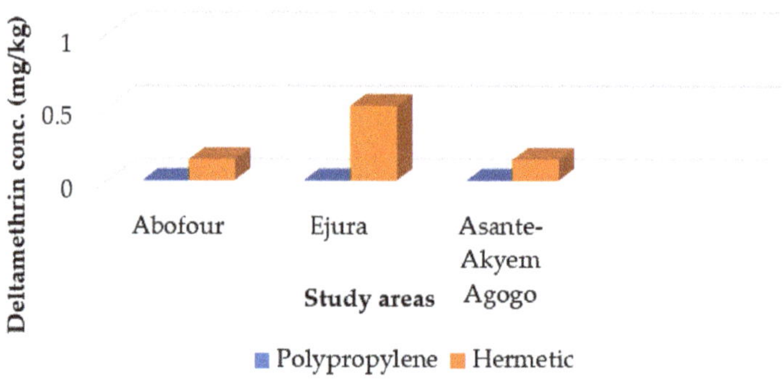



**Figure 3.** Deltamethrin residue in hermetic and polypropylene bags among study locations.

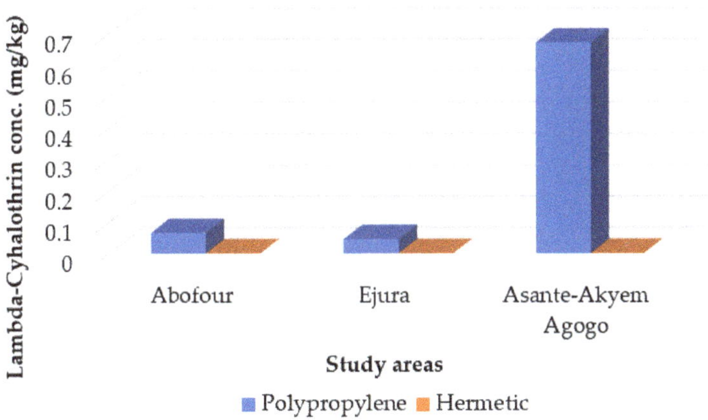



**Figure 4.** Lambda-cyhalothrin residue in hermetic and polypropylene bags among study locations.

## 4. Discussion

### 4.1. Grain Quality of Maize Samples

Generally, the recorded moisture content of the samples suggests adequate drying of the grains before storage by the warehouse operators; however, the relatively higher moisture content of maize samples stored in the polypropylene bags could be due to the gaseous exchange between the maize samples and the immediate environment in the storage area. Polypropylene bags have been reported to be porous in nature and therefore allows for moisture absorption or loss unlike hermetic bags that have a restrictive gaseous interchange barrier [9]. Moisture content beyond 13% encourages microbial growth and favors mycotoxin development, implying that maize grains stored in polypropylene bags will be susceptible to microbial and aflatoxin contamination compared to those stored in hermetic bags [8]. Gasparin et al. [10] and Bewley et al. [11] reported that control of maize grain moisture is the surest way of sustaining its viability, quality and safety throughout storage.

The variability in the organic and inorganic matter and stained and other grain qualities assessed could be attributed to decreased metabolic respiration/activity of mold in the hermetic bags compared to the polypropylene bags. Inorganic matter constitutes the presence of inanimate objects like stones, metals, plastics, cloth, etc. whereas organic matter takes into account wood, cobs, leaves, sticks, etc. in maize grains [15]. Both organic and inorganic matter are given keen attention by stakeholders in the maize value chain as they pose food safety threats to humans and animals aside from increasing the cleaning costs of processing industries. The decomposition of organic matter in maize adds to filth, stain and discolor of maize grains. The presence of diseased grains was significantly higher in the polypropylene bags than the hermetic bags. According to [16] this could be attributed to bacterial or fungal infections due to the presence of insect activities in the bag.

Discolored maize is grain that has an alteration in its regular (white or yellow) coloration to red, brown or a dark smear, which is usually influenced by excessive heat and/or excessive respiration [17]. The percentage of discolored grains in the polypropylene bags (3.70%) was significantly ($p < 0.05$) higher than that of the hermetic bags (0.79%). The observed discoloration could be attributed to respiration from insect and fungi activity within polypropylene bags and is an indication of a higher population of insects and fungi present in maize stored in the polypropylene bags compared to the grains in the hermetic bags. Fungi that occur in maize storage include members of the genera Aspergillus and Penicillium; their adaptation leads to the colonization of the embryo, which causes discoloration and rotting due to increased fatty acid content, oil rancidity and heating of the seed mass [18]. The grain quality analysis conducted on the stored maize samples revealed a significant ($p < 0.05$) concentration of broken/chipped maize in polypropylene bags (7.85%)

as compared to grains in hermetic bags (6.04%) in all the three study communities. The difference could be ascribed to the high moisture content of grains and poor post-harvest handling practices such as shelling, cleaning and winnowing. High moisture content has been found to significantly contribute to breakage of grains during shelling [19]. A similar study by Adu et al. [3] established that traditional shelling where maize cobs are packed in sacks and beaten with sticks usually results in an uncontrolled breakage of maize grains.

Mutungi et al. [7] reported that hefty sums of broken grain facilitate insects and microbial development and hence are undesirable in grain lots projected for longstanding storage. Maize grain processors lay emphasis on the amount of broken/chipped maize in their decision to accept or reject raw materials (maize) since it has the tendency to increase percentage grain losses in cleaning processes.

Maize samples stored in polypropylene bags recorded a higher amount of insect damaged grains (9.75%) in comparison to 1.96% recorded of sampled grains from the hermetic bags. Invasion of insects such as the maize weevil and the larger grain borer is responsible for such observation and was very profound in polypropylene bags due to its porous nature, permitting influx of oxygen for insect activity. These insects feed on maize endosperm leading to reduction in grain weight and end-product yield [8]. The lower insect damaged grains recorded in hermetic bags shows that it is better to store maize grains in hermetic bags than in polypropylene bags. The results further inform stakeholders along the maize value chain of the benefits of hermetic storage in the quest to reduce post-harvest losses in maize due to insect activity, which contributes to about 90% of post-harvest losses of maize globally according to [19].

Shriveled maize grains are underdeveloped, thin and papery in appearance, potentially resulting from a couple of factors such as soil and nutrient condition, moisture deficiency, drought and incidence of diseases [20,21]. Limiting growth factors that affect biomass and photosynthetic potential hinder the development of the reproductive organs of maize and consequently affect grain sizes. Results from the comparative analysis revealed that the percentages of shriveled grains in the polypropylene bags (3.52%) were not significantly different ($p > 0.05$) from those of the hermetic bags (1.45%), informing that the parameter is independent of the method of storage.

Maize grain quality assessment between the two methods of storage revealed a significant difference between the overall or total defective grains found in the polypropylene (30.49%) and hermetic bags (9.58%). The lower percentage of total defective grains in hermetic bags makes it a better option over polypropylene bags for maize storage.

### 4.2. Aflatoxin Content of Maize Samples

Aflatoxin contamination in maize usually occurs in two different phases: pre-harvest and post-harvest contamination. High humidity, insufficient grain drying, high temperatures and poor storage surroundings are typical causes of aflatoxin development [9]. Efficient post-harvest management of maize is an important factor in mitigating post-harvest storage-related losses. Aflatoxin in maize significantly affects the market value of maize and threatens consumer health and food security. A study by Bakoye et al. [22] revealed that aflatoxin contamination is not directly correlated to moisture content but emphasized moldy grains, foreign materials, and the prevalence of insects as a function of aflatoxin contamination in grains.

Mutambuki et al. [23] explained that hermetically sealed containers operate on a phenomenon of restricting $O_2$ availability to microbes and insects already in cereal grains upon storage. The elimination of oxygen is primarily achieved through the exchange of gases between cereals, insects and microbes inside airtight containers; respiration within the airtight container leads to a reduction in oxygen volumes with an increase in carbon dioxide volumes, causing suffocation and subsequent death of insects and microbes. Gaseous exchange within polypropylene sacks is unrestricted, as the porosity of the bags allows free movement of oxygen and carbon dioxide in and out of the bag, ensuring balance in respiration among maize grains, insects and microbes [24]. The accessibility of oxygen by microbes (fungi) supports their growth and consequently increases aflatoxin levels in the maize samples stored in the polypropylene sacks. The results obtained from the present study show that hermetic storage bags have competitive advantages over polypropylene bags in terms of aflatoxin prevention.

*4.3. Pesticide Residue Concentration in Maize Samples*

The detection of lambda-cyhalothrin and deltamethrin, all belonging to the synthetic pyrethroid class of pesticides, in sampled maize grains could be as a result of the substitution of organochlorines with a more biodegradable option of synthetic pyrethroids. Results from this study corroborate a study by Bempah et al. [25] that detected pyrethroid residues in fruits and vegetables, which emphasizes a signal of a paradigm shift in the usage of pesticides in Ghana from organochlorine to less toxic, biodegradable pyrethroid pesticides. Dziembowska et al. [26] reported that pyrethroids have a high efficacy of about 2250 counts and are particularly lethal to insects compared to advanced animals. The detection of lambda-cyhalothrin above its stipulated maximum residue limit of 0.02 mg/kg suggests the possibility of misapplication and abuse of the insecticide. The detection could also originate from environmental contamination as a result of previous agricultural activities (such as chemical spraying against weeds and insects) in the growing communities. Pyrethroids usually exhibit low toxicity with respect to humans, characterized by a speedy breakdown in adults, as they do not bio-accumulate in adult tissues and are expelled out of the body through urine [26]. Since pyrethroid insecticide residues have shown some form of toxicity to humans, bioaccumulation along the food chain may subject an exposed population to harmful long-term health hazards.

## 5. Conclusions

The study discovered that maize grains stored in hermetic bags recorded lower aflatoxin and pesticide residue concentrations and higher grain quality than those stored in polypropylene bags with respect to diseased, discolored, broken/chipped, insect-damaged, stained, germinated, shriveled, other grains, total defective, inorganic and organic matter. Only lambda-cyhalothrin and deltamethrin were detected in maize stored in polypropylene and hermetic bags, respectively. Lambda-cyhalothrin showed residue levels higher than its maximum residue limit (MRL) of 0.02 mg/kg and this poses safety issues for consumers, whilst maize samples stored in hermetic bags had deltamethrin residues below the MRL of 2.00 mg/kg. The findings point to the many benefits of the use of hermetic bags over polypropylene bags in maize grain storage and the urgent need to establish reliable monitoring programs for pesticides so that any exceedance in concentration over quality standards can be detected with appropriate actions taken.

Further research could focus on evaluating pesticide residue concentrations of maize from production through to the point of entry into the market to establish at what point(s) along the maize supply chain pesticides are being introduced.

**Author Contributions:** The following authors contributed to the work: Conceptualization, J.O.A. and F.D.W.-M.; data curation, S.K.N. and Y.G.A.; formal analysis, S.K.N. and Y.G.A.; funding acquisition, J.O.A.; investigation, S.K.N.; methodology, S.K.N.; project administration, F.D.W.-M.; resources, J.O.A. and F.D.W.-M.; Supervision, F.D.W.-M.; validation, Y.G.A.; visualization, Y.G.A.; writing—original draft, S.K.N.; writing—review & editing, Y.G.A., J.O.A. and F.D.W.-M. All authors have read and agreed to the published version of the manuscript.

**Funding:** Funding for this study was provided under grants from The Rockefeller Foundation (grant 2018 FOD 004) and the Foundation for Food and Agriculture Research (grant DFs-18-0000000008).

**Institutional Review Board Statement:** Not applicable.

**Informed Consent Statement:** Not applicable.

**Data Availability Statement:** Not applicable.

**Conflicts of Interest:** The authors declare no conflict of interest.

## References


1. Ragasa, C.; Chapoto, A.; Kolavalli, S. Maize productivity in Ghana. *Int. Food Policy Res. Inst.* **2014**, *5*, 1–3.
2. Opit, G.P.; Campbell, J.; Arthur, F.; Armstrong, P.; Osekre, E.; Washburn, S.; Baban, O.; McNeill, S.; Mbata, G.; Ayobami, I.; et al. Assessment of maize postharvest losses in the middle belt of Ghana. In Proceedings of the 11th International Working Conference on Stored Product Protection, Chiang Mai, Thailand, 24–28 November 2014; pp. 860–868.
3. Adu, G.B.; Abdulai, M.S.; Alidu, H.; Nustugah, S.K.; Buah, S.S.; Kombiok, J.M.; Obeng-Antwi, K.; Abudulai, M.P.M.E. Recommended Production Practices for Maize in Ghana. Available online: http://sari.csir.org.gh/wp-content/uploads/2020/01/Recommended-practices-for-maize-production-in-Ghana-web-1.pdf (accessed on 20 March 2021).
4. Likhayo, P.; Bruce, A.Y.; Tefera, T.; Mueke, J. Maize Grain Stored in Hermetic Bags: Effect of Moisture and Pest Infestation on Grain Quality. *J. Food Qual.* **2018**, *2018*, 2515698. [CrossRef]
5. Food and Agriculture Organization of the United Nations (FAO). *FAOSTAT Online Statistical Service*; FAO: Rome, Italy, 2012; Available online: https://faostat.fao.org/ (accessed on 15 January 2021).
6. Fianko, J.R.; Donkor, A.; Lowor, S.T.; Yeboah, P.O.; Glover, E.T. Health Risk Associated with Pesticide Contamination of Fish from the Densu River Basin in Ghana. *J. Environ. Prot.* **2011**, *2*, 115–123. [CrossRef]
7. Mutungi, C.; Muthoni, F.; Bekunda, M.; Gaspar, A.; Kabula, E. Physical quality of maize grain harvested and stored by smallholder farmers in the Northern highlands of Tanzania: Effects of harvesting and pre-storage handling practices in two marginally contrasting. *J. Stored Prod. Res.* **2019**, *84*, 1–12. [CrossRef]
8. Tefera, T.; Teshome, A.; Singano, C. Effectiveness of Improved Hermetic Storage Structures Against Maize Storage Insect Pests Sitophilus zeamais and Prostephanus truncatus. *J. Agric. Sci.* **2018**, *10*, 100–106. [CrossRef]
9. Walker, S.; Jaime, R.; Kagot, V.; Probst, C. Comparative effects of hermetic and traditional storage devices on maize grain: Mycotoxin development, insect infestation and grain quality. *J. Stored Prod. Res.* **2018**, *77*, 34–44. [CrossRef]
10. Gasparin, E.; Araujo, M.M.; Tolfo, C.V.; Foltz, D.R.B.; Magistrali, P.R. Substrates for germination and physiological quality of storage seeds of *Parapiptadenia rigida* (Benth.) Brenan. *J. Seed Sci.* **2013**, *35*, 77–85. [CrossRef]
11. Bewley, J.D.; Bradford, K.J.; Hilhorst, H.W.; Nonogaki, H. Longevity, storage, and deterioration. In *Seeds*; Springer: New York, NY, USA, 2013; pp. 341–376.
12. Omari, R.; Tetteh, E.K.; Baah-Tuahene, S.; Karbo, R.; Adams, A.; Asante, I.K. Aflatoxins and their Management in Ghana: A Situational Analysis. *FARA Res. Rep.* **2020**, *5*, 1–80.
13. Blair, A.; Ritz, B.; Wesseling, C.; Freeman, L.B. Pesticides and human health. *Occup. Environ. Med.* **2014**, *72*, 81–82. [CrossRef] [PubMed]
14. Milne, M. *Pesticide Residues: Maximum Residue Limits.* 2013. Available online: http://extwprlegs1.fao.org/docs/pdf/tha161066.pdf (accessed on 14 March 2021).
15. Ghana Standards Authority. *National Aflatoxin Sensitization and Management (NASAM) Project*; Ghana Standards Authority: Accra, Ghana, 2013.
16. Asare-Bediako, E.; Kvarnheden, A.; Van der Puije, G.C.; Taah, K.J.; Agyei-Frimpong, K.; Amenorpe, G.; Appiah-Kubi, A.; Lamptey, J.N.L.; Oppong, A.; Mochiah, M.B.; et al. Spatio-temporal variations in the incidence and severity of maize streak disease in the Volta region of Ghana. *J. Plant Pathol. Microbiol.* **2017**, *8*, 1–7.
17. Fernandez, M.; Wang, H.; Singh, A. Impact of seed discolouration on emergence and early plant growth of durum wheat at different soil gravimetric water contents. *Can. J. Plant Pathol.* **2014**, *36*, 509–516. [CrossRef]
18. Thiessen, L.D.; Woodward, J.E. Diseases of peanut caused by soil borne pathogens in the southwestern United States. *ISRN Agron.* **2012**, *2012*, 517905. [CrossRef]
19. Darfour, B.; Rosentrater, K.A. Maize in Ghana: Maize in Ghana: An Overview of Cultivation to Processing. In Proceedings of the ASABE International Meeting, Orlando, FL, USA, 17–20 July 2016; pp. 1–16.
20. Asfaw, A.; Almekinders, C.J.; Struik, P.C.; Blair, M.W. Farmers' common bean variety and seed management in the face of drought and climate instability in southern Ethiopia. *Sci. Res. Essays* **2013**, *8*, 1022–1037.
21. Gao, Z.; Liu, H.; Wang, H.; Li, N.; Wang, D.; Song, Y.; Song, C. Generation of the genetic mutant population for the screening and characterization of the mutants in response to drought in maize. *Chin. Sci. Bull.* **2014**, *59*, 766–775. [CrossRef]
22. Bakoye, O.N.; Baoua, I.B.; Seyni, H.; Amadou, L.; Murdock, L.L.; Baributsa, D. Quality of maize for sale in markets in Benin and Niger. *J. Stored Prod.* **2017**, *71*, 99–105. [CrossRef] [PubMed]
23. Mutambuki, K.; Affognon, H.; Likhayo, P.B.D. Evaluation of Purdue Improved Crop Storage Triple Layer Hermetic Storage Bag against Prostephanus. *Insects* **2019**, *10*, 204. [CrossRef] [PubMed]
24. Ndegwa, M.; De Groote, H.; Gitonga Zachary, B.A. Effectiveness and Economics of Hermetic Bags for Maize Storage: Results of a Randomized Controlled Trial in Kenya. *Int. Conf. Agric. Econ.* **2015**, *307*, 3–6. [CrossRef]
25. Bempah, C.K.; Asomaning, J.B.J. Market basket survey for some pesticides residues in fruits. *J. Microbiol. Biotechnol. Food Sci.* **2012**, *2*, 851–855.
26. Dziembowska, I.; Bogusiewicz, J. Current Research on the Safety of Pyrethroids Used as Insecticides. *Medicina* **2018**, *54*, 61.

Article

# Quantitative Evaluation of Color, Firmness, and Soluble Solid Content of Korla Fragrant Pears via IRIV and LS-SVM

Yuanyuan Liu [1,2,*], Tongzhao Wang [1,2], Rong Su [1,2], Can Hu [1,2], Fei Chen [1,2] and Junhu Cheng [3]

[1] College of Mechanicaland Electrical Engineering, Tarim University, Alar 843300, China; Wtz080921@163.com (T.W.); 10757202211@stumail.taru.edu.cn (R.S.); 120140004@taru.edu.cn (C.H.); 10757192126@stumail.taru.edu.cn (F.C.)

[2] Agricultural Engineering Key Laboratory at Universities of Education Department of Xinjiang Uygur Autonomous Region, Tarim University, Alar 843300, China

[3] College of Light Industry and Food Sciences, South China University of Technology, Guangzhou 510641, China; fechengjh@scut.edu.cn

* Correspondence: 120080015@taru.edu.cn

**Abstract:** Customers pay significant attention to the organoleptic and physicochemical attributes of their food with the improvement of their living standards. In this work, near infrared hyperspectral technology was used to evaluate the one-color parameter, a*, firmness, and soluble solid content (SSC) of Korla fragrant pears. Moreover, iteratively retaining informative variables (IRIV) and least square support vector machine (LS-SVM) were applied together to construct evaluating models for their quality parameters. A set of 200 samples was chosen and its hyperspectral data were acquired by using a hyperspectral imaging system. Optimal spectral preprocessing methods were selected to obtain out partial least square regression models (PLSRs). The results show that the combination of multiplicative scatter correction (MSC) and Savitsky-Golay (S-G) is the most effective spectral preprocessing method to evaluate the quality parameters of the fruit. Different characteristic wavelengths were selected to evaluate the a* value, the firmness, and the SSC of the Korla fragrant pears, respectively, after the 6 iterations. These values were obtained via IRIV and the reverse elimination method. The correlation coefficients of the validation set of the a* value, the firmness, and the SSC measure 0.927, 0.948, and 0.953, respectively. Furthermore, the values of the regression error weight, $\gamma$, and the kernel function parameter, $\sigma^2$, for the same parameters measure ($8.67 \times 10^4$, $1.21 \times 10^3$), ($1.45 \times 10^4$, $2.93 \times 10^4$), and ($2.37 \times 10^5$, $3.80 \times 10^3$), respectively. This study demonstrates that the combination of LS SVM and IRIV can be used to evaluate the a* value, the firmness, and the SSC of Korla fragrant pears to define their grade.

**Keywords:** IRIV; LS-SVM; Korla fragrant pear; quality parameter; evaluation

**Citation:** Liu, Y.; Wang, T.; Su, R.; Hu, C.; Chen, F.; Cheng, J. Quantitative Evaluation of Color, Firmness, and Soluble Solid Content of Korla Fragrant Pears via IRIV and LS-SVM. *Agriculture* **2021**, *11*, 731. https://doi.org/10.3390/agriculture11080731

Academic Editor: Isabel Lara

Received: 21 June 2021
Accepted: 26 July 2021
Published: 31 July 2021



## 1. Introduction

Korla fragrant pears are very popular among customers due to their thin skin, juicy, sweaty taste, and delicate flesh [1,2]. Nowadays, customers pay significant attention to both the organoleptic and physicochemical attributes of fruits with the improvement of their living standards. The organoleptic parameter, color of skin, is related to maturity of Korla fragrant pear. The sunward side of most mature Korla fragrant pears has blush which is also distinctive in all kinds of pears. However, only physicochemical parameters are used as quality evaluation attributes to grade Korla fragrant pears.

Several non-destructive studies have been carried out to evaluate the soluble solid content (SSC) of Korla fragrant pears [3,4]. Zhu et al. [5] used hyperspectral imaging and support vector regression to define this parameter. The correlation coefficient ($R_C$) and the root mean square error ($RMSE_C$) in their calibration set measured 0.986 and 0.186%, respectively. In their validation set the correlation coefficient ($R_V$) and the root mean square error ($RMSE_V$) measured 0.946 and 0.403%. Zhan et al. [6] quantitatively determined the

SSC of Korla fragrant pears via least square support vector machine (LS-SVM) and partial least square regression (PLSR). The $R_V$ and $RMSE_V$ reported in this study measure 0.851 and 0.291%, respectively.

Other researchers investigated the firmness of Korla fragrant pears vie quantitative predictions. For instance, Sheng et al. [7] used near-Infrared (NIR) spectroscopy together with different variable selecting methods to construct a set of partial least square models to describe firmness. Yu et al. [8] predicted both the firmness and the SSC by developing a deep learning method based on Vis/NIR hyperspectral reflectance imaging. Their combination model of a series of stacked auto-encoders and a fully connected neural network achieved a reasonable prediction performance with $R_V$ and $RMSE_V$ values of 0.9434 and 1.81 N, respectively.

However, no investigation reported results on the simultaneous measurement of the organoleptic and physicochemical attributes on Korla fragrant pears. According to the requirements of the latest group standard on Korla fragrant pears [9], organoleptic and physicochemical attributes appear to have the same importance in the grade definition. The skin color of Korla fragrant pears changes from green to red-yellow, as the fruit ripes. The a* value represents the color change from red to green in chromatic aberration data. Therefore, the organoleptic quality of the samples can be defined according to their a* values. The firmness and the SSC are the most significant edible quality parameters in Korla fragrant pears, and they are directly related to consumers' satisfaction [10]. Thus, the three parameters, a*, firmness, and SSC, must be carefully evaluated to determine the influence of the postharvest storage period on the fruit quality control process.

Both quality and safety parameters can be accurately evaluated via hyperspectral imaging [11–13], although the hyperspectral approach requires expensive equipment and complex data analysis Compared with other nondestructive testing methods [14]. However, in order to define simple predicting models and improve their prediction efficiency, a set of wavebands have to be selected. These wavebands can be related with several important chemical bonds, which can be used to discriminate the samples based on their quality and safety parameters. Successive projection algorithms (SPAs) [15–18], competitive adaptive reweighting sampling (CARS) [19,20], and uninformative variable elimination (UVE) [21,22] have been used by to choose such wavebands. Despite these selection methods are quite effective, they do not account for the combination effects among the wavebands. The iteratively retaining informative variables (IRIV) method ensures that each variable has the same probability to take part into the selection process and increases filtering speed by using a set of binary mixing filters [23,24].

To this date, the combination of IRIV and LS-SVM has not been investigated to quantitatively predict the quality parameters of Korla fragrant pears. In this work, IRIV-LS-SVM is used to (1) obtain the a* value, the firmness, and the SSC of Korla fragrant pears, (2) analyze the spectral features of Korla fragrant pears in the 945–1670 nm wavelength range, (3) select the optimal wavebands related to the C-H, N-H, and O-H chemical bonds, and (4) construct a set of predicting models to define the quality parameters for Korla fragrant pears.

## 2. Materials and Methods

### 2.1. Korla Fragrant Pears and Pretreatment

Korla fragrant pears were collected from a plantation located near Tarim University (80°30′–81°58′ E, 40°22′–40°57′ N) from September 11th to September 15th 2019. A set of 200 samples with a uniform shape, a single fruit weight of 120 ± 10 g, and intact epidermis was selected. The side of each Korla fragrant pear, which was exposed to the sunlight, was labeled.

The samples were sprayed with a special fruit cleaning agent (Almawin, Germany), soaked in water for about 30 s, and then rinsed with distilled water twice. The cleaned pears were dried at room temperature (20 °C), and then stored in a preservation box at

4 °C. The samples were placed on the desk at room temperature for 30 min to eliminate the influence of the temperature change before the hyperspectral image data acquisition.

*2.2. Hyperspectral Imaging System and Diffuse Reflectance Spectrum Data Acquisition*

The hyperspectral imaging system used in this study is shown in Figure 1. It consists of a push-broom scanning system composed of a spectrograph (N17E, Spectral Imaging Ltd., Oulu, Finland), an enhanced near-infrared hyperspectral camera (Xeva-1.7-320, Xenics Infrared Solutions, Leuven, Belgium), four halogen light sources with a maximum power of 150 W each, a stepper-motor-driving stage, a dark box, and a computer.

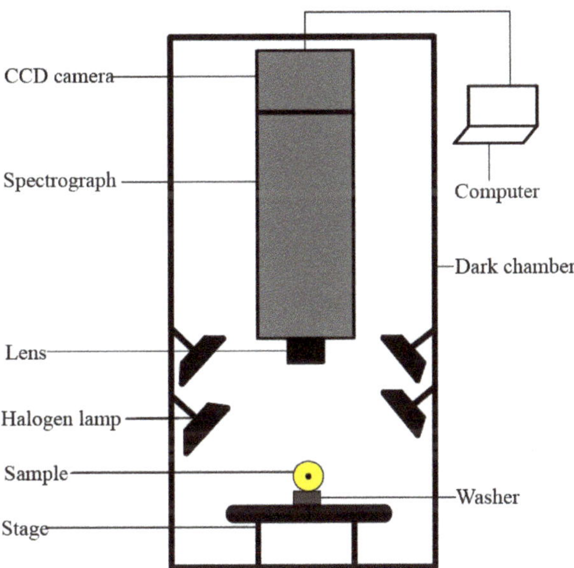


**Figure 1.** Schematic diagram of the hyperspectral imaging system.

Before data acquisition, the system was preheated for about 30 min to ensure its temperature stability. A sample with a uniform shape and a moderate weight was placed on the stage to adjust the calibration parameters of the instrument. In order to ensure the hyperspectral image integrity of the samples, the distance between the bottom of the spectrograph and the stage was set to 310 mm and maintained fixed. Moreover, the focal length was adjusted by rotating the aperture until sharp reflection peaks appeared. The moving speed of the platform and the exposure time of the camera were set to 18 mm/s and 20 ms, respectively.

A strip-shaped standard reflecting whiteboard was placed under the spectrograph to obtain the white and the black references separately by opening and closing the lens cover. The long axis of a Korla fragrant pear sample was positioned along the moving direction of the stage to ensure a uniform irradiation. The sunward side of each Korla fragrant pear was placed upside in order to reduce the influences on spectral data and measured quality parameter values of the sunward side and nightside. The sample hyperspectral image data were corrected by taking into account the black and white references to eliminate the influence of the light source intensity differences and the camera dark current noise, as described in Formula (1).

$$I = (I_o - I_b)/(I_w - I_b) \tag{1}$$

Here, $I_o$ corresponds to the original hyperspectral image data, $I_b$ to the black reference data collected when the lens cover is closed, $I_w$ refers to the white reference image data of the strip-shaped standard reflecting whiteboard when the lens cover is open.

The spectral data of the region of interest (ROI) were extracted by using the ENVI 5.1 software (Exelis Visual Information Solutions, Boulder, Colorado, USA). The shape of the ROI was rectangular, and its center was located near the intersection between the long axis and the equator of the pear. The corresponding pixel numbers of each ROI were 90 along the long axis and 70 along the equator.

### 2.3. Measurement of the Sensory and Physicochemical Parameters

The sensory and physicochemical parameters of the Korla pears were measured after the hyperspectral image data acquisition. The sensory parameter a* was obtained by employing a precision chromatic aberration meter (HP-C220, Shenzhen HanPu Testing Instrument Co., Ltd., Shenzhen, China). Each measurement consisted of an average of five points randomly selected on the ROI surface.

The firmness was obtained by averaging the values collected at five different locations of the pears. They were set at 12 mm center distance between two adjacent ROIs and were measured by a firmness tester (GY-4 with a probe diameter of 7.9 mm, Top instrument). The SSC was measured by using a digital refractometer (PAL-1, ATAGO, Tokyo, Japan). Before the measurement, the refractometer was calibrated with distilled water. Three small pieces of pulp of about 5 g each were cut out from the ROI. Their liquid content was dropped into a sample tank by manual extrusion. The average value of the solid content of the three pulp samples was taken as the measurement value.

### 2.4. Spectral Preprocessing

The standard normal variable transformation (SNV) is a normalization, which is sometimes employed in near infrared spectroscopy [25,26]. This preprocessing algorithm can center and scale each spectrum. Multiple scatter correction (MSC) is used to compensate for the non-uniform scattering effects in spectral data, when heterogenous sample sizes, irregular distributions, and other physical effects are present [27]. Whereas the Savitsky-Golay (S-G) algorithm can be used to improve smoothness of spectral curves. The different preprocessing effects obtained with MSC, SNV, MSC-SG, and SNV-SG were compared to evaluate the characteristics of the PLSR models.

### 2.5. Division Calibration Set and Validation Set

The sample set partitioning method based on the joint x-y distance algorithm (SPXY) was proposed by Galvão et al. [28]. This algorithm considers the reflection spectrum distribution and the standard value distribution equally important in the data characterization process by increasing the representativity of both the calibration and validation set. In this study, the calibration set and the validation set were grouped by SPXY with a 3:1 ratio.

### 2.6. Selection of Important Wavelengths

The principal components were determined by using the partial least square regression (PLSR) models established via the 5-fold cross-validation method to select the most significant wavelengths in different iterations. The process of selecting the important wavelengths for one quality parameter in Round I is shown in Figure 2.

Here, $Spec_{(I-1)in}$ corresponds to the matrix of the spectral data, which is composed of the set of wavelengths selected during the last iteration, $Y_k$ refers to the measurement value matrix of the $k^{th}$ quality parameter. $N_I$, and $C_I$ in the figure correspond to the binary matrix lines in the $I^{th}$ iteration, and the optimal wavelengths selected in the $(i-1)^{th}$ iteration, respectively, $Num\_total_I$ is the total number of uninformative and interfering wavelengths.

According to the number of wavelengths selected in the $(i-1)^{th}$ iteration, a binary matrix shuffler filter, $M_{Iin}$, with $C_I$ columns and $N_I$ rows for Round I was generated. The value of $M_{Iin}(i,j)$ indicates that Wavelength i is used to construct the predicting quality model j. The root mean square error $RMSECV_{Iin}(:,j)$ for the $N_I$ possible wavelength combinations was calculated separately. Each $RMSECV_{Iin}(:,j)$ value was set as $RMSECV_{Iin}(i,j)$. The binary matrix $M_{Iex}$ was obtained by inverting the elements of $M_{Iin}$, implying a change in the including state of the sample spectrum for its corresponding wavelength. A new PLSR prediction model and its corresponding root mean square error $RMSECV_{Iex}(i,j)$ was calculated when the inclusion state of wavelength j changed into the i[th] wavelength.

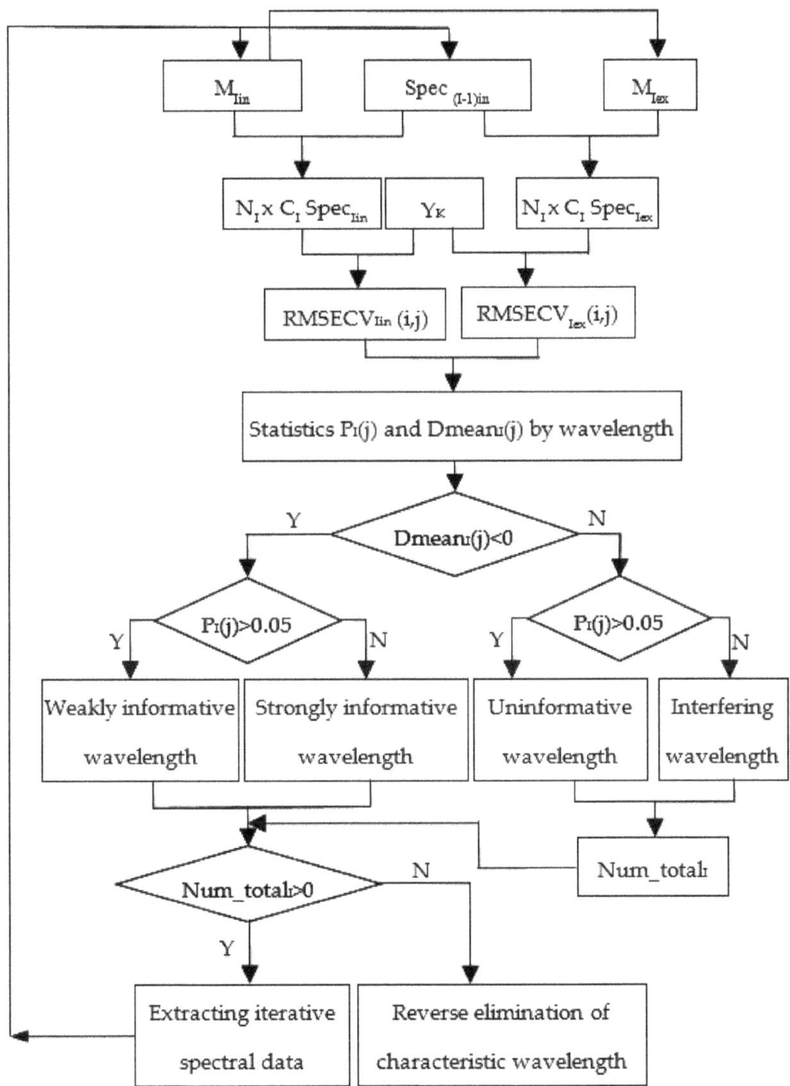



**Figure 2.** Iteration process of Round I.

The values of $RMSECV_{Iin}(i,j)$ and $RMSECV_{Iex}(i,j)$ of the i[th] wavelength combination with and without including the wavelength j were calculated according to the $M_{Iin}$ and $M_{Iex}$ values. $RMSECV_{Iex}(i,j)$ and $RMSECV_{Iin}(i,j)$ were tested via the Mann–Whitney U test with a significance level of 0.05. The difference between the two values of the wavelength j was defined as $Dmean_I(j)$. The wavelengths were classified into four types with the test level $P_I(j)$ and $Dmean_I(j)$, as shown in Table 1. Strongly informative wavelengths can be used in to drive prediction models, contrarily to weakly informative wavelengths. Interfering wavelengths create noise inside the model and lower significantly its performance, whereas uninformative wavelengths play the same role of interfering wavelengths but have a lower effect on the model performance.

**Table 1.** Variable classification rules.

| Wavelength Type | Classification Rules |
| --- | --- |
| Strongly informative wavelength | Dmean(j) < 0, P(j) < 0.05 |
| Weakly informative wavelength | Dmean(j) < 0, P(j) > 0.05 |
| Uninformative wavelength | Dmean(j) > 0, P(j) > 0.05 |
| Interfering wavelength | Dmean(j) > 0, P(j) < 0.05 |

When DmeanI(j) was smaller than 0, its corresponding wavelength was entered into a new iteration. When the number of uninformative and interfering wavelengths (Num_totalI) was smaller than 0, the iteration stopped and the RMSECV value was calculated using the spectra with strongly and weakly informative wavelengths together with their quality values.

Reverse elimination was then performed. When either a strongly informative wavelength or a weakly one was eliminated, a new set of PLSR models was established and the corresponding RMSECV' values were obtained. If the RMSECV' was smaller than the RMSECV, the corresponding wavelength was eliminated and remaining wavelengths were defined as important ones.

### 2.7. Modeling Algorithm

The least square support vector machine (LS-SVM) is an improved SVM algorithm proposed by Suykens [29]. Its operation speed can be significantly improved by solving a set of linear equations instead of the complex quadratic programming problem of the SVM. In this work, the radial basis function (RBF) was used as the kernel function, and the combination of the regression error weight, $\gamma$, and the kernel function parameter, $\sigma^2$, were optimized via grid search based on the cross-validation model. The quality parameters of the LS-SVM models were evaluated by using the $RMSE_C$, $R_C$, $RMSE_V$, and $R_V$ values. The results show that the model performs better when $RMSE_C$ and $RMSE_V$ are small and $R_C$ and $R_V$ are large.

## 3. Results

### 3.1. Statistics and Analysis of the Sensory and Physicochemical Values

The statistic values of a*, the firmness, and the SSC of Korla fragrant pears are shown in Table 2. The value of a* lies in the −7.108–3.254 range. When a* is positive the color of the tested area is red, whereas when a* is negative is green. The firmness lies in the $10.4 \times 10^5$–$14.1 \times 10^5$ Pa range. This value is larger than that measured in other studies [30,31] probably because, in this work, the skin of the pears was not removed. This method was preferred since it meets the most common eating habits of the customers, who generally eat the pears with the skin to increase their uptake of vitamin C. The SSC lies in the 10.0–13.4 °Brix range. Such range is narrower that the one defined by Yu X J et al. and Li J B et al. probably due to the differences in planting locations. On the other hand, the value ranges in the calibration set include those in the other set: Both sets, in fact, are representative since the mean values and dispersion degree of the two sets are similar.

**Table 2.** Statistics of the quality parameters in the calibration and validation sets.

| Quality Parameters | Group | Min | Max | Mean Value | Standard Deviation |
|---|---|---|---|---|---|
| a* | Correction set | −7.108 | 3.254 | −3.459 | 0.987 |
| | Verification set | −5.794 | 2.282 | −3.989 | 0.997 |
| Firmness ($10^5$ Pa) | Correction set | 10.4 | 14.1 | 12.1 | 0.760 |
| | Verification set | 10.8 | 13.4 | 11.3 | 0.637 |
| SSC (°Brix) | Correction set | 10.0 | 13.4 | 12.1 | 0.693 |
| | Verification set | 11.5 | 13.2 | 12.2 | 0.443 |

The a* color space method recommended by the International Commission on illumination (CIE) used "*" in the expression of three parameters.

### 3.2. Spectrum Data Processing

#### 3.2.1. Spectral Curves

The spectral curves with the largest distance in most wavelengths are shown in Figure 3. The measured values of a*, of the firmness, and of the SSC of sample 1 and sample 2 are 3.194, $13.9 \times 10^5$ Pa, and 12.0 °Brix and −6.934, $10.4 \times 10^5$ Pa, and 10.1 °Brix, respectively. Three reflection valleys can be observed near 1140 nm, 1440 nm, and 1640 nm, whereas two reflection peaks are located at 960 nm and 1270 nm. A water absorption band exists near 960 nm [32]. The reflection valleys near 1140 nm and 1640 nm may correspond to the first and second overtones of the C-H group, respectively [33]. The strong reflection valley at 1440 nm can be assigned to the first overtone of the O-H and N-H bonds [34]. The reflection peaks near 1270 nm may be related to the second overtones of the O-H and C-H bonds, respectively [35].

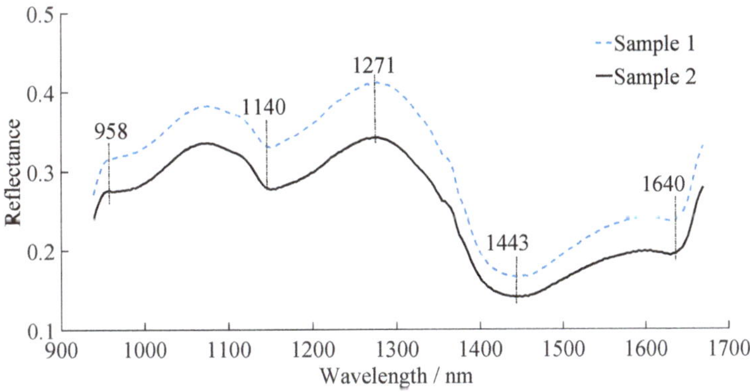


**Figure 3.** Reflective spectral curves.

#### 3.2.2. PLSR Models for the Quality Parameters and Optimization of the Principal Components Based on the Full Spectral Analysis

The PLSR models for the a* value, the firmness, and the SSC of Korla fragrant pears were obtained by analyzing the spectral data after different spectral pre-processing processes. The spectra after pretreatment with MSC-SG are shown in Figure 4. The selection process of the numbers of principal components is shown in Figure 5. The principal components to determine the a* value, the firmness, and the SSC are 10, 8, and 9, respectively. The prediction results are listed in Table 3. The results show that the PLSR models with MSC-SG pretreatment exhibit the highest evaluating ability. The $R_C$ and $RMSE_C$ values obtained for a* measure 0.907 and 0.448, respectively, in the case of the calibration set, whereas $R_V$ and $RMSE_V$ measure 0.894 and 0.402 when the validation set is used. The $R_C$ and $RMSE_C$ values of the firmness are 0.914 and $0.352 \times 10^5$ Pa, respectively, for the calibration set and the $R_V$ and $RMSE_V$ values of 0.903 and $0.317 \times 10^5$ Pa, respectively, are obtained from the validation set. The $R_C$ and $RMSE_C$ of the SSC measure 0.925 and 0.314 °Brix, respectively, when the calibration set is considered, whereas $R_V$ and $RMSE_V$ measure 0.912 and 0.301 °Brix, respectively, in the case of the validation set.

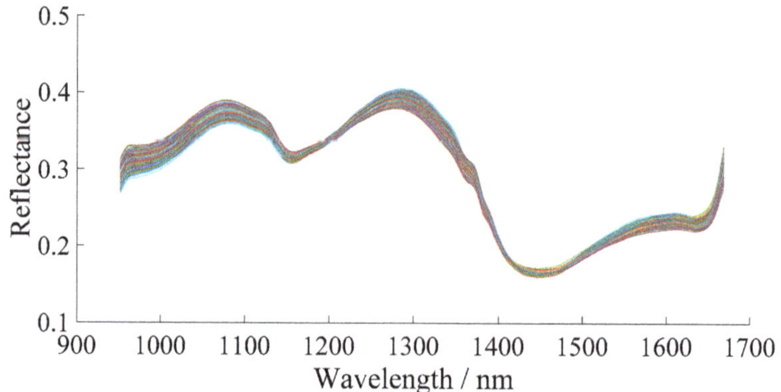

**Figure 4.** Spectra after the MSC-SG preprocessing.

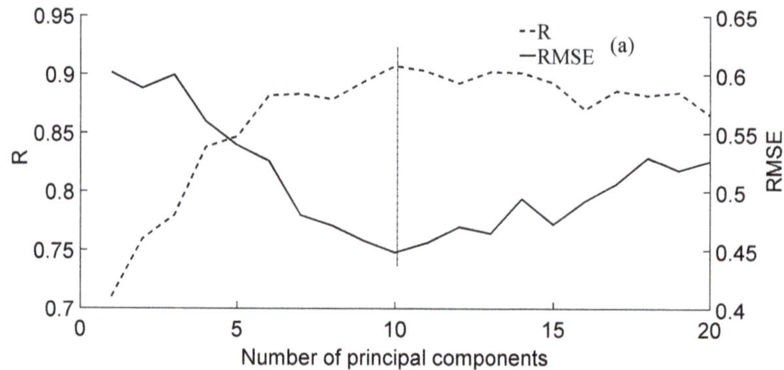

**(a)**

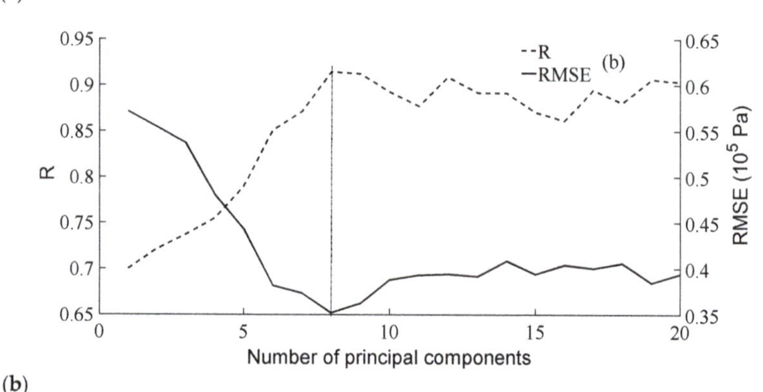

**(b)**

**Figure 5.** *Cont.*

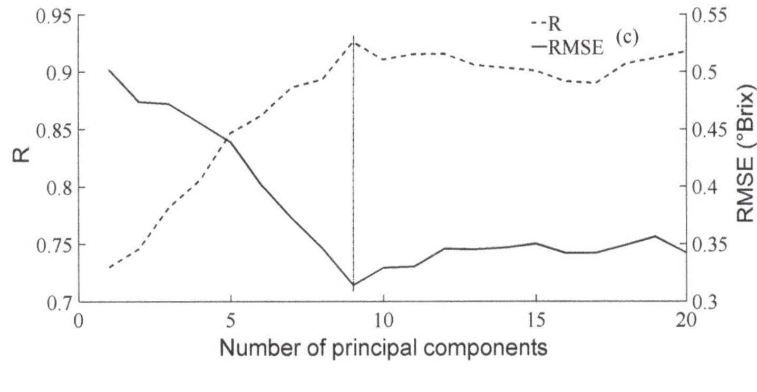


(c)

**Figure 5.** Selection process of principal components based on MSC-SG. (**a**) Selecting process to estimate the a* value. (**b**). Selecting process to estimate the firmness. (**c**) Selecting process to estimate the SSC.

**Table 3.** Modeling results to estimate the quality parameters for Korla pears.

| Quality Parameters | Preprocessing Algorithm | $R_C$ | $RMSE_C$ | $R_V$ | $RMSE_V$ |
|---|---|---|---|---|---|
| a* | MSC | 0.875 | 0.552 | 0.867 | 0.522 |
| | SNV | 0.872 | 0.551 | 0.873 | 0.568 |
| | MSC + S − G | 0.907 | 0.448 | 0.894 | 0.402 |
| | SNV + S − G | 0.896 | 0.437 | 0.882 | 0.484 |
| Firmness ($10^5$ Pa) | MSC | 0.892 | 0.357 | 0.898 | 0.338 |
| | SNV | 0.898 | 0.399 | 0.881 | 0.322 |
| | MSC + S − G | 0.914 | 0.352 | 0.903 | 0.317 |
| | SNV + S − G | 0.906 | 0.397 | 0.894 | 0.379 |
| SSC (°Brix) | MSC | 0.914 | 0.410 | 0.903 | 0.480 |
| | SNV | 0.894 | 0.415 | 0.883 | 0.482 |
| | MSC + S − G | 0.925 | 0.314 | 0.912 | 0.301 |
| | SNV + S − G | 0.915 | 0.339 | 0.902 | 0.322 |

The a* color space method recommended by the International Commission on illumination (CIE) used "*" in the expression of three parameters.

### 3.2.3. Visualization of the Iterative Process and Selection of the Important Wavelengths

In an iterative process, wavelengths can be classified into different groups according to their P and Dmean values. Figure 6 shows the distribution of the P and D-means values for each wavelength obtained in the second iteration. The strongly informative wavelengths, weakly informative wavelengths, uninformative wavelengths, and interfering wavelengths are 7, 57, 37, and 14 to estimate the a* value, 15, 37, 47, and 7 to define the firmness, and 8, 59, 34, and 13 to calculate the SSC, respectively.

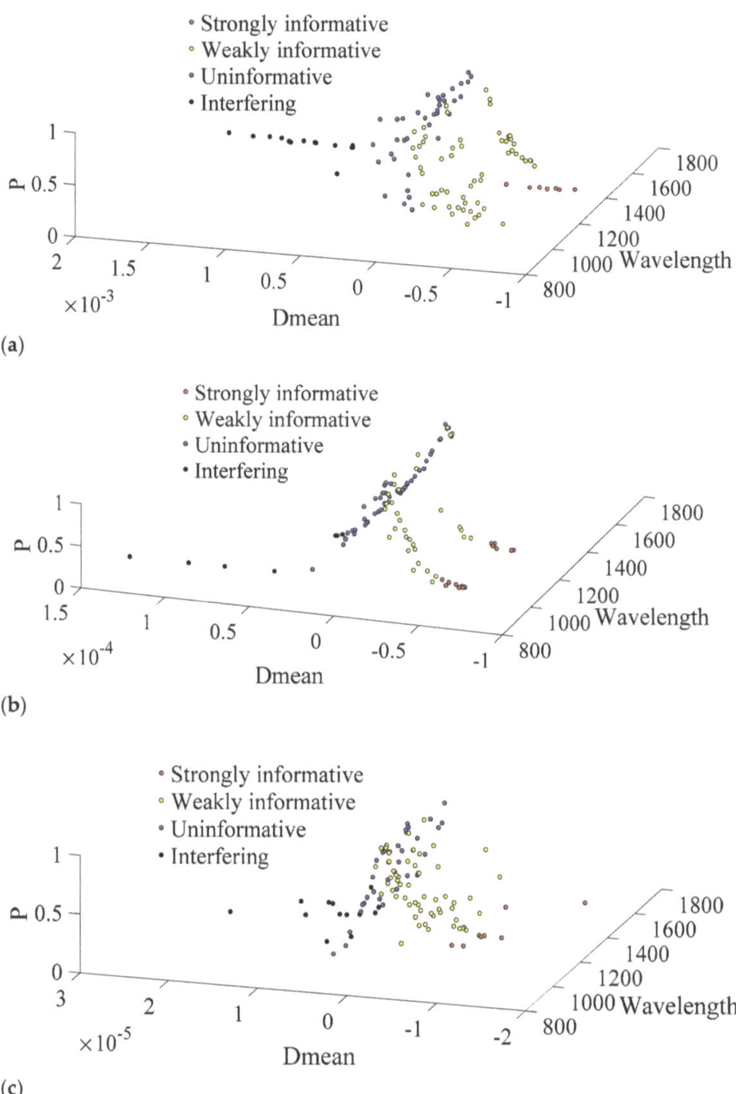



**Figure 6.** Distribution of wavelengths for different parameters obtained in the second iteration. (a) Wavelengths to estimate the a* value. (b) Wavelengths to estimate the firmness. (c) Wavelengths to estimate the SSC.

The number of wavelengths selected for a*, the firmness, and the SSC in different iterations are shown in Figure 7. Their number in the first three rounds initially decreases rapidly and then slows down. Both the irrelevant wavelengths and the interference wavelengths are completely removed after the 6th iteration. The important wavelengths, which were missed during the process, were selected after reverse elimination. To estimate the a* value, the firmness, and the SSC, 8, 11, and 16 important wavelengths are necessary. Selected wavelengths for each parameter are shown in Table 4. The number of important wavelengths of different quality parameters accounts for 3.9%, 5.4%, 7.9% of the valid wavelengths, respectively.

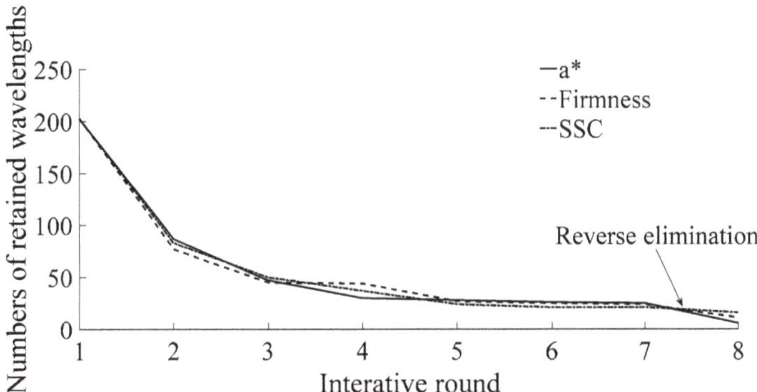


**Figure 7.** Number of retained wavelengths in each IRIV iteration.

**Table 4.** Important wavelengths for different parameters.

| Quality Parameters | Important Wavelengths |
|---|---|
| a* | 1078.70 nm, 1130.32 nm, 1238.28 nm, 1321.41 nm, 1453.38 nm, 1508.33 nm, 1535.98 nm, and 1605.63 nm |
| Firmness | 1114.14 nm, 1185.69 nm, 1254.82 nm, 1341.53 nm, 1392.12 nm, 1405.68 nm, 1453.38 nm, 1477.36 nm, 1529.06 nm, 1570.71 nm, and 1616.14 nm |
| SSC | 1046.67 nm, 1053.06 nm, 1179.15 nm, 1211.93 nm, 1234.98 nm, 1241.59 nm, 1304.69 nm, 1385.35 nm, 1415.87 nm, 1463.65 nm, 1487.67 nm, 1491.11 nm, 1508.33 nm, 1518.68 nm, 1581.17 nm, and 1630.18 nm |

The a* color space method recommended by the International Commission on illumination (CIE) used "*" in the expression of three parameters.

### 3.2.4. Evaluation of the Quality Parameters Based on the LS-SVM Model

In this study, several evaluation models were established based on the LS-SVM and the PLSR methods for a set of selected wavelengths. The optimal combinations of the regression error weight, $\gamma$, and the kernel function parameter, $\sigma^2$, are $(8.67 \times 10^4, 1.21 \times 10^3)$, $(1.45 \times 10^4, 2.93 \times 10^4)$, and $(2.37 \times 10^5, 3.80 \times 10^3)$ for the a* value, the firmness, and the SSC, respectively. Figure 8a–c shows the results on the 3 quality parameters obtained via the IRIV-LS-SVM model. The $R_C$ and $R_V$ values measure 0.932 and 0.927, respectively, in the case of the a* value; They are 0.954 and 0.948 for the firmness, and 0.955 and 0.953 for the SSC. The $RMSE_C$ and $RMSE_V$ value measure 0.426 and 0.475, respectively, for the a* value, $0.310 \times 10^5$ Pa and $0.345 \times 10^5$ Pa for the firmness, and 0.319 °Brix and 0.346 °Brix for the SSC.

The principal components used in the PLSR models to estimate a*, the firmness, and the SSC are 8, 8, and 9, respectively. Figure 8d–f shows the results obtained by using the IRIV-PLSR model. The $R_C$ and $R_V$ values of a* measure 0.921 and 0.915, respectively, in the case of the firmness, these values are 0.940 and 0.933, respectively, whereas for the SSC, the measure 0.951 and 0.942. The $RMSE_C$ and the $RMSE_V$ of the a* are 0.447 and 0.406, in the case of the firmness they measure $0.330 \times 10^5$ Pa and $0.395 \times 10^5$ Pa, whereas for the SSC 0.346 °Brix and 0.340 °Brix, respectively.

These results show that the IRIV-LS-SVM model provides more accurate results than the IRIV-PLSR one.

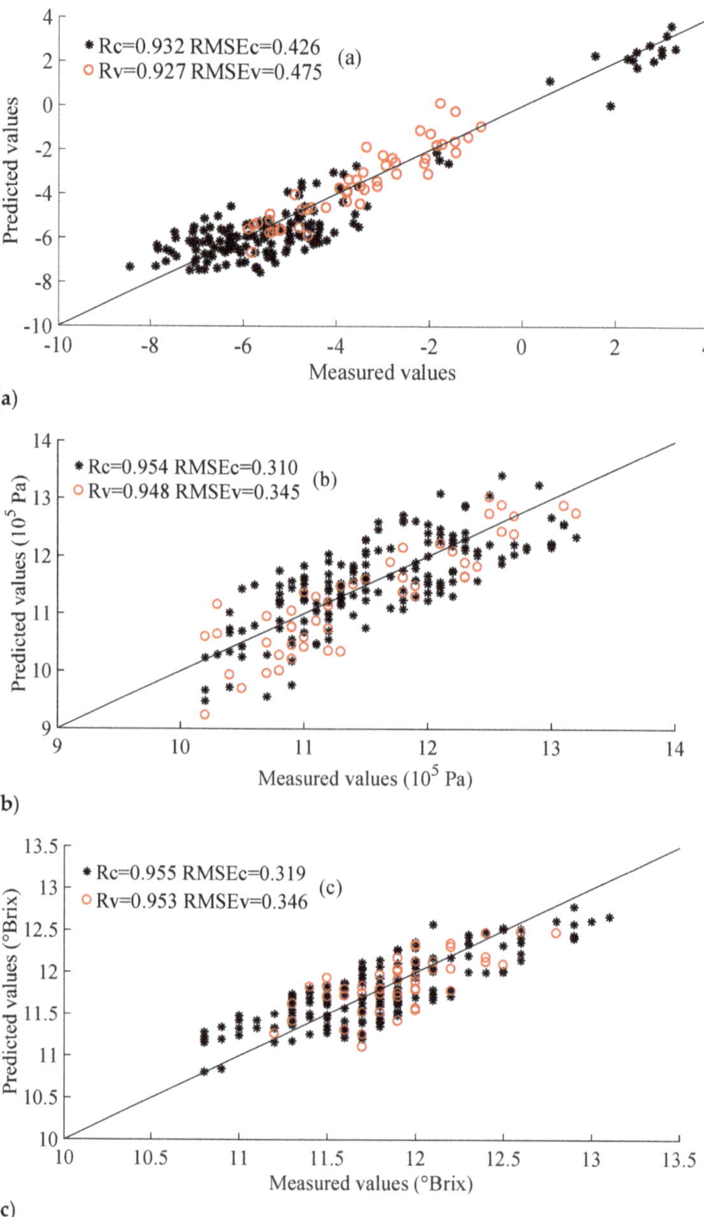


**Figure 8.** *Cont.*

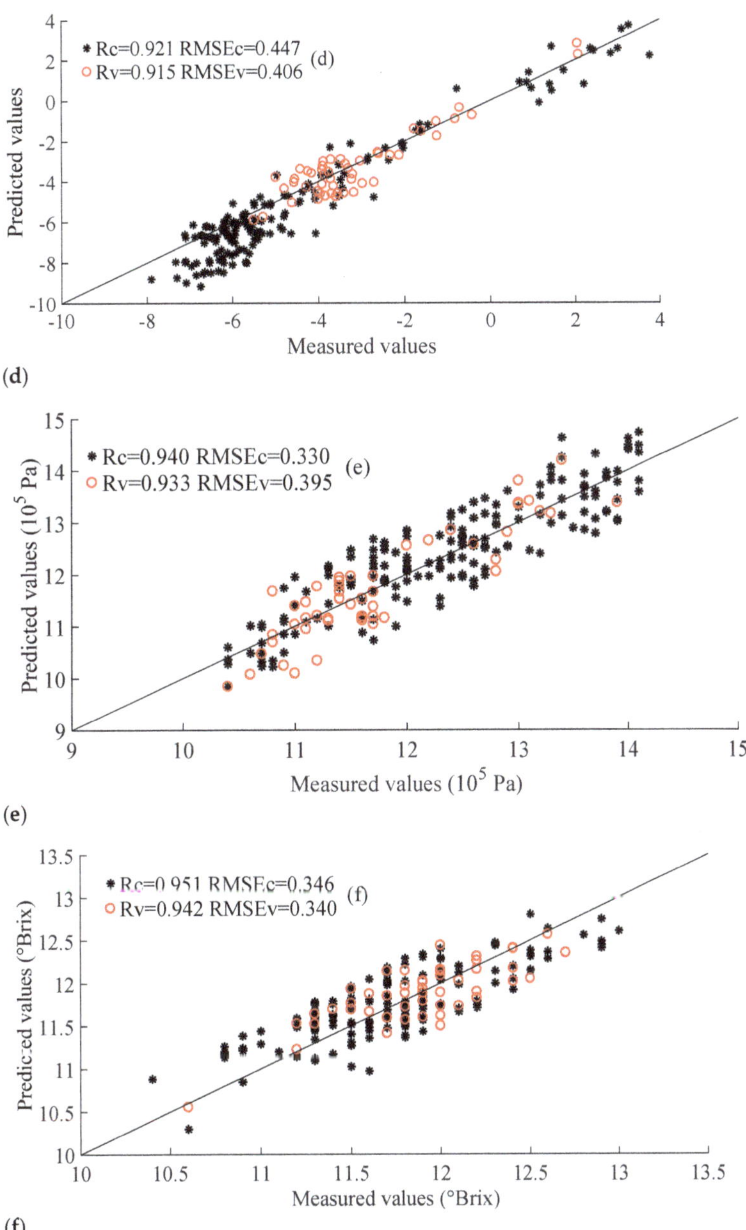



(**d**)

(**e**)

(**f**)

**Figure 8.** Scatter plots of the calibration set (*) and prediction set (o) for each quality parameter. (**a**) Scatter plots of the LS-SVM mold of the a* value. (**b**) Scatter plots of the LS-SVM mold of the firmness. (**c**) Scatter plots of the LS-SVM mold of the SSC. (**d**) Scatter plots of the PLSR mold of the a* value. (**e**) Scatter plots of the PLSR mold of the firmness. (**f**) Scatter plots of the PLSR mold of the SSC.

## 4. Discussion

This work demonstrates that hyperspectral imaging can be used to quantitatively analyze the a* value, the firmness, and the SSC of Korla fragrant pears. Both the PLSR and the LS-SVM models were implemented in combination with the IRIV algorithm to select the important wavelengths. The optimal ($\gamma$ and $\sigma^2$) combinations found in this study are ($8.67 \times 10^4$, $1.21 \times 10^3$), ($1.45 \times 10^4$, $2.93 \times 10^4$), and ($2.37 \times 10^5$, $3.80 \times 10^3$) for the a* value, the firmness, and the SSC, respectively. In the LS-SVM model, the combination of the $R_C$ and $RMSE_C$ values for a*, the firmness, and the SSC measures (0.892, 0.726), (0.914, 0.410), and (0.925, 0.319), respectively. These combinations are (0.883, 0.775), (0.908, 0.548), and (0.916, 0.346), respectively, when the validation set is considered. These results show that the IRIV-LS-SVM model can efficiently evaluate the main important parameters of Korla fragrant pears, which can be used for the quantitative evaluation and grading of fruit.

## 5. Conclusions

Compared with traditional detection methods, multiple parameter detection based on hyperspectral imaging technology has the technical advantages of being nondestructive, real-time and accurate.

There were two ways to reduce the spectral influences caused by different optical path lengths of ROI of Korla fragrant pear. Firstly, there were four halogen light sources at the same vertical plane in the irreflexive hyperspectral imaging system. The center of the four lights was in the center of the moving stage. Secondly, some spectral preprocessing algorithms were used in order to reduce the effects. The combination of MSC and SG exhibited the highest evaluating ability.

Most previous studies predicted only one or two parameters of fruits by non-destructive technologies. Three quality parameters related to the maturity and grading were predicted at the same time in this paper. Both the PLSR and the LS-SVM models were implemented in combination with the IRIV algorithm to select the important wavelengths. Both the irrelevant wavelengths and the interference wavelengths are completely removed after the 6th iteration. 8, 11, and 16 important wavelengths are selected to estimate the a* value, the firmness, and the SSC. The optimal ($\gamma$ and $\sigma^2$) combinations found in this study are ($8.67 \times 10^4$, $1.21 \times 10^3$), ($1.45 \times 10^4$, $2.93 \times 10^4$), and ($2.37 \times 10^5$, $3.80 \times 10^3$) for the a* value, the firmness, and the SSC, respectively. In the LS-SVM model, the combination of the $R_C$ and $RMSE_C$ values for a*, the firmness, and the SSC measures (0.892, 0.726), (0.914, 0.410), and (0.925, 0.319), respectively. These combinations are (0.883, 0.775), (0.908, 0.548), and (0.916, 0.346), respectively, when the validation set is considered. These results show that the IRIV-LS-SVM model can efficiently evaluate the main important parameters of Korla fragrant pears, which can be used for a quantitative evaluation and grading of the fruit. At the same time, this study also has a certain guiding significance for the qualitative detection of other fruits.

There are some research demands in the future. Firstly, a large number of experiments are needed to extend this method to more fruit detection fields through the adjustment of key parameters and the development of supporting equipment. Secondly, the number of Korla fragrant pears can be increased, so as to guarantee the grading quality and realize the industrial upgrading. Thirdly, this research mainly used spectral data to quantitatively predict the quality parameters of Korla fragrant pear although hyperspectral imaging technology has the characteristics of atlas integration. The image processing technology can be introduced to identify the kind of defects, defect level, maturities, et al. of Korla fragrant pear according to more organoleptic attributes.

**Author Contributions:** Resources, F.C.; data curation, R.S.; writing—original draft preparation, T.W.; writing—review and editing, Y.L.; visualization, C.H.; supervision, J.C.; project administration, Y.L. All authors have read and agreed to the published version of the manuscript.

**Funding:** This work was supported by the National Natural Science Foundation of China, grant number 31960498, the Open Project Program of the Key Laboratory of Colleges & Universities under the Department of Education of Xinjiang Uygur Autonomous Region, grant number TDNG20200102, and the Science and Technology Planning Project of 1st Division of Xinjiang Production and Construction Corps, grant number 2019XX02.

**Institutional Review Board Statement:** Not applicable.

**Informed Consent Statement:** Not applicable.

**Data Availability Statement:** The data presented in this study are available on request from the corresponding author. The data are not publicly available due to the request of funding scientific research projects.

**Acknowledgments:** This work was supported in part by the National Natural Science Foundation of China (Project No. 31960498), the Open Project Program of the Key Laboratory of Colleges & Universities under the Department of Education of Xinjiang Uygur Autonomous Region (No. TDNG20150101 and No. TDNG20200102), the Science and Technology Planning Project of 1st Division of Xinjiang Production and Construction Corps (No. 2019XX02). The authors are grateful to anonymous reviewers for their comments.

**Conflicts of Interest:** The authors declare that they have no known competing financial interest or personal relationship that could have appeared to influence the work reported in this paper.

## References

1.  Lan, H.; Jia, F.; Tang, Y.; Zhang, Q.; Han, Y.; Liu, Y. Quantity evaluation method of maturity for Korla fragrant pear. *Trans. CSAE* **2015**, *31*, 325–330.
2.  Wei, J.; Ma, J.; Chen, J.; Wang, X.; Ren, X. Quality differences and comprehensive evaluation of Korla fragrant pear from different habitats. *Food Sci.* **2017**, *38*, 87–91.
3.  Li, J.; Zhang, H.; Zhan, B.; Wang, Z.; Jiang, Y. Determination of SSC in pears by establishing the multi-cultivar models based on visible-NIR spectroscopy. *Infrared Phys. Technol.* **2019**, *102*, 1–10. [CrossRef]
4.  Tian, X.; Wang, Q.; Li, J.; Peng, F.; Huang, W. Non-destructive prediction of soluble solids content of pear based on fruit surface feature classification and multivariate regression analysis. *Infrared Phys. Technol.* **2018**, *92*, 336–344. [CrossRef]
5.  Zhu, X.; Li, G.; Zhang, M. Prediction of soluble solids content of Korla pears based on CARS-MIV. *Spectrosc. Spectr. Anal.* **2019**, *39*, 3547–3552.
6.  Zhan, B.; Ni, J.; Li, J. Hyperspectral technology combined with CARS algorithm to quantitatively determine the SSC in Korla fragrant pear. *Spectrosc. Spectr. Anal.* **2014**, *34*, 2752–2757.
7.  Sheng, X.; Li, Z.; Li, Z.; Zhang, F.; Zhu, T.; Wang, J.; Yin, J.; Song, Q. Determination of Korla fragrant pear firmness based on Near-Infrared Spectroscopy. *Spectrosc. Spectr. Anal.* **2019**, *39*, 2818–2822.
8.  Yu, X.; Lu, H.; Wu, D. Development of deep learning method for predicting firmness and soluble solid content of postharvest Korla fragrant pear using Vis/NIR hyperspectral reflectance imaging. *Postharvest Biol. Technol.* **2018**, *141*, 39–49. [CrossRef]
9.  T/XLXH001-2019, Korla Pear Group Standard. Korla Fragrant Pear Association of Bayingolin Mongolian Autonomous Prefecture. 2019, p. 29. Available online: www.ttbz.org.cn/Home/Show/8587?tdsourcetag=s_pcqq_aiomsg (accessed on 1 August 2019).
10.  Wang, J.H.; Wang, J.; Chen, Z.; Han, D.H. Development of multi-cultivar models for predicting the soluble solid content and firmness of European pear (*Pyrus communis* L.) using portable vis-NIR spectroscopy. *Postharvest Biol. Technol.* **2017**, *129*, 143–151. [CrossRef]
11.  Lu, Y.Z.; Saeys, W.; Kim, M.; Peng, Y.K.; Lu, R.F. Hyperspectral imaging technology for quality and safety evaluation of horticultural products: A review and celebration of the past 20-year progress. *Postharvest Biol. Technol.* **2020**, *170*, 111318. [CrossRef]
12.  Achata, E.; Oliveira, M.; Esquerre, C.; Tiwari, B.; O'Donnell, C. Visible and NIR hyperspectral imaging and chemometrics for prediction of microbial quality of beef Longissimus dorsi muscle under simulated normal and abuse storage conditions. *J. Food Sci. Technol.* **2020**, *128*, 109463. [CrossRef]
13.  Bergsträsser, S.; Fanourakis, D.; Schmittgen, S.; Cendrero-Mateo, M.; Jansen, M.; Scharr, H.; Rascher, U. HyperART: Non-invasive quantification of leaf traits using hyperspectral absorption-reflectance-transmittance imaging. *Plant Methods* **2015**, *11*, 1. [CrossRef]
14.  Fanoourakis, D.; Kazakos, F.; Nektarios, P. Allometric individual leaf area estimation in Chrysanthemum. *Agronomy* **2021**, *11*, 795. [CrossRef]
15.  Tian, X.; Li, J.; Yi, S.; Jin, G.; Qiu, X.; Li, Y. Nondestructive determining the soluble solids content of citrus using near infrared transmittance technology combined with the variable selection algorithm. *Artif. Intell. Agric.* **2020**, *4*, 48–57. [CrossRef]
16.  Guo, Z.; Wang, M.; Agyekum, A.; Wu, J.; Chen, Q.; Zuo, M.; El-Seedi, R.; Tao, F.; Shi, J.; Ouyang, Q.; et al. Quantitative detection of apple watercore and soluble solids content by near infrared transmittance spectroscopy. *J. Food Eng.* **2020**, *279*, 109955. [CrossRef]

17. Almeida, V.; Araújo, G.; Sousa, F.; Goicoechea, H.; Gaolvão, R.; Araújo, M. Vis-NIR spectrometric determination of Brix and sucrose in sugar production samples using kernel partial least squares with interval selection based on the successive projections algorithm. *Talanta* **2018**, *181*, 38–43. [CrossRef] [PubMed]

18. Sun, J.; Zhou, X.; Hu, Y.; Wu, X.; Zhang, X.; Wang, P. Visualizing distribution of moisture content in tea leaves using optimization algorithms and NIR hyperspectral imaging. *Comput. Electron. Agric.* **2019**, *160*, 153–159. [CrossRef]

19. Fan, S.; Zhang, B.; Li, J.; Liu, C.; Huang, W.; Tian, X. Prediction of soluble solids content of apple using the combination of spectra and textural features of hyperspectral reflectance imaging data. *Postharvest Biol. Technol.* **2016**, *121*, 51–61. [CrossRef]

20. Wang, H.; Yang, G.; Zhang, Y.; Bao, Y.; He, Y. Detection of fungal disease on tomato leaves with competitive adaptive reweighted sampling and correlation analysis methods. *Spectrosc. Spectr. Anal.* **2017**, *27*, 2115–2119.

21. Hu, M.; Dong, Q.; Liu, B.; Opara, U. Prediction of mechanical properties of blueberry using hyperspectral interactance imaging. *Postharvest Biol. Technol.* **2016**, *115*, 122–131. [CrossRef]

22. Jie, D.; Xie, L.; Rao, X.; Ying, Y. Using visible and near infrared diffuse transmittance technique to predict soluble solids content of watermelon in an on-line detection system. *Postharvest Biol. Technol.* **2014**, *90*, 1–6. [CrossRef]

23. Yun, Y.; Wang, W.; Tan, M.; Liang, Y.; Li, H.; Cao, D.; Lu, H.; Xu, Q. A strategy that iteratively retains informative variables for selecting optimal variable subset in multivariate calibration. *Anal. Chim. Acta* **2014**, *807*, 36–43. [CrossRef]

24. Ren, G.; Ning, J.; Zhang, Z. Intelligent assessment of tea quality employing visible-near infrared spectra combined with a hybrid variable selection strategy. *Microchem. J.* **2020**, *157*, 1–8. [CrossRef]

25. Syvilay, D.; Wilkie-Chancellier, N.; Trichereau, B.; Texier, A.; Martinez, L.; Serfaty, S.; Detalle, V. Evaluation of the standard normal variate method for laser-induced breakdown spectroscopy data treatment applied to the discrimination of painting layers. *Spectrochim. Acta Part B* **2015**, *114*, 38–45. [CrossRef]

26. Bi, Y.; Yuan, K.; Xiao, W.; Wu, J.; Shi, C.; Xia, J.; Chu, G.; Zhang, G.; Zhou, G. A local pre-processing method for near-infrared spectra, combined with spectral segmentation and standard normal variate transformation. *Anal. Chim. Acta* **2016**, *909*, 30–40. [CrossRef]

27. Silalahi, D.; Midi, H.; Arasan, J.; Mustafa, M.; Caliman, J. Robust generalized multiplicative scatter correction algorithm on pretreatment of near infrared spectral data. *Vib. Spectrosc.* **2018**, *97*, 55–65. [CrossRef]

28. Galvão, R.; Araujo, M.; José, G.; Pontes, M.; Silva, E.; Saldanha, T. A method for calibration and validation subset partitioning. *Talanta* **2005**, *67*, 736–740. [CrossRef]

29. Sukens, J.; Vandewalle, J. Least squares support vector machine classifiers. *Neural Process. Lett.* **1999**, *9*, 293–300. [CrossRef]

30. Jiang, Y.; Wang, Y.; Mao, H.; Lv, Y.; Chen, G. Delaying the aging process of pears by maintain cuticular waxes under high humidity storage conditions. *Trans. CSAE* **2020**, *36*, 287–295.

31. Wang, Z.; Wu, J.; Zhao, Z.; Zhang, H.; Mei, W. Nondestructive testing of pear firmness based on acoustic vibration response method. *Trans. CSAE* **2016**, *32*, 277–283.

32. Gomez, A.; He, Y.; Pereira, A. Non-destructive measurement of acidity, soluble solids and firmness of Satsuma mandarin using Vis/NIR-spectroscopy techniques. *J. Food Eng.* **2006**, *77*, 313–319. [CrossRef]

33. Hu, L.; Yin, C.; Ma, S.; Liu, Z. Rapid detection of three quality parameters and classification of wine based on Vis-NIR spectroscopy with wavelength selection by ACO and CARS algorithms. *Spectrochim. Acta Part A Mol. Biomol. Spectrosc.* **2018**, *205*, 574–581. [CrossRef] [PubMed]

34. Arendse, E.; Fawole, O.; Magwaza, L.; Nieuwoudt, H.; Opara, U. Fourier transform near infrared diffuse reflectance spectroscopy and two spectral acquisition modes for evaluation of external and internal quality of intact pomegranate fruit. *Postharvest Biol. Technol.* **2018**, *138*, 91–98. [CrossRef]

35. Louw, E.D.; Theron, K.I. Robust prediction models for quality parameters in Japanese plums (*Prunus salicina* L.) using NIR spectroscopy. *Postharvest Biol. Technol.* **2010**, *60*, 174–176. [CrossRef]

*Article*

# Mango Postharvest Technologies: An Observational Study of the Yieldwise Initiative in Kenya

**Hory Chikez** [1,*]**, Dirk Maier** [1] **and Steve Sonka** [1,2]

[1] Agricultural and Biosystems Engineering, Iowa State University, 1340 Elings Hall, 605 Bissell Road, Ames, IA 50011, USA; dmaier@iastate.edu (D.M.); ssonka@illinois.edu (S.S.)

[2] Ed Snider Center for Enterprise and Markets, University of Maryland, College Park, MD 20742, USA

* Correspondence: horych@iastate.edu

**Abstract:** Several studies have evaluated the effects of postharvest technologies on postharvest loss (PHL) incurred at a single stage of a food value chain. However, very few studies have assessed the effect of multiple technologies on PHL incurred at various stages of a food value chain. This study evaluated the effect of five technologies (harvesting tools, cold stores, plastic crates, fruit fly traps, and ground tarps) promoted by the Rockefeller Foundation Yieldwise Initiative (YWI) in Kenya on PHL incurred at three mango value chain stages (harvest, transportation, and point of sale). After extensive screening of the YWI data, the Kruskal–Wallis statistical test was used to compare each YWI promoted technology to smallholder farmers (SHF) traditional practices. Results indicated that plastic crates used to transport or store mangos and fruit fly traps used to attract and kill fruit flies were statistically significant ($p < 0.05$) in reducing PHL at the point of sale. Meanwhile, no statistical evidence of PHL reduction was observed from SHF using harvesting tools, cold stores, and ground tarps. Cold stores were the least adopted of the promoted technologies due to their high costs of implementation and utilization. While this study asserts that increased technology adoption is associated with PHL reduction, further research is needed to identify additional factors that favor technologies' efficacy in reducing PHL in similar food value chains.

**Keywords:** postharvest technologies; mango postharvest loss; Yieldwise Initiative

**Citation:** Chikez, H.; Maier, D.; Sonka, S. Mango Postharvest Technologies: An Observational Study of the Yieldwise Initiative in Kenya. *Agriculture* **2021**, *11*, 623. https://doi.org/10.3390/agriculture11070623

Academic Editor: Isabel Lara Ayala

Received: 12 June 2021
Accepted: 29 June 2021
Published: 1 July 2021



## 1. Introduction

Rising incomes in low-income countries are driving changes in dietary patterns and increasing the demand for safe and nutritious food [1]. However, to equate future demand and supply of safe and healthy agricultural food, global food production will need to increase at a rate of 1.3 percent every year [2]. Sustainably achieving such a growth rate will require increasing plant-based food production. Such an effort will promote long-term food security without sacrificing nutrition [3] and will provide increased employment opportunities for farm workers [4].

The two commonly documented approaches for increasing plant-based food production are agricultural intensification and cropland expansion [5]. While both have contributed to global food security substantially, several limitations have also been reported. For example, the former has been challenging to achieve in geographic areas affected by climate change, especially as it pertains to increasing crop yield [6]. Meanwhile, the latter constitutes a potential threat to biodiversity by driving habitat loss. Additionally, cropland expansion impacts carbon storage through the loss of biomass and soil carbon [7].

Given these limitations, numerous studies have suggested postharvest loss (PHL) reduction as an essential and complementary approach to meeting the increasing demand for safe and nutritious food [8]. PHL can be defined as a measurable reduction in agricultural products that arise from changes these products undergo during postharvest handling [9]. Therefore, PHL reduction efforts, especially in sub-Saharan Africa (SSA), could be a catalyst for increasing profit for food value chain actors while at the same time improving food security [10]. Given the importance of PHL reduction, several PHL mitigation studies have been initiated over the last decade, focusing on improving food security in SSA, which remains the most food-insecure region in the world [11].

For example, notable PHL mitigation studies in SSA include introducing the Purdue Improved Crop Storage (PICS) hermetic bags, which prevent storage losses due to insects in maize and other grains without chemical pesticides [12]. The commercialization of this technology, funded by the Bill and Melinda Gates Foundation, led at least five other manufacturers to introduce hermetic storage bag technology products [13]. In 2016, the Rockefeller Foundation launched the Yieldwise Initiative (YWI), intending to provide smallholder farmers (SHF) access to markets, technologies, training, and financing [14] to reduce PHL of mangos in Kenya, maize in Tanzania, and tomatoes in Nigeria. More recently, the Consortium for Innovation in Postharvest Loss and Food Waste Reduction launched as a collaborative effort between the Foundation for Food and Agriculture Research (FFAR), the Rockefeller Foundation, Iowa State University (ISU), and several other academic and research institutions around the world (reducePHL.com (accessed on 2 June 2021)) to address social, economic, and environmental impacts from food loss and waste.

Over time several additional PHL mitigation projects have emerged [15], with a focus on either quantifying PHL by stages of a food value chain [16,17] or comparing the effect of postharvest interventions on PHL incurred at a single stage of a food value chain. However, relatively few PHL mitigation projects have compared the effect of several postharvest technologies on PHL incurred at several stages of a food value chain. Therefore, this study analyzed the YWI dataset generated within the Kenyan mango value chain to evaluate the effect of five YWI promoted technologies (harvesting tools, cold stores, plastic crates, fruit fly traps, and ground tarps) on PHL incurred at three value chain stages (harvest, transportation, and point of sale).

Over the past decades, mango farming in Kenya has expanded considerably, involving several value chain actors such as non-governmental organizations, farmer cooperative groups, aggregation centers, financial institutions, mango processors, and others [18]. Additionally, annual mango production in Kenya is estimated at 1,024,500 metric tons, with approximately 80% being sold to local markets [18]. Thus, mango farming is considered a major income earner for many SHF households in Kenya [19]. However, mango production is accompanied by major PHL estimated at 40–50%, which are mainly the result of a lack of suitable technologies for the postharvest handling and processing into a wide range of value-added mango products [18]. Therefore, comparing YWI promoted technologies and identifying the value chain stage at which they are most effective, is a key step in reducing PHL along the entire value chain and improving SHF livelihoods.

## 2. Materials and Methods

### 2.1. Data Collection

Following the launch of the YWI, the Rockefeller Foundation contracted Technoserve Kenya for implementation of the mango value chain study, whereby they conducted in-person surveys and collected field data from participating farmers and other value chain actors between June and July 2018 (The authors of this paper were neither involved in the survey design nor the data collection process.). Technoserve collected data from 920 SHF (*row entries*) who provided answers based on September 2017 to March 2018 mango harvesting season. For each respondent farmer, there were 697 recorded variables (*column entries*) grouped into 12 sections, including geography and socio-demographics, farm demographics, inputs and input costs, labor costs, production, production and PHL practices, harvesting, sales, grading and storage, training, top five sources of household income, and credit access. Finally, the YWI was performed in a quasi-experimental design. Its interventions were not randomly assigned to farmers, and farmers who benefited from the interventions were not randomly selected.

## 2.2. Data Review

Review of the mango dataset began by separating the dependent variables from the independent variables, also referred to as factors in this study. Thus, all numerical variables within the dataset are expressed in the unit of mango fruit, such as mangos consumed, mangos sold, and mangos losses in different ways, and were designated as potential dependent variables. Twenty-five (25) such potential dependent variables were determined from the dataset's 697 variables (total). The remaining 672 variables were designated as factors that potentially affect the dependent variables.

### 2.2.1. Independent Variables

The 672 potential factors were sorted by removing factors with one or more missing entries, except for the "production and PHL practices" factor. Following the removal of factors with missing entries, the resulting dataset was reduced to 61 factors.

Then, factors containing the respondent farmers' identification information, such as name, contact information, and survey starting and ending times were removed. Additionally, all factors containing "true and false" entries were removed from the dataset. Furthermore, several numerical factors were positively correlated, such as the "total number of mango trees" and "number of productive mango trees" owned by a farmer. In such cases, one (number of productive mango trees) of the two was removed to avoid collinearity [20].

Finally, a listwise deletion of rows within the factor "production and PHL practices" was performed. As mentioned in the first step, this factor was the only one that was not entirely removed from the dataset despite missing entries. The reason being that Technoserve experts suggested the "fruit fly traps," a subset of the "production and PHL practices" factor, played a crucial role in reducing insect infestations of mangos before harvest. Hence, by retaining this factor in the dataset, the importance of "fruit fly traps" in reducing insect infestations of mangos before harvest could be compared to its importance in preserving quality and reducing loss after harvest. The listwise deletion of rows was applied to remove any randomly missing entries of this factor. Although the listwise deletion of rows is a commonly used technique for handling missing data [21], it was only applied to the "production and PHL practices" factor and not to the entire dataset. Using such an approach to the entire raw dataset would have resulted in a 100% loss of information due to multiple missing entries.

The final dataset of factors consisted of nine sections and 21 factors (Table 1), where 19 factors were categorical (each containing at least two subsets), and two were numerical. Therefore, harvest methods, type of storage used after harvest, type of package for sale, and production PHL practices are the four identified factors that contain various technology subsets as specified in Table 1. Their effect on mango PHL will be evaluated in this study. Additionally, certain factors and subsets were renamed to provide more clarity, and some subsets were combined into fewer to facilitate the evaluation of their effect on PHL.

Following factor review and summarization, the four factors that contained postharvest technologies are listed in Table 2, along with their subsets, subset descriptors, and descriptions.

### 2.2.2. Dependent Variables

The 25 potential dependent variables were also sorted to identify the various types of mango losses along the value chain. The first step consisted of removing variables or columns with at least one missing entry. The second step consisted of identifying all mango PHL along the value chain. Though all 25 potential dependent variables were numerical data representing quantities of mango fruit sold, given to family, used as payment-in-kind, consumed by farmers, and lost along the value chain, not all were PHL variables. PHL variables are the hotspots of loss that form the entire PHL [22]. Therefore, in this study, mango losses that occurred during harvest and losses that occurred after harvest were the only types of losses considered to be PHL variables.

**Table 1.** A summary of the dataset showing sections, factors, subsets of factors, and respondent farmers: Column (a) lists the nine sections to which each factor belongs. Column (b) lists all 21 factors, including the 19 categorical [C], two numerical [N], and four containing postharvest technologies [T]. Column (c) expands each factor into subsets. Subsets with the superscript [PHT] are identified as postharvest technologies. Subsets with the superscript [PRHT] are identified as pre-harvest technologies. Numerical factors consist of numerical values estimated by each respondent farmer. Column (d) renames subsets and combines them into fewer categories to facilitate subsequent analysis. Subset descriptors with the superscript [YWI] are identified as technologies promoted by the YWI. Column (e) indicates the number of respondent farmers belonging to each subset. For each factor, respondent farmers who reported more than one subset were assigned the subset Other **.

| (a) Sections | (b) Factors | (c) Subsets of Factors | (d) Subset Descriptors | (e) # Observations (Respondent Farmers) |
|---|---|---|---|---|
| A. Geography and socio-demographics | 1. county [C] | Embu | eastern | 159 |
| | | Garissa | north eastern | 6 |
| | | Kilifi | coast | 1 |
| | | Kirinyaga | central | 1 |
| | | Lamu | coast | 12 |
| | | Machakos | eastern | 49 |
| | | Makueni | eastern | 88 |
| | | Meru | eastern | 86 |
| | | Muranga | central | 12 |
| | | Tana river | coast | 332 |
| | | Tharaka nithi | eastern | 7 |
| | 2. treatment control [C] | control | non beneficiary | 282 |
| | | treatment | yieldwise beneficiary | 471 |
| | 3. farm ownership [C] | no | no | 135 |
| | | yes | yes | 618 |
| B. Labor costs | 4. labor costs [C] | no | no | 468 |
| | | yes | yes | 285 |
| C. Harvesting | 5. who harvested mango [C] | both | farmer and buyer | 133 |
| | | buyer only | buyer | 411 |
| | | self-family | farmer | 181 |
| | | other ** | other | 28 |
| | 6. inform when to harvest [C] | days after blooming | days after blooming | 5 |
| | | fruit color | fruit color | 165 |
| | | fruit size or shape | fruit size or shape | 49 |
| | | test for maturity | test for maturity | 13 |
| | | other ** | other | 521 |
| | 7. frequency of harvest [C] | daily | daily | 53 |
| | | fortnightly | fortnightly | 231 |
| | | monthly | monthly | 52 |
| | | weekly | weekly | 308 |
| | | other ** | other | 109 |
| | 8. methods of harvest [C, T] | handpicking | traditional practices | 276 |
| | | harvesting tools [PHT] | harvesting tools [YWI] | 48 |
| | | poles | traditional practices | 67 |
| | | shaking trees or branches | traditional practices | 11 |
| | | other ** | other | 350 |
| D. Sales | 9. how farmer identified buyer [C] | brokers | brokers | 407 |
| | | farmer-based organization (FBO) | fbo | 12 |
| | | own effort neighbor family or friend | own effort | 253 |
| | | other ** | other | 81 |

**Table 1.** *Cont.*

| (a) Sections | (b) Factors | (c) Subsets of Factors | (d) Subset Descriptors | (e) # Observations (Respondent Farmers) |
|---|---|---|---|---|
| E. Grading and storage | 10. harvested mango graded [C] | no | no | 346 |
| | | yes | yes | 407 |
| | 11. market destination [C] | export | export | 106 |
| | | local market | local market | 362 |
| | | processing | processing | 41 |
| | | supermarket | supermarket | 5 |
| | | other ** | other | 239 |
| | 12. storage after harvesting [C, T] | cold store [PHT] | cold store [YWI] | 18 |
| | | did not store | traditional practices | 386 |
| | | shade | traditional practices | 212 |
| | | store shed [PHT] | traditional practices | 88 |
| | | other ** | other | 49 |
| | 13. package for sale [C, T] | in crates cartons [PHT] | plastic crates [YWI] | 320 |
| | | in sacks [PHT] | traditional practices | 119 |
| | | other ** | other | 314 |
| | 14. mango price [N] | Ksh per mango | Ksh per mango | 753 |
| F. Training | 15. receive production training [C] | no | no | 534 |
| | | yes | yes | 219 |
| G. Credit access | 16. have bank account [C] | no | no | 374 |
| | | yes | yes | 379 |
| | 17. have mobile money account [C] | no | no | 95 |
| | | yes | yes | 658 |
| | 18. receive remittances [C] | no | no | 467 |
| | | yes | yes | 286 |
| | 19. taken loan for farm [C] | no | no | 695 |
| | | yes | yes | 58 |
| H. Production and phl practices | 20. production PHL practices [C, T] | fruit fly traps [PRHT] | fruit fly traps [YWI] | 125 |
| | | none | traditional practices | 218 |
| | | scouting fruit fly | traditional practices | 91 |
| | | tarp [PHT] | tarp [YWI] | 115 |
| | | other ** | other | 204 |
| I. Farm demographics | 21. total trees [N] | # of trees | # of trees | 753 |

Following the selection of mango PHL variables, the resulting dependent variables consisted of nine types of mango PHL (Table 3) from the raw dataset's initial 25 potential dependent variables. The nine types of mango PHL were subsequently grouped based on the stages of the value chain at which they occurred (Table 3).

The third step consisted of identifying and removing outliers [23] from dependent variables. To identify outliers, mango gross production per farmer was calculated for each farmer. The calculation consisted of summing all variables that contributed to mango gross production, including mangos sold, given to family, used as payment-in-kind, consumed by farmers, and all PHL variables shown in Table 3. It was then observed that the calculated mango gross production distribution was skewed with outliers. Hence, removing the rows containing mango gross production outliers resulted in eliminating outliers from PHL distributions at each value chain stage.

The last step consisted of expressing mango PHL at all three value chain stages as percentages of gross production (Figure 1) for all 753 respondent farmers.

**Table 2.** Summarizing and describing factors containing postharvest technology subsets: Column (a) shows the four factors containing postharvest technologies. Column (b) shows the subset descriptors, which are renamed subsets; these were determined to reduce the raw data into fewer categories, to facilitate subsequent analysis. Column (c) shows the subsets of each factor as initially recorded in the raw data. Column (d) describes the purpose of each subset. The superscripts [YWI] in Columns (b) and (c) refer to technologies that the YWI promoted.

| (a) Factors | (b) Subset Descriptors | (c) Subsets | (d) Description |
|---|---|---|---|
| methods of harvest | harvesting tools [YWI] | harvesting tools [YWI] | Tools that reduce/eliminate the need for harvesting by hand and catch mangos in a soft fabric sack, thereby preventing bruising that may occur due to hard grips or when mangos fall on hard surfaces |
| | traditional practices | shaking trees or branches | Harvesting practice consisting of the farmer shaking the mango tree or branches, causing it to detach from the tree and fall on the ground |
| | traditional practices | handpicking | Not specified in the data |
| | traditional practices | poles | Not specified in the data |
| storage after harvesting | cold store [YWI] | cold store [YWI] | Cold stores consist of charcoal evaporative coolers, brick evaporative coolers, insulated air-conditioned containers powered by photovoltaic cells or by the electrical grid |
| | traditional practices | did not store | Not specified in the data |
| | traditional practices | shade | Trees shade |
| | traditional practices | store shed | Shed built to store mango |
| package for sale | plastic crates [YWI] | in crates [YWI] | Plastic rectangular containers that protect/preserve quality by reducing impact damage during transport, and each crate can hold up to 50 mangos |
| | traditional practices | in sacks | Not specified in the data |
| production phl practices | fruit fly traps [YWI] | fruit fly traps [YWI] | A container with chemicals like bactrolure or metarhizium anisopliae ICIPE 69 that attract fruit flies and eventually kills them, either directly by chemical exposure or through secondary transmission from other fruit flies |
| | tarp [YWI] | tarp [YWI] | Large plastic covers/surfaces mainly used to prevent bruising of mango during harvest by reducing the impact of mango. Mangos harvested by hand are thrown down on the tarp which acts as a cushion to reduce the mechanical impact force on the fruits. Tarps are also used after harvest to protect mangos from weather effects, including rain, moisture, or direct sunlight |
| | traditional practices | none | Not specified in the data |
| | traditional practices | scouting fruit fly | Not specified in the data |

**Table 3.** Types of mango PHL within the dataset of dependent variables.

| (A) Mango Value Chain Stages | (b) Types of Mango PHL (Dependent Variables) | (c) Description |
| --- | --- | --- |
| Harvest | PHL during harvest | Mango fruit (quantity) discarded by the farmer as a result of bruises or injuries caused to the fruit during harvesting activities |
| | PHL during harvest other ways | Not specified in the data (unclear) |
| Transportation to the point of sale or aggregation site | PHL during transportation | Mango fruit (quantity) discarded by the farmer as a result of unspecified quality issues during transportation |
| Point of sale (off-takers, wholesaler or brokers) | PHL due to mangos being rejected by buyers | Mango fruit (quantity) discarded by the buyer as a result of unspecified quality issues |
| | PHL due to mangos being overripe | Mango fruit (quantity) discarded by the farmer as a result of the fruit being too overripe for sale |
| | PHL due to mangos physical damage | Mango fruit (quantity) discarded by the farmer as a result of bruises or injuries caused to the fruit after harvest |
| | PHL due to mangos being rotten | Mango fruit (quantity) discarded by the farmer as a result of the fruit being rotten |
| | PHL due to low-quality mangos being fed to livestock | Mango fruit (quantity) discarded by the farmer and fed to livestock as a result of the fruit being unfit for human consumption |
| | PHL other ways | Not specified in the data (unclear) |

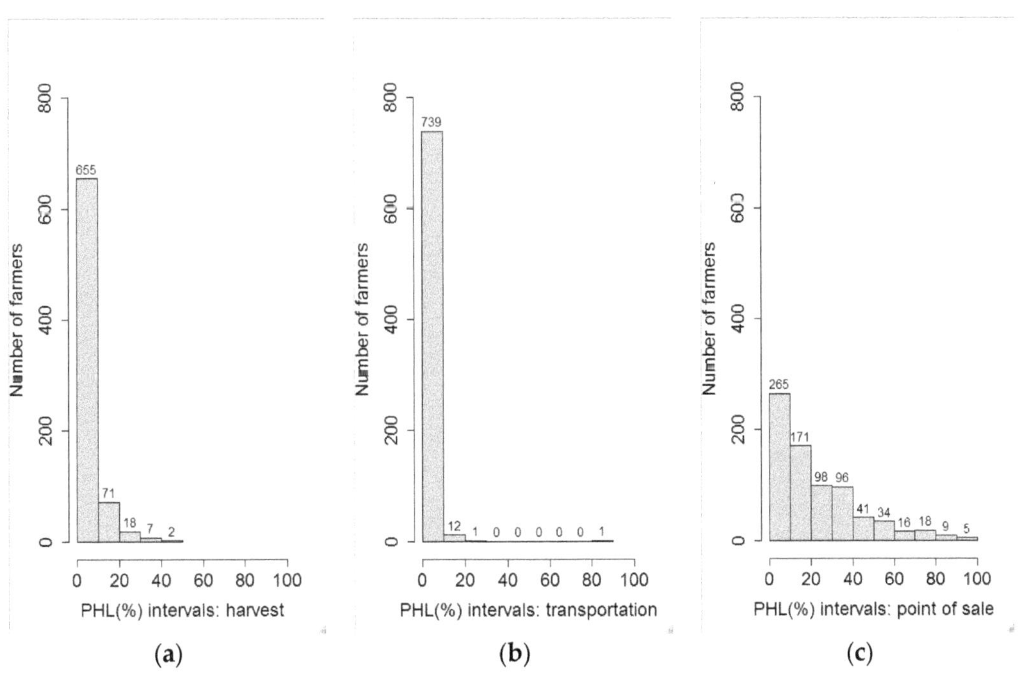

**Figure 1.** Distributions of mango PHL (%) during harvest (**a**), transportation (**b**), and at point of sale (**c**).

The PHL data summarized in Figure 1 were subsequently combined with the factors listed in Table 1. This combination resulted in creating the YWI mango dataset (summarized in Table 4) that formed the basis for the analysis and results presented in this study.

**Table 4.** Summary of all factors, subsets of factors, respondent farmers, and the seven types of mango PHL: Column (a) lists all 21 factors including the 19 categorical [C], two numerical [N], and four containing postharvest technologies [T]. Column (b) expands each factor into subsets that were previously referred to as subset descriptors in Table 1. The superscript [YWI] is used to identify technologies promoted by the YWI. "Other" refers to the combination of multiple subsets as reported by respondent farmers. Column (c) indicates the number of respondent farmers belonging to each subset. Column (d) encompasses mango PHL at harvest, during transportation, at point of sale, and as a total of all three value chain stages. PHL averages cannot be categorized by numerical factors, hence the n/a notation.

| (a) Factors | (b) Subsets of Factors | (c) Observations (Respondent Farmers *n*) | (d) Average PHL (%) Per Farmer Per Value Chain Stage | | | |
|---|---|---|---|---|---|---|
| | | | Harvest | Transportation | Point of Sale | Entire Value Chain |
| 1. county [C] | central | 13 | 4 | 1 | 20 | 25 |
| | coast | 345 | 6 | 2 | 25 | 32 |
| | eastern | 389 | 4 | 1 | 20 | 25 |
| | north eastern | 6 | 3 | 0 | 15 | 18 |
| 2. treatment control [C] | non beneficiary | 282 | 4 | 1 | 25 | 31 |
| | yieldwise beneficiary | 471 | 5 | 1 | 21 | 27 |
| 3. farm ownership [C] | no | 135 | 5 | 1 | 22 | 28 |
| | yes | 618 | 5 | 1 | 23 | 28 |
| 4. labor costs [C] | no | 468 | 5 | 1 | 25 | 30 |
| | yes | 285 | 5 | 1 | 19 | 25 |
| 5. who harvested mango [C] | buyer | 411 | 4 | 0 | 22 | 27 |
| | farmer | 181 | 6 | 2 | 25 | 33 |
| | farmer and buyer | 133 | 5 | 1 | 21 | 27 |
| | other | 28 | 4 | 1 | 21 | 27 |
| 6. inform when to harvest [C] | days after blooming | 5 | 6 | 2 | 5 | 14 |
| | fruit color | 165 | 5 | 1 | 24 | 30 |
| | fruit size or shape | 49 | 8 | 2 | 34 | 43 |
| | test for maturity | 13 | 3 | 2 | 24 | 29 |
| | other | 521 | 5 | 1 | 21 | 27 |
| 7. frequency of harvest [C] | daily | 53 | 6 | 1 | 25 | 31 |
| | fortnightly | 231 | 5 | 1 | 23 | 29 |
| | monthly | 52 | 4 | 1 | 29 | 34 |
| | weekly | 308 | 5 | 1 | 20 | 26 |
| | other | 109 | 4 | 1 | 24 | 28 |
| 8. methods of harvest [C, T] | harvesting tools [YWI] | 49 | 6 | 1 | 19 | 25 |
| | traditional practices | 544 | 5 | 1 | 24 | 30 |
| | other | 160 | 4 | 2 | 19 | 25 |
| 9. how farmer identified buyer [C] | brokers | 407 | 5 | 1 | 23 | 29 |
| | farmer based organization | 12 | 2 | 0 | 9 | 12 |
| | own effort | 253 | 4 | 2 | 20 | 26 |
| | other | 81 | 4 | 1 | 29 | 33 |
| 10. harvested mango graded [C] | no | 346 | 4 | 1 | 25 | 30 |
| | yes | 407 | 6 | 1 | 20 | 27 |

**Table 4.** *Cont.*

| (a) Factors | (b) Subsets of Factors | (c) Observations (Respondent Farmers *n*) | (d) Average PHL (%) Per Farmer Per Value Chain Stage | | | |
|---|---|---|---|---|---|---|
| | | | Harvest | Transportation | Point of Sale | Entire Value Chain |
| 11. market destination [C] | export | 106 | 3 | 0 | 15 | 19 |
| | local market | 362 | 5 | 1 | 23 | 29 |
| | processing | 41 | 7 | 2 | 26 | 34 |
| | supermarket | 5 | 6 | 0 | 20 | 25 |
| | other | 239 | 5 | 2 | 24 | 31 |
| 12. storage after harvesting [C, T] | cold store [YWI] | 18 | 8 | 1 | 16 | 25 |
| | traditional practices | 686 | 5 | 1 | 22 | 28 |
| | other | 49 | 4 | 1 | 25 | 29 |
| 13. package for sale [C, T] | plastic crates [YWI] | 320 | 5 | 1 | 18 | 24 |
| | traditional practices | 119 | 6 | 2 | 26 | 34 |
| | other | 314 | 5 | 1 | 25 | 31 |
| 14. mango price [N] | Ksh per mango | 753 | n/a | n/a | n/a | n/a |
| 15. receive production training [C] | no | 534 | 5 | 1 | 25 | 31 |
| | yes | 219 | 5 | 1 | 15 | 21 |
| 16. have bank account [C] | no | 374 | 5 | 1 | 24 | 31 |
| | yes | 379 | 5 | 1 | 21 | 26 |
| 17. have mobile money account [C] | no | 95 | 6 | 2 | 24 | 32 |
| | yes | 658 | 5 | 1 | 22 | 28 |
| 18. receive remittances [C] | no | 467 | 5 | 1 | 22 | 28 |
| | yes | 286 | 5 | 1 | 23 | 29 |
| 19. taken loan for farm [C] | no | 695 | 5 | 1 | 23 | 29 |
| | yes | 58 | 3 | 1 | 21 | 25 |
| 20. production phl practices [C, T] | fruit fly traps [YWI] | 125 | 4 | 0 | 20 | 25 |
| | tarp [YWI] | 115 | 7 | 2 | 25 | 33 |
| | traditional practices | 310 | 4 | 1 | 24 | 29 |
| | other | 203 | 5 | 1 | 21 | 27 |
| 21. total trees [N] | # of trees | 753 | n/a | n/a | n/a | n/a |

In addition to summarizing the YWI mango dataset in Table 4, each stage's PHL was expressed as a proportion of the total PHL (Figure 2) by dividing each stage's average by the average PHL of the entire value chain. Furthermore, an online interactive mango PHL dashboard was created (https://phldashboard.shinyapps.io/phldashboard/ (accessed on 2 June 2021)) to explore average mango PHL as a function of each factor in Table 4 Column (a) and as a function of a selected combination of factors.

*2.3. Statistical Analysis*

Identification of the four factors containing postharvest technology subsets (Table 2) and the subsequent quantification of mango losses associated with each subset (Table 4) provided a basis for comparing PHL averages per subset and quantifying the effect size among postharvest technology subsets. However, to ensure that the PHL averages are significantly different among subsets or technologies, a preliminary analysis of the subsets' data was conducted to identify an appropriate statistical tool for comparing means. The initial analysis consisted of verifying the main mathematical assumptions of normality, homogeneity of variance, and independence [24] required to use parametric statistical tools.

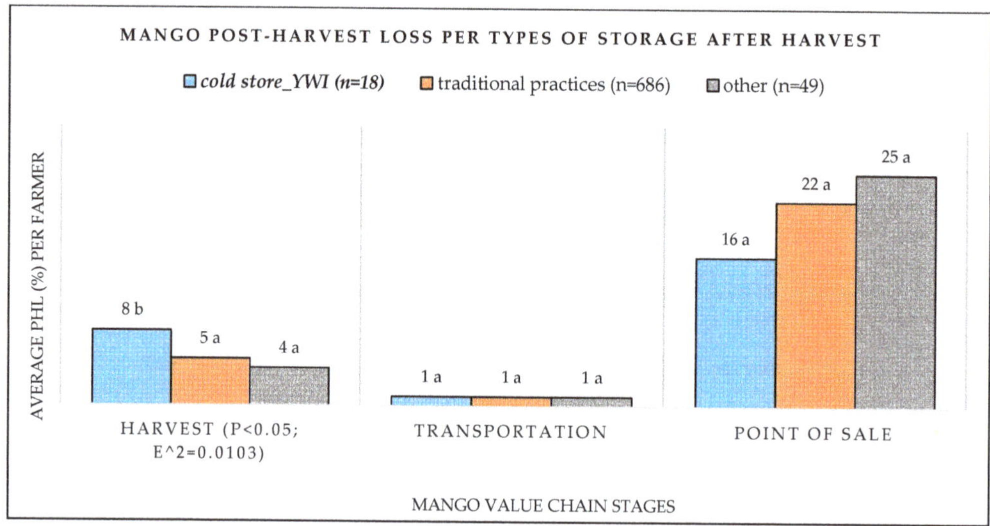


**Figure 4.** Comparing cold stores to alternative storage types after harvest. Values with different letters are significantly different at $p < 0.05$ from the Kruskal–Wallis analysis, Dunn test, and Benjamini–Hochberg adjustment. E^2 = Epsilon-squared value for effect size. YWI refers to the technology that the Yieldwise Initiative promoted. (*n*) refers to the number of farmers who reported using a given practice or technology. 'Other' refers to practices that combined both YWI promoted technologies and traditional practices.

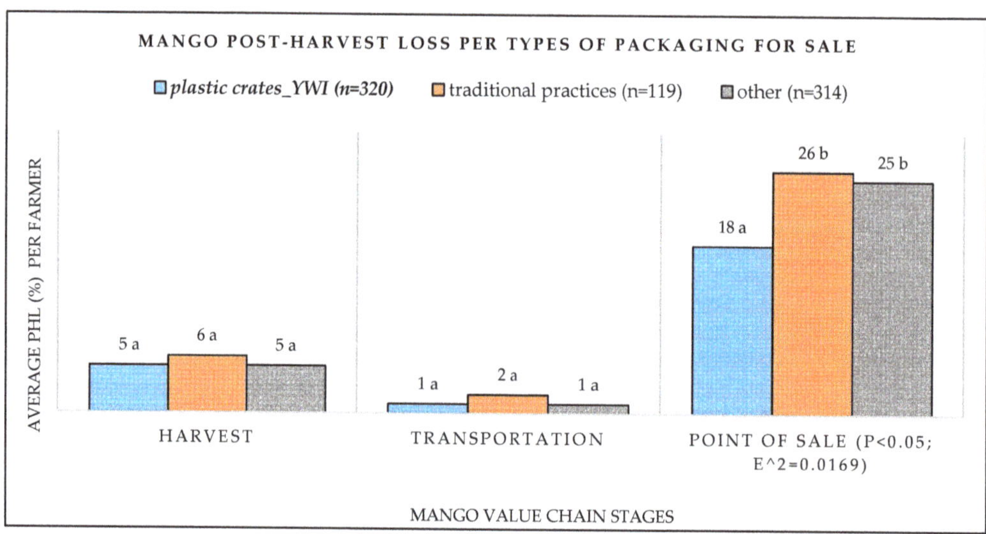


**Figure 5.** Comparing plastic crates to traditional practices. Values with different letters are significantly different at $p < 0.05$ from the Kruskal–Wallis analysis, Dunn test, and Benjamini–Hochberg adjustment. E^2 = Epsilon-squared value for effect size. YWI refers to the technology that the Yieldwise Initiative promoted. (*n*) refers to the number of farmers who reported using a given practice or technology. 'Other' refers to practices that combined both YWI promoted technologies and traditional practices.

### 3.4. Fruit Fly Traps and Ground Tarps

Fruit fly traps were statistically significant ($p < 0.05$) in reducing PHL incurred at the point of sale (Figure 6), although the effect size of the reduction was weak (Epsilon-squared = 0.017). Additionally, PHL reduction was detected during transportation due to SHF using fruit fly traps over traditional production practices. However, this reduction was not statistically significant ($p < 0.05$). Moreover, no PHL reduction was detected during harvest from SHF using fruit fly traps over traditional production practices (Figure 6). Meanwhile, moderate PHL increases during harvest ($p < 0.05$, Epsilon-squared = 0.04) and during transportation ($p < 0.05$, Epsilon-squared = 0.06), and a weak PHL increase ($p < 0.05$, Epsilon-squared = 0.017) at the point of sale were detected from SHF using ground tarps over any other harvest practice (Figure 6).

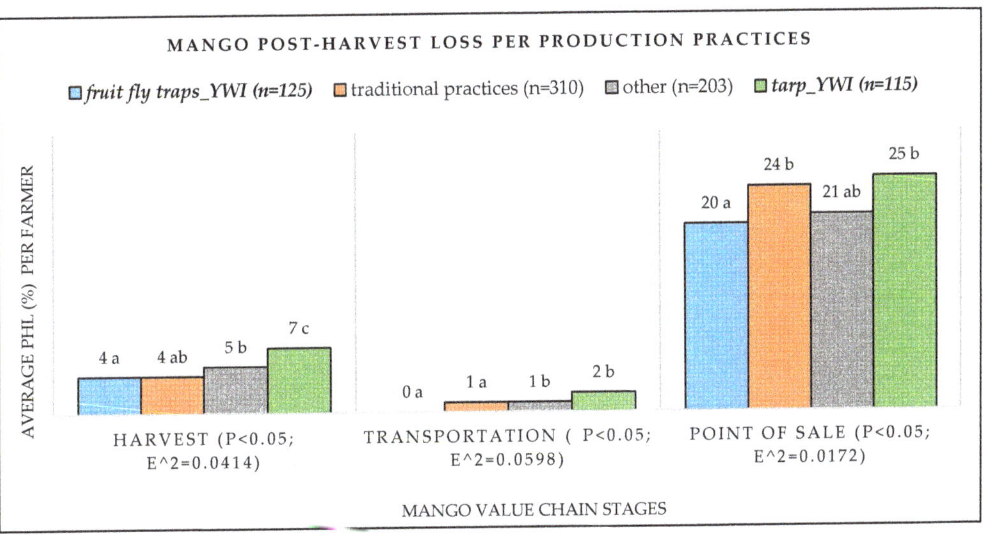


**Figure 6.** Comparing fruit fly traps to alternative production practices. Values with different letters are significantly different at $p < 0.05$ from the Kruskal–Wallis analysis, Dunn test, and Benjamini–Hochberg adjustment. E^2 = Epsilon-squared value for effect size. YWI refers to the technology that the Yieldwise Initiative promoted. (*n*) refers to the number of farmers who reported using a given practice or technology. 'Other' refers to practices that combined both YWI promoted technologies and traditional practices.

## 4. Discussion

While mango SHF have reported seeing a PHL reduction due to using harvesting tools [28,29], traditional harvesting practices such as handpicking can also reduce damage caused during harvest [30–32]. Hence, increasing the adoption of correct mango handpicking practices could be effective, if not more effective, than harvesting tools (Figure 3).

Cold stores utilized by SHF (photovoltaic-powered coolers, charcoal evaporative coolers, and brick evaporative coolers) effectively preserve mangos [33,34]. However, they are costly for individual farmers to own. Hence, most mango cold stores are owned by farmers' cooperatives [33], requiring farmers to inspect mangos during harvest and only store fruits that can be well preserved in the cold stores. Therefore, PHL increase during harvest from SHF using cold stores (Figure 4) can be attributed to large quantities of poor fruit quality set aside during the inspection process before storage.

Packaging mangos in plastic crates instead of sacks allows adequate packaging and storage of mangos [35] needed to preserve quality and provide greater wholesale value for the fruit [36,37]. Packaging mangos in crates can also reduce damage caused to the fruit during transportation (Figure 5), and by extension, reduces PHL at the point of sale

(Figure 5). Furthermore, plastic crates had the highest adoption ($n = 320$, Figure 5) of all the YWI promoted technologies as SHF and value chain actors saw value in using them.

Fruit fly traps were statistically significant in reducing PHL at the point of sale as their adoption was relatively higher ($n = 125$, Figure 6) than the other YWI promoted technologies, except for plastic crates. Two major factors were reported related to slowing the adoption of fruit fly traps. First, farmers' beliefs that fruit fly traps attract fruit flies from other farms caused the farmers to remove traps, leaving mangos susceptible to infestation and diminishing fruit fly traps' efficacy over traditional production practices [35]. Second, although farmers reported having fruit fly trap containers, without adequate financing, they could not refill the fruit fly trap containers with bait refills frequently enough for the traps to be effective [35]. Thus, overcoming these challenges could result in higher adoption of fruit fly traps.

Although essential, increased adoption and access to the technologies can be difficult to achieve. Cold stores, for example, are too expensive for SHF to own or utilize, especially without access to affordable credit [28,29]. Hence their adoption within the YWI was relatively low ($n = 18$, Figure 4). On the other hand, technologies easily accessible to farmers, such as plastic crates and fruit fly traps [38], had a relatively higher adoption rate. Hence, providing SHF easier access to affordable credit through innovative financing [35] or lower discount rates [39] could be an essential and initial step toward enabling increased adoption of preferred technologies. Alternatively, facilitating access to postharvest technologies through innovative subsidy programs could also increase the adoption of preferred technologies [40,41].

Lastly, discussions with Kenyan SHF revealed that buyers mainly do the harvesting and thus due to the informal and often hierarchal relationships between the two groups, farmers cannot intervene with the harvesting. Therefore, they do not have a say about whether or not ground tarps are used, increasing the chances of experiencing PHL during harvesting and, by extension, several other PHL types along the value chain (Figure 6). Moreover, training and promotion of ground tarps delivered through the YWI may lose their impact over time, and refresher training will be necessary [36].

## 5. Conclusions

This study quantitatively compared postharvest technologies and their effects on mango PHL in Kenya via the Rockefeller Foundation's YWI. Five YWI promoted technologies were compared to Kenyan SHF's traditional practices at three value chain stages. Subsequently, the following conclusions were inferred from analyzing the YWI mango dataset:

Efforts to reduce PHL in the mango value chain should prioritize adopting plastic crates and fruit fly traps. These technologies were statistically significant in reducing PHL incurred at the point of sale. In addition to preserving quality, plastic crates and fruit fly traps can be easily accessed and adopted by SHF compared to harvesting tools, cold stores, and ground tarps.

Harvesting tools as a YWI promoted technology and handpicking as a traditional practice to harvest mangos are similar in that both require careful handling of the fruit when picking. Therefore, PHL reduction from SHF using harvesting tools was not statistically significant because handpicking can effectively reduce mango PHL when done correctly. Further research is needed to determine factors other than increased adoption that increase the effectiveness of harvesting tools in reducing PHL.

PHL reduction from SHF using cold stores was not statistically significant. While several factors can contribute to this lack of statistical significance, this study posits that the low adoption of cold stores among SHF is due to their high cost of ownership or utilization.

The benefits of ground tarps should be further investigated because SHF are not always involved in the harvest and do not have a say about whether or not ground tarps are used, resulting in increased PHL. Additionally, training and promotion of technologies delivered through the YWI may lose their impact over time, and refresher training is recommended.

While this study asserts that increased technology adoption is necessary to obtaining better PHL reduction efficacy, further research is needed to identify additional factors of importance that favor technologies' efficacy in reducing PHL in similar food value chains.

**Author Contributions:** Conceptualization, H.C., D.M. and S.S.; data curation, H.C.; formal analysis, H.C.; investigation, H.C., D.M. and S.S.; methodology, H.C., D.M. and S.S.; software, H.C.; visualization, H.C. and D.M.; roles/writing—original draft, H.C.; writing—review and editing, H.C. and D.M.; funding acquisition, D.M. and S.S.; project administration, D.M. and S.S.; resources, D.M. and S.S.; supervision, D.M. and S.S.; validation, D.M. and S.S. All authors have read and agreed to the published version of the manuscript.

**Funding:** This research was funded by The Rockefeller Foundation (Grant 2018 FOD 004), the Foundation for Food and Agriculture Research (Grant DFs-18-0000000008), and the Iowa Agriculture and Home Economics Experiment Station.

**Data Availability Statement:** An online interactive mango PHL dashboard was created at (https://phldashboard.shinyapps.io/phldashboard/ (accessed on 2 June 2021)) to support this study's results and to further explore average mango PHL as a function of several factors and combinations thereof.

**Acknowledgments:** Funding for this study was provided by The Rockefeller Foundation (2018 FOD 004), the Foundation for Food and Agriculture Research (DFs-18-0000000008), and the Iowa Agriculture and Home Economics Experiment Station. The authors would also like to thank the following individuals and their respective organizations for assistance in better understanding the YWI and data collection procedures: Amos Kioko at The Rockefeller Foundation (formally), Paul Ngugi, Isaiah Kirema, and Onesmus Kinene (formally) at Technoserve.

**Conflicts of Interest:** The funders had no role in the design of this study, analyses, or interpretation of data; in the writing of the manuscript, or in the decision to publish the results.

## References

1. FAO. *The State of Food and Agriculture 2019. Moving Forward on Food Loss and Waste Reduction;* FAO: Rome, Italy, 2019.
2. Alexandratos, N.; Bruinsma, J. *World Agriculture Towards 2030/2050: The 2012 Revision;* FAO: Rome, Italy, 2012.
3. De Boer, J.; Aiking, H. On the merits of plant-based proteins for global food security: Marrying macro and micro perspectives. *Ecol. Econ.* **2011**, *70*, 1259–1265. [CrossRef]
4. Joosten, F.; Dijkxhoorn, Y.; Sertse, Y.; Ruben, R. *How does the Fruit and Vegetable Sector Contribute to Food and Nutrition Security?* LEI Wageningen UR: The Hague, The Netherlands, 2015.
5. Wu, W.; Yu, Q.; You, L.; Chen, K.; Tang, H.; Liu, J. Global cropping intensity gaps: Increasing food production without cropland expansion. *Land Use Policy* **2018**, *76*, 515–525. [CrossRef]
6. Pugh, T.A.M.; Müller, C.; Elliott, J.; Deryng, D.; Folberth, C.; Olin, S.; Schmid, E.; Arneth, A. Climate analogues suggest limited potential for intensification of production on current croplands under climate change. *Nat. Commun.* **2016**, *7*, 1–8. [CrossRef]
7. Molotoks, A.; Stehfest, E.; Doelman, J.; Albanito, F.; Fitton, N.; Dawson, T.P.; Smith, P. Global projections of future cropland expansion to 2050 and direct impacts on biodiversity and carbon storage. *Glob. Chang. Biol.* **2018**, *24*, 5895–5908. [CrossRef] [PubMed]
8. Bradford, K.J.; Dahal, P.; Van Asbrouck, J.; Kunusoth, K.; Bello, P.; Thompson, J.; Wu, F. The dry chain: Reducing postharvest losses and improving food safety in humid climates. *Trends Food Sci. Technol.* **2018**, *71*, 84–93. [CrossRef]
9. Grolleaud, M. *Post-Harvest Losses: Discovering the Full Story;* Post-Harvest System and Food Losses; FAO: Rome, Italy, 1997; Chapter 2; Available online: http://www.fao.org/3/ac301e/AC301e03.htm (accessed on 4 August 2020).
10. Sheahan, M.; Barrett, C.B. Food loss and waste in Sub-Saharan Africa: A critical review. *Food Policy* **2017**, *70*, 1–12. [CrossRef] [PubMed]
11. Xie, H.; Perez, N.; Anderson, W.; Ringler, C.; You, L. Can Sub-Saharan Africa feed itself? The role of irrigation development in the region's drylands for food security. *Water Int.* **2018**, *43*, 796–814. [CrossRef]
12. Williams, S.B.; Murdock, L.L.; Baributsa, D. Sorghum seed storage in Purdue Improved Crop Storage (PICS) bags and improvised containers. *J. Stored Prod. Res.* **2017**, *72*, 138–142. [CrossRef]
13. Baributsa, D.; Ignacio, M.C. Developments in the use of hermetic bags for grain storage. In *Advances in Postharvest Management of Cereals and Grains;* Burleigh Dodds Science Publishing: London, UK, 2020; pp. 171–198.
14. Flanagan, K.; Robertson, K.; Hanson, C. *Reducing Food Loss Setting a Global Action Agenda;* World Resources Institute (WRI): Washington, DC, USA, 2019.
15. Stathers, T.; Holcroft, D.; Kitinoja, L.; Mvumi, B.M.; English, A.; Omotilewa, O.; Kocher, M.; Ault, J.; Torero, M. A scoping review of interventions for crop postharvest loss reduction in sub-Saharan Africa and South Asia. *Nat. Sustain.* **2020**, *3*, 821–835. [CrossRef]
16. APHLIS. APHLIS+. 2021. Available online: https://www.aphlis.net/en (accessed on 4 May 2021).

17. FAO. Food Loss and Waste Database | FAO | Food and Agriculture Organization of the United Nations. 2021. Available online: http://www.fao.org/platform-food-loss-waste/flw-data/en/ (accessed on 4 May 2021).

18. Engineering for Change. *Landscape Analysis of Post-Harvest Technologies for Mango Production in East Africa*; Engineering for Change: New York, NY, USA, 2020.

19. FSD. *Kenya Opportunities for Financing the Mango Value Chain: A Case Study of Lower Eastern Kenya*; FSD Kenya: Nairobi, Kenya, 2015; Volume 52.

20. Kuhn, M.; Johnson, K. *Applied Predictive Modeling*; Springer: New York, NY, USA, 2013; ISBN 9781461468493.

21. Cheema, J.R. Some general guidelines for choosing missing data handling methods in educational research. *J. Mod. Appl. Stat. Methods* **2014**, *13*, 53–75. [CrossRef]

22. Affognon, H.; Mutungi, C.; Sanginga, P.; Borgemeister, C. Unpacking postharvest losses in sub-Saharan Africa: A Meta-Analysis. *World Dev.* **2015**, *66*, 49–68. [CrossRef]

23. Schwertman, N.C.; Owens, M.A.; Adnan, R. A simple more general boxplot method for identifying outliers. *Comput. Stat. Data Anal.* **2004**, *47*, 165–174. [CrossRef]

24. Ramsey, F.; Schafer, D. *The Statistical Sleuth: A Course in Methods of Data Analysis*, 3rd ed.; Cengage Learning: Boston, MA, USA, 2012; ISBN 9781133490678.

25. Hecke, T. Van Power study of anova versus Kruskal-Wallis test. *J. Stat. Manag. Syst.* **2012**, *15*, 241–247.

26. Tomczak, M.; Tomczak, E. The need to report effect size estimates revisited. An overview of some recommended measures of effect size. *Trends Sport Sci.* **2014**, *1*, 19–25.

27. Rea, L.M.; Parker, R.A. *Designing and Conducting Survey Research: A Comprehensive Guide*, 4th ed.; Jossey-Bass: Hoboken, NJ, USA, 2014; ISBN 9781118767030.

28. Anonymous. *Yieldwise: Kenya Mango Quarterly Report-Q1 (Jan-Mar 2018)*; Technoserve: Nairobi, Kenya, 2018.

29. Ran, Y.; Annebäck, J.; Widmark, E.; Osborne, M. *Boosting Technology Uptake: Some Ideas for Improving Small-Scale Mango Farming in Kenya*; Stockholm Environment Institute: Stockholm, Sweden, 2018.

30. Bally, I.S.E.; Kulkarni, V.J.; Johnson, P.R. Mango production in Australia. *Acta Hortic.* **2000**, *509*, 79–85. [CrossRef]

31. Baloch, M.K.; Bibi, F. Effect of harvesting and storage conditions on the post harvest quality and shelf life of mango (Mangifera indica L.) fruit. *S. Afr. J. Bot.* **2012**, *83*, 109–116. [CrossRef]

32. Singh, V.; Kumar, A.; Kumar, R.; Malik, A.; Kumar, R.; Kumar, A. Adoption of post-harvest management practices by Mango growers of Haryana. *Studies* **2020**, *10*, 158–161.

33. Ambuko, J.L. Tackling Postharvest Losses in Mango among Resource- Poor Farmers in Kenya. *Chron. Hortic.* **2020**, *60*, 28–29.

34. Ntsoane, M.L.; Zude-Sasse, M.; Mahajan, P.; Sivakumar, D. Quality assesment and postharvest technology of mango: A review of its current status and future perspectives. *Sci. Hortic.* **2019**, *249*, 77–85. [CrossRef]

35. Verrinder, N. *End of Project Evaluation for the Scale-Up Phase of the YieldWise Initiative (Mango, Kenya) Final Evaluation Report*; Genesis Analytics: Johannesburg, South Africa, 2018.

36. Chonhenchob, V.; Singh, S.P. Testing and Comparison of Various Packages for Mango Distribution. *J. Test. Eval.* **2004**, *32*, 69–72. [CrossRef]

37. Kitinoja, L.; Saran, S.; Roy, S.K.; Kader, A.A. Postharvest technology for developing countries: Challenges and opportunities in research, outreach and advocacy. *J. Sci. Food Agric.* **2011**, *91*, 597–603. [CrossRef] [PubMed]

38. Sonka, S.T. Measuring to Manage. Reduction and of Food Loss Waste. A Cooperation between the Pontifical Academy of Science and the Rockefeller Foundation. In Proceedings of the Pontifical Academy of Science, Casina Pio IV, Vatican City, 11–12 November 2019; p. 240, ISBN 9788877611154.

39. Mujuka, E.; Mburu, J.; Ogutu, A.; Ambuko, J. Returns to investment in postharvest loss reduction technologies among mango farmers in Embu County, Kenya. *Food Energy Secur.* **2020**, *9*, 1–9. [CrossRef]

40. Carter, M.R.; Laajaj, R.; Yang, D. *Susidies and the Persistence of Technology Adoption: Field Experimental Evidence from Mozambique*; National Bureau of Economic Research: Cambridge, UK, 2014.

41. Omotilewa, O.J.; Ricker-Gilbert, J.; Ainembabazi, J.H. Subsidies for Agricultural Technology Adoption: Evidence from a Randomized Experiment with Improved Grain Storage Bags in Uganda. *Am. J. Agric. Econ.* **2019**, *101*, 753–772. [CrossRef] [PubMed]

 *agriculture*

*Article*

# Techno-Economic Analysis of a Crossflow Column Dryer for Maize Drying in Ghana

George Obeng-Akrofi [1,*], Joseph O. Akowuah [2], Dirk E. Maier [1] and Ahmad Addo [2]

[1] Department of Agricultural and Biosystems Engineering, Iowa State University, Ames, IA 50011, USA; dmaier@iastate.edu

[2] Department of Agricultural and Biosystems Engineering, Kwame Nkrumah University of Science and Technology, Kumasi, Ghana; akowuahjoe@yahoo.co.uk (J.O.A.); ahmadaddo@gmail.com (A.A.)

* Correspondence: georgeo@iastate.edu

**Abstract:** In Ghana, smallholder maize farmers continue to serve as the primary contributor to maize production. These farmers, however, still face challenges of access to appropriate, effective, and efficient drying systems. They continue to depend on open sun drying, which leads to high post-harvest losses. In this study, a 500 kg portable column dryer with a biomass burner heat source was evaluated using maize. Indicators such as drying rate, drying efficiency, and moisture extraction rate were used to assess its technical performance. The economic performance of the drying system was appraised using Net Present Value (NPV), Internal Rate of Return (IRR), Benefit-Cost Ratio (BCR), and Payback Period (PBP). The results showed that the moisture content of maize was reduced from 22.3% to 13.4 ± 2.6% in 5 h at an average drying rate of 1.81%/h and drying efficiency of 64.7%. Utilization of the column dryer for the provision of drying services in a maize-growing community over a 10-year utilization period proved viable with an NPV and IRR of $1633 and 71%, respectively, PBP of less than two years, and BCR of 2.82. Adoption of such low-capacity mobile grain dryers in sub-Saharan Africa would be beneficial in providing timely drying services and improve the socio-economic status of smallholder maize farmers in the region.

**Keywords:** biomass utilization; economic analysis; grain dryer; maize drying; technical performance

**Citation:** Obeng-Akrofi, G.; Akowuah, J.O.; Maier, D.E.; Addo, A. Techno-Economic Analysis of a Crossflow Column Dryer for Maize Drying in Ghana. *Agriculture* **2021**, *11*, 568. https://doi.org/10.3390/agriculture11060568

Academic Editor: Johannes Sauer

Received: 15 May 2021
Accepted: 18 June 2021
Published: 21 June 2021



## 1. Introduction

The importance of maize cannot be overlooked due to its significant role in fighting hunger and improving the socio-economic comfort of the people in sub-Saharan Africa [1]. In Ghana and many countries in sub-Saharan Africa, the crop is the most produced and consumed staple [2]. It is harvested at high moisture content, and as such, it is required that the moisture content be reduced to 12–14% to ensure safe storage for future use in humid and warm countries like Ghana [3,4].

Drying as a post-harvest activity is the most attractive method for conditioning food grains by removing moisture to a safe moisture level. This is because the drying process has proven reliable and flexible for removing moisture from food grains [5]. Although drying of food products is widely applied in various industries globally, it has been a challenge for the smallholder farmer in Ghana and other parts of sub-Saharan Africa.

In Ghana, drying of harvested maize is usually done using traditional drying methods where farmers leave the crop to dry in the field or the open sun next to farmers' homes or along roadsides, either on bare ground or on tarpaulins [2]. This reduces the quality of the dried maize grain and leads to contamination of dried food grains [6]. The situation becomes challenging when harvesting of food grains coincides with unfavorable drying weather conditions such as the rainy season, during which the drying process can take up to 5 days. Drying under such adverse conditions leads to the growth of molds [7], resulting in a considerable loss of food grains in terms of quality.

In attempts to improve the process of crop drying by reducing the drying period, lower drying cost, reliability and accessibility of drying systems, and environmental issues associated with drying, there has been the introduction of varieties of drying system such as solar dryers [8,9], biomass assisted hybrid dryers [10], and other mechanical drying systems. However, most farmers have not widely adopted these interventions, and they continue to dry their harvested produce using the unreliable and inefficient open sun drying method [4]. According to Kaaya and Kyamuhangire [11], such drying technologies are capital intensive to install and operate, making their operation expensive for the smallholder farmer to patronize.

According to Chua and Chou [12], low-cost drying systems are more suitable for smallholder farmers in developing countries. They highlighted that such drying technologies should have low initial capital cost, easy to operate with no complicated electronic and/mechanical protocol, effective in promoting better drying kinetics. The authors also reported that low-cost drying systems should also be easily constructed with available local materials and be run on renewable energy.

The economic and technical appraisal of such drying technologies is vital for their adoption by smallholder farmers. Successful assessment of these low-cost technologies drives their scale-up from research laboratories to commercialization and adoption. This study sought to assess the economic and technical performance of a portable locally fabricated half-tonne capacity column drying system with a biomass burner for maize drying in Ghana.

## 2. Materials and Methods

### 2.1. Technical Performance Study

#### 2.1.1. Study Site

The drying experiment was conducted at the Department of Agricultural and Biosystems Engineering of Kwame Nkrumah University of Science and Technology (KNUST) in the Ashanti Region of Ghana. It is located at 06°41′5.67″ N 01°34′13.87″ W with average temperature and rainfall being 26 °C and 1448 mm, respectively. During the major maize harvesting period, average temperature and relative humidity conditions varies between 31 °C to 32 °C and 74% to 75%, respectively.

#### 2.1.2. Dryer Description

The crossflow column dryer, shown in Figure 1, was fabricated at the Department of Agricultural and Biosystems Engineering, KNUST. It is a mobile drying system that can be transported from one place to another. The dryer consists of three main parts: a cylindrical drying bin, a portable biomass burner, and a fan (blower). The drying bin of 1.20 m height is made up of an inner and an outer wall cylinders that holds grains in the annular space of 0.25 m thickness. The inner and outer wall cylinders with radii 0.15 m and 0.40 m, respectively, make up the plenum and drying chamber of the dryer, respectively. Both the inner and outer wall cylinders were constructed with a perforated metal sheet to allow hot air movement across the inner bin, through the grains, and exit of moist air through the outer bin. The biomass burner serves as the primary heat-generating component of the dryer, and it is made up of heat exchangers. The burner is designed to accommodate a variety of biomass such as corn cobs, wood chippings, and rice husk. From a preliminary experiment conducted on just the biomass burner, corn cobs were fed into to burner at a feed-rate of 12 kg/h. After combustion of biomass, ashes fall through a grate in the combustion chamber for easy collection. The blower sucks air from the biomass burner through the heat exchangers and then forces the drying air through an air delivery tube to the drying bin. At the dryer's plenum, drying air is forced to pass through the drying chamber radially by restricting the movement of the drying air in the plenum by using a stopper.

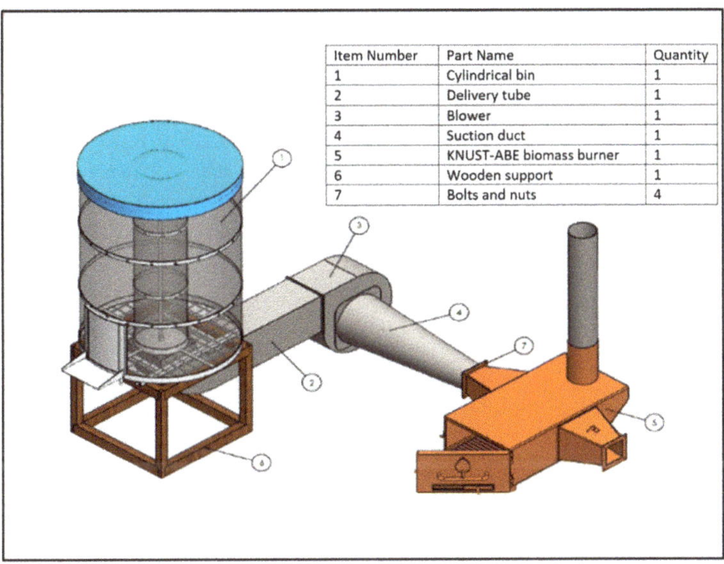

**Figure 1.** CAD model of the crossflow column dryer showing all of its parts.

| Item Number | Part Name | Quantity |
| --- | --- | --- |
| 1 | Cylindrical bin | 1 |
| 2 | Delivery tube | 1 |
| 3 | Blower | 1 |
| 4 | Suction duct | 1 |
| 5 | KNUST-ABE biomass burner | 1 |
| 6 | Wooden support | 1 |
| 7 | Bolts and nuts | 4 |

### 2.1.3. Experimental Procedure

Freshly harvested maize from a local farm was used to evaluate the performance of the dryer. The initial and final moisture contents of the sample was determined using a pre-calibrated John Deere (JD) moisture meter manufactured by AgraTronix™ (Moisture Check Plus™), (SW08120, Moline, IL, USA). Temperature distribution in the dryer was monitored by temperature sensors positioned in the dryer, as shown in Figure 2. From the base of the drying bin, temperature sensors were placed at 15 cm, 30 cm, and 45 cm representing Level 1 (L1), L2, and L3, respectively. At each level (L1, L2, and L3) in the drying chamber, three different sensors were distributed radially at every level. With the use of an anemometer thermo-anemometer (Extech, Melrose, MA, USA) at the suction end of the blower, the airflow rate during the experiment was measured to be 10 m$^3$/s.

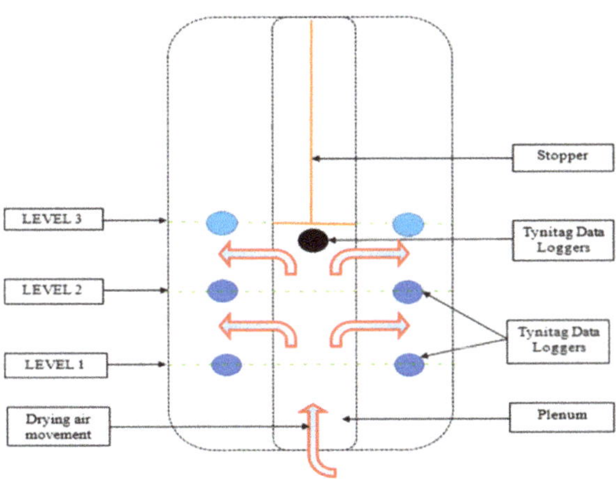


**Figure 2.** Longitudinal cross-section showing points of data collection.

In the process of monitoring moisture loss in the drying bin, a sampling rod was used to take maize samples from the inner and outer sections of the drying bin, at each of the drying levels as shown in Figure 3, and 70 g was taken from the sample lot for analyses. Samples were taken at all levels; L1, L2, and L3, at three different points, P1, P2, and P3. Furthermore, samples were taken from both inner and outer sections to check for moisture reduction in maize at every given point. Representative moisture content and temperature data were analyzed based on the average of the data taken at various points at each level.

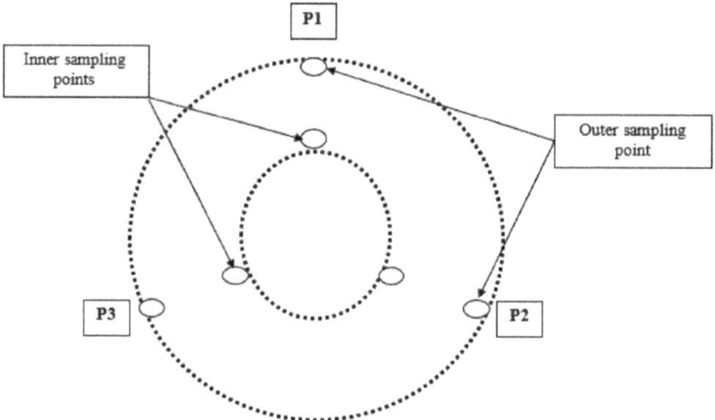


**Figure 3.** The transversal section of the drying chamber showing sampling points for moisture determination.

### 2.1.4. Dryer Performance Indices

Dryer performance indices such as drying rate, moisture extraction rate, and drying efficiency were considered for the performance assessment of the crossflow column dryer. Equations (1)–(3) show the expressions that were used to determine the performance indices.

Drying Rate, DR

The drying rate, DR, was determined using Equation (1).

$$DR = \frac{M_i - M_d}{t}.$$ (1)

where $M_i$ = initial moisture content (% w.b.), $M_d$ = final moisture content (% w.b.), and $t$ = drying time (h).

Moisture Extraction Rate, MER

Moisture extraction rate was determined using Equation (2).

$$M_{ER} = (W_i \times (\frac{M_i - M_d}{100 - M_d}))/t$$ (2)

where $M_{ER}$ = moisture extraction rate (kg/hr), $W_i$ = initial mass of grain dried (kg), $M_i$ = initial moisture content (% w.b.) and $M_d$ = final moisture content (% w.b.) and $t$ = drying time (h).

Drying Efficiency, $\eta$

The drying efficiency, which gives the ratio of the energy used to evaporate moisture from the product to the energy provided by the drying air, was determined using Equation (3).

$$\eta = \frac{M_{ER} L_v}{M_{air} C p_{air} \Delta T} \times 100$$ (3)

where $\eta$ = drying efficiency (%), $M_{ER}$ = rate of moisture evaporation (kg/hr), $L_v$ = latent heat of vaporization of water (kJ/kg), $M_{air}$ = mass flow rate of air (kg/hr), $Cp_{air}$ = specific heat capacity of air (kJ/kg. °C) and $\Delta T$ = change in temperature between the ambient and drying air (°C).

The drying efficiency was converted in specific energy values in MJoules per kilogram of moisture removed using Equation (4).

$$Specific\ Energy\ Consumption = \frac{M_{Biomass} \times H_{Biomass}}{M_w}. \tag{4}$$

where $M_{Biomass}$ = mass of biomass combusted during drying (kg), $H_v$ = heat value of biomass (kJ/kg), $M_w$ = mass of moisture removed from maize during drying (kg).

### 2.2. Economic Performance Study

The economic assessment on the column drying system was appraised from the perspective of a smallholder maize farmer using the discounted method where the time value of money is considered.

#### 2.2.1. Case Study Scenario

The following assumptions were made for the scenario considered for the study:

1. A smallholder farmer owns the column dryer for drying maize.
2. The farmer owns/cultivates maize on a 2-ha farmland.
3. The farmer harvests 1.5 t/ha of maize as estimated by the Ministry of Agriculture [13] in Ghana.
4. The farmer uses the column dryer to dry all his/her maize.
5. The farmer provides drying services to other maize farmers in his community.

The model scenario outlay is shown in Figure 4.

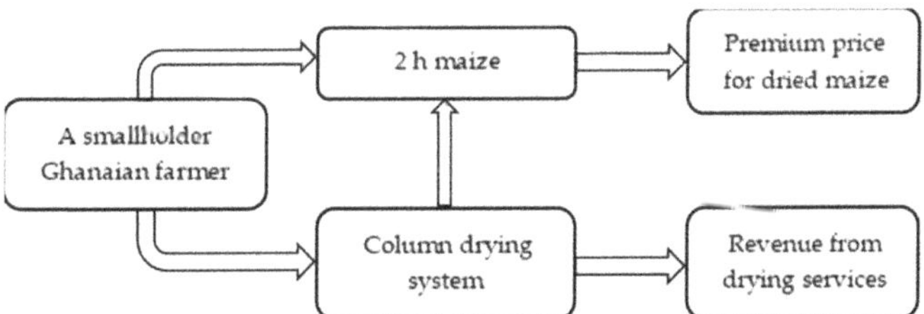


**Figure 4.** Schematic description of the model scenario considered in the study.

#### 2.2.2. Estimation of Cost and Revenue

The cost component was made up of the investment cost and cost of operation and maintenance. The investment cost consisted of all the expenses required to set up the complete drying system. This included the cost associated with the fabrication of the drying column, biomass burner, and an electric blower. The cost of electricity for operating the drying system during operation and a flat rate of 2% of equipment and machinery cost was assumed to be operation and maintenance costs, respectively. The cost of fuel, which comprised of the cost of corn cobs, was not considered since the biomass residue is anticipated to be readily available in the study area. The revenue generation stream was sourced from the price charged for providing drying services to other farmers in the community. The drying charge and quantity of maize anticipated to be dried were presented in the economic model to determine the annual total revenue generated.

#### 2.2.3. Economic Appraisal

Net Present Value (NPV), Internal Rate of Return (IRR), Benefit-Cost Ratio (BCR), and Payback Period were used to evaluate the economic performance of the column dryer.

NPV uses a discounting method for evaluating the economic viability of the investment and gives the value of all future cash flows in today's currency. This provides a true measure of an investment's economic feasibility. It presents the present value of cash in and outflows [14]. A positive NPV indicates an economically viable investment or project, while a negative one shows that it is not economically feasible to carry out such investment or project [15]. Equation (5) was used to calculate the NPV.

$$NPV = \sum_{t=0}^{N} a_t S_t \tag{5}$$

where $S_t$ = net cash flow at a specific time ($t$), $N$ = number of years (10 years), and $a_t$ = financial discount factor, which was calculated using Equation (6).

$$a_t = \frac{1}{(1+i)^t} \tag{6}$$

where $t$ = time from 0 and 10 years and $i$ = the discount rate (%).

IRR is the discount rate that makes the net present value of all cash flows from a particular investment equal to zero. Generally, the higher the IRR, the more desirable it is to undertake the project [16]. IRR was determined using Equation (7).

$$NPV = \sum \frac{S_t}{(1 + IRR)^t} = 0 \tag{7}$$

PBP is the number of years it takes to recover an investment's initial cost. It provides a simple way to assess the economic merit of investments. Equation (8) was used to calculate the PBP.

$$PBP = \frac{C_i}{S} \tag{8}$$

where $C_i$ = initial investment cost and $S$ = net cash flow

BCR is the ratio of total discounted benefit to total discounted cost. Projects with a benefit-cost ratio greater than 1 have greater benefits than costs; hence, they have positive net benefits. The higher the ratio, the greater the benefits relative to the costs. It was calculated using Equation (9) [17].

$$BCR = \sum \left( B_i / (1+d)^i \right) \div \sum \left( C_i / (1+d)^i \right) \tag{9}$$

where $B_i$ = benefit of the project in year $i$ ($i$ = 0 to 10 years), $C_i$ = cost of the project in year $i$, and $d$ = discount rate.

### 2.2.4. Financial Assumptions

The following financial assumptions were made during the assessment:

1. Cash flows were discounted over a ten-year period based on the expected useable lifetime of the Crossflow Column Dryer.
2. An operation period of three months per year is considered since the major harvest season starts in June/July and ends in August/September. This harvesting period coincides with the rainy season, making maize drying a challenge for a typical maize farmer in Ghana.
3. The dryer is expected to be operated at 500 kg full capacity.
4. The dryer will be operated by the farmer, and as such, no cost for labor is expected for operating the dryer.
5. A discount rate of 14%, which is Ghana's discount rate of February 2019 [18], was used for the analysis.
6. A percentage of 2% of the investment cost was assumed to be maintenance cost in the financial analysis.

2.2.5. Sensitivity Analysis

Sensitivity analysis was carried out by varying one parameter of the economic model and determining the effect of that change on the economic indicators. Analysis of such sort is needed to measure the effect of changes in critical variables of investments on the economic indicators [14,19]. The critical variables considered for the sensitivity analysis included:

1. Discount rate variation: Discount rate is one of the key variables that determine the NPV of investments. The analysis was made using a 7%, 14%, 21%, and 28% discount rate. The basis for considering these variations was based on the variation of discount rate in Ghana since 2000 [18], which has witnessed minimum and maximum values of 12.5% and 27.5%, respectively.

2. Drying prices of $0.75, $0.94, $1.13, and $1.32 per bag of maize were considered for the analysis. The drying price was varied from $0.75 to $1.32 per bag of maize. These drying prices are lower than the least price charged for drying maize by other installed mechanical drying facilities in most farming communities in Ghana, which is about $ 2.80.

3. The investment cost, which consists of setting up the complete drying system, was also varied to determine its effect on the viability of the case scenario. This was done because it is anticipated that any investor who may deal in the manufacture and distribution of the column drying system would want to gain profit by selling the dryer at a higher price than the estimated investment price. Increments were made at 20%, 50%, and 80% more of the base investment cost.

## 3. Results

### 3.1. Technical Performance Evaluation

3.1.1. Temperature Variation during Drying

Temperature variation in the plenum and the drying chamber, compared to the ambient during the drying process, is shown in Figure 5. The plenum temperature increased steadily from 38 °C to a maximum of 58 °C within the 5 h drying period. This resulted in a corresponding increase in the air temperature in the drying chamber from 35 °C to a maximum of 44 °C during the same period. Average temperatures of 51.5 ± 4.8 °C and 38.5 ± 2.8 °C were recorded at the plenum and drying chamber of the column dryer, respectively. The mean temperature in the drying chamber was 9 °C higher than the ambient temperature.

As a common phenomenon with column drying systems, there was not much difference in the drying air temperatures at different levels, L1, L2, and L3, as shown in Figure 6. In the drying chamber, average temperatures of 39.4 ± 2.3, 36.1 ± 1.2 and 39.9 ± 2.5 °C were recorded at L1, L2, and L3, respectively. A similar observation was made by Alam et al. [20] and Kumar et al. [21], who worked on the performance of a column drying system where there were no substantial differences in temperatures between the top, middle, and bottom sections of the drying chamber.

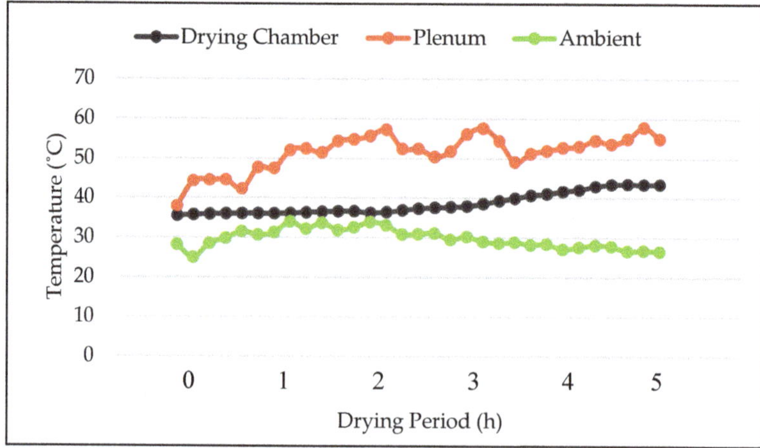

**Figure 5.** Temperature variations in the plenum and drying chamber during the experiment.

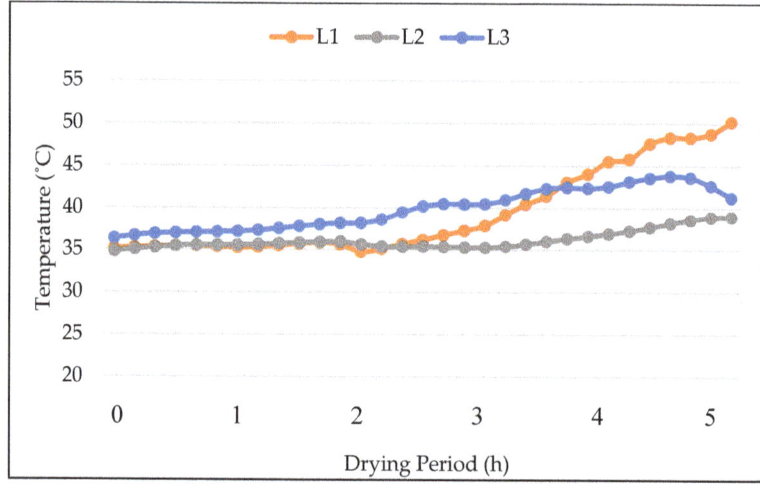

**Figure 6.** Temperature variations at L1, L2, and L3 in the drying chamber.

### 3.1.2. Moisture Content Variation During Drying

The variation in moisture content at the inner and the outer sections of the column dryer during the drying experiment is shown in Figure 7. The grain moisture content decreased with drying time, with grains closer to the plenum (inner section) reaching a lower moisture content after 5h of drying compared to grains close to the outer edge of the drying column (outer section). The drying process occurred in the falling rate period, as shown in Figure 7, where the moisture content of maize decreased from 22.3% wet basis (w.b.) to 11.6 ± 0.3% and 15.4 ± 0.3% for grains at the inner and outer sections, respectively, within the 5 h drying period.

Variations in the moisture content of grains sampled transversally across the drying chamber could be attributed to variation in the drying air temperature. It is forced through the drying bed with all the grains not fully exposed to the same drying air condition which is established by plenum. As demonstrated in Figure 8, the grain mass at the inner section of the dryer is exposed to drying air of high temperature compared to the grain mass at the outer section. As drying air moves from the plenum across the mass of maize, moisture is lost from the grains to the drying air, thereby increasing the humidity of the drying air along the transversal depth of the maize grains towards the outer section. According to Chakraverty and Singh [22], this is a common phenomenon in deep bed dryers, which leads to grains at the inner section drying faster compared to grains at the outer section.

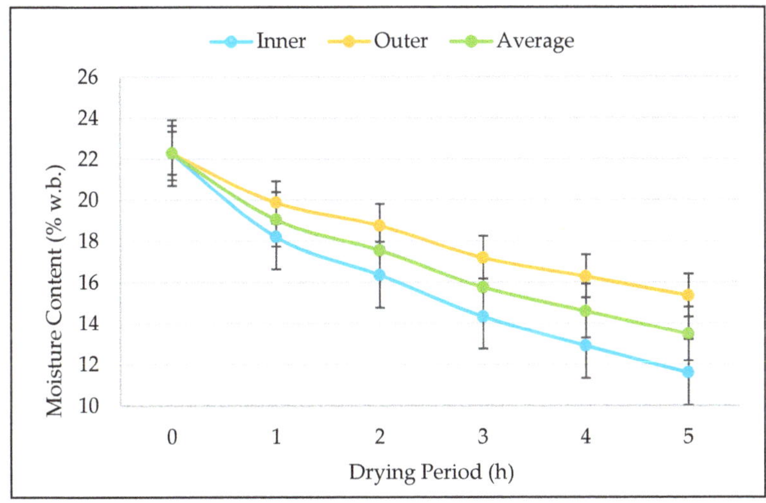


**Figure 7.** Moisture content variation with time at the inner and outer sections of the dryer.

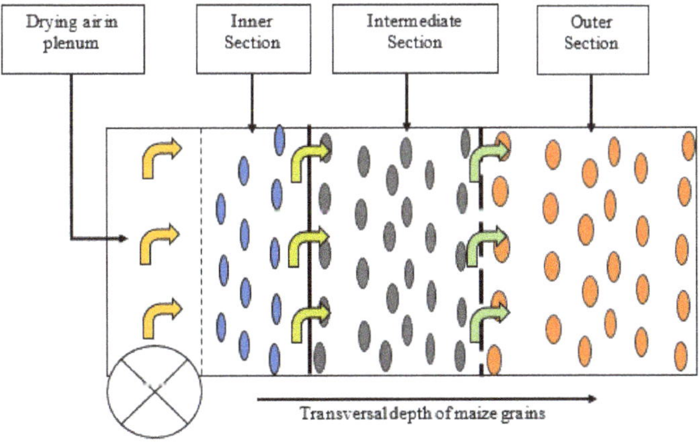


**Figure 8.** Schematics of the deep bed drying principle.

The findings of this study are corroborated by Kumar et al. [21]. They made similar observations in moisture content variations of wheat and maize grains dried in a similar deep bed dryer. The final moisture content of dried samples was 10.76% and 10.84% for the inner and outer sections, respectively.

The moisture contents of maize grains at different levels L1, L2, and L3, as shown in Figure 2, along the longitudinal depth at the inner and outer sections of the dryer was analyzed. The analysis of MC of grains across the longitudinal depth at both the inner and outer section of the dryer did not vary significantly, as shown in Figure 9. Grains at different levels at both the inner and outer sections for the drying chamber reached $11.6 \pm 0.3\%$ and $15.4 \pm 0.3\%$ moisture contents, respectively.

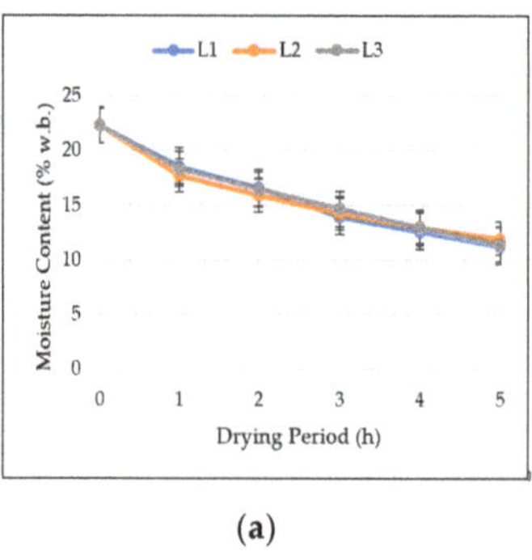
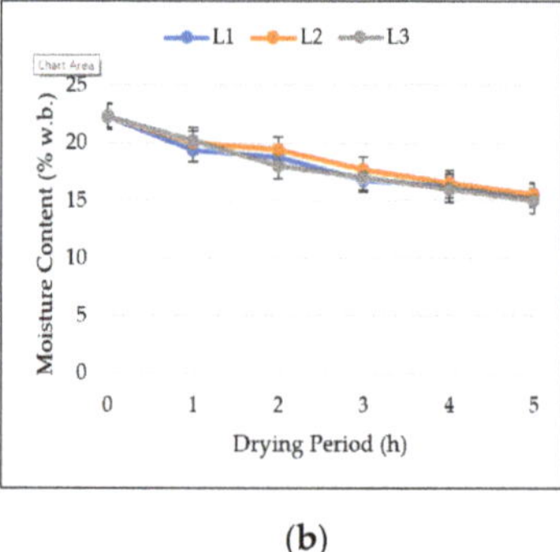

**Figure 9.** Variation of moisture contents at L1, L2, and L3 at the (**a**) inner section of the drying chamber; and (**b**) outer sections of the drying chamber.

To ensure that maize dried in a column dryer reaches the safe moisture content of about 13% (w.b.) before storage in tropical weather conditions like Ghana, Kaaya and Kyamuhangire [11] recommend that thorough mixing of grains close to the inner and outer sections during unloading should be encouraged.

### 3.1.3. Dryer Performance Specification

Table 1 shows the column dryer's technical performance, which satisfies the drying needs of smallholder maize farmers. The average temperature distribution of 38.5 ± 2.8 °C in the drying chamber was not too much of a drying temperature that can result in the loss of seed viability [2,23,24]. This is an essential consideration for adopting grain dryers as about 80% of smallholder grain farmers rely on their seed stock from their previous harvest [2,25]. Hence, using dryers that tend to reduce the seed viability of their harvest, usually due to high drying temperatures of 70–100 °C, should be avoided. More so, the designed capacity of the grain dryer matches the harvesting rate of grain farmers, making the column dryer suitable for adoption by grain farmers [12].

**Table 1.** Summary of dryer technical performance.

| Parameter | Value |
|---|---|
| Dryer | |
| Initial mass of maize | 250 kg |
| Initial moisture content | 22.3% |
| Final moisture content | 13.3 ± 2.6% |
| Average drying time | 5h |
| Average drying rate | 1.8%/h |
| Drying Efficiency | 64.7% |
| Specific energy consumption | 9.23 MJ/kg of moisture |
| Average Drying Temperature | 38.5 ± 2.8 °C |
| Average MER | 5.1 kg/h |

### 3.2. Economic Performance Evaluation

### 3.2.1. Technical and Financial Analysis of the Drying System

Table 2 presents the financial and technical parameters considered for the operation of the dryer for the case scenario. Based on a drying capacity of 500 kg of maize per batch, it is expected that two batches of drying could be achieved per day. Performance study of the drying system shows that a farmer can dry his maize from an initial moisture content of about 22% (w.b.) to a safe moisture content of 13% (w.b.) within a period of 5 h. This translates to a 720-h operational period of three months from June/July to August/September. This period happens to be the time when over 60% of Ghana's maize produced in the major production season by smallholder farmers along the transition belt of Ghana is harvested. Drying services are critically needed during this period as the harvesting period normally coincides with the onset of rains used for the minor season maize cultivation. Based on the dryer's specified capacity and the drying time, it is estimated that 72 tonnes of maize (554 bags) are expected to be processed within the operational period of three months in a year.

**Table 2.** Technical and financial parameters considered for the business model proposed in the study.

| Parameter | Value |
| --- | --- |
| Capacity of drier (kg) | 500 |
| Number of batches per day | 2 |
| Number of hours required per batch of drying | 5 |
| Number of operational days per week | 6 |
| Number of operational hours per week | 60 |
| Number of operational months per year | 3 |
| Operational hours per year (h) | 720 |
| Quantity of maize per bag (kg) | 130 |
| Number of bags of maize dried per day | 8 |
| Number of bags dried of maize per week | 46 |
| Quantity of maize dried per year (t) | 72 |
| Number of bags of produce processed per year | 554 |
| Estimated amount of crop produced per year in the district (t) | 20,000 |
| Number of dyers required to process the total available maize | 278 |
| Lifespan of drying system (years) | 10 |
| Price charged for drying a bag of maize ($) | 0.94 |

Furthermore, according to MoFA-SRID [13], most smallholder maize farmers in Ghana cultivate an estimated average farmland size of 2 ha at an average yield of 1.5t/ha, correspond to 3 tonnes of maize produced by a farmer in a cropping season. This quantity of maize is projected to be dried within three operation days of the dryer. This means the dryer would be available to other smallholder maize farmers who otherwise will use the unreliable open-sun method for drying their maize. In that regard, about 24 smallholder maize farmers (72 tonnes of maize ÷ (2 ha/smallholderfarmer × 1.5 tonnes of maize/ha)), therefore, could rely on the dryer for their drying services within the operational period used for the case scenario.

### 3.2.2. Cost and Returns on Investment

The initial capital cost for the complete drying system is presented in Table 3. The main cost component, as shown in Table 3, is the fan cost, estimated to be 46.9% of the total investment cost.

**Table 3.** Capital cost of the drying system.

| Investment | Cost Value (USD) | % of Total Investment Cost |
| --- | --- | --- |
| Column dryer plus auxiliary units (air delivery ducts) | 189.00 | 31.3 |
| Biomass Burner | 132.00 | 21.9 |
| Blower (Fan) | 283.00 | 46.9 |
| Total Fixed Cost | 604.00 | 100.0 |

The costs associated with the operation and maintenance of the dryer are presented in Table 4. An amount of $12 representing 2% of the total investment cost is allocated for maintenance and overhead expenses. With a fan of a motor rating of 0.75 kW, a total power of 540 kWh required per an operation cycle, an amount of $82 is estimated as the cost of electricity for the operation of the drying system. The cost of electricity was estimated at $0.15 per kWh in Ghana [26].

**Table 4.** Operation and maintenance cost of running the drying system.

| Operations and Maintenance | Cost Value (USD)/Operation Cycle |
|---|---|
| Maintenance and overhead expenses (2% of investment cost) | 12.00 |
| Cost of electricity | 82.00 |
| Total running/variable Cost | 94.00 |

At a projected six days of operation per week for the three-month operational period per year, it is expected that the column dryer will be used to dry 72,000 kg of maize per year. At a drying price of $0.94 charged per bag (130 kg) of maize to be dried, total revenue of $523 is anticipated since 554 bags of maize will be dried during the operation cycle of three months in a 12-month year.

### 3.2.3. Economic Appraisal of the Business Model

For the case scenario considered in the study to be financially viable, Abbood et al. [15] reported that an NPV of a positive value and an IRR greater than the present interest rate (14% for the case study) should be targeted. The variation of the NPV and IRR over the operation period is presented in Figure 10. Economic analysis of the case scenario revealed that, at a discount rate of 14% over a projected 10-year lifespan of the drying system, an NPV and IRR of $1633 and 71%, respectively, can be achieved at a payback period of 1.41 years after operations begin. The economic indicators' values prove the viability of the case scenario where a farmer can invest in owning and running the column dryer as a business in the study area. The study results agree with studies by Adams et al. [14] and Mensah et al. [27], who worked on the financial feasibility of a mango-chip processing and small-scale meat production, respectively, in Ghana. In their studies, the authors reported the economic viability of their case studies in Ghana, where there were similar trends in NPV and IRR for the operational period of the individual startups.

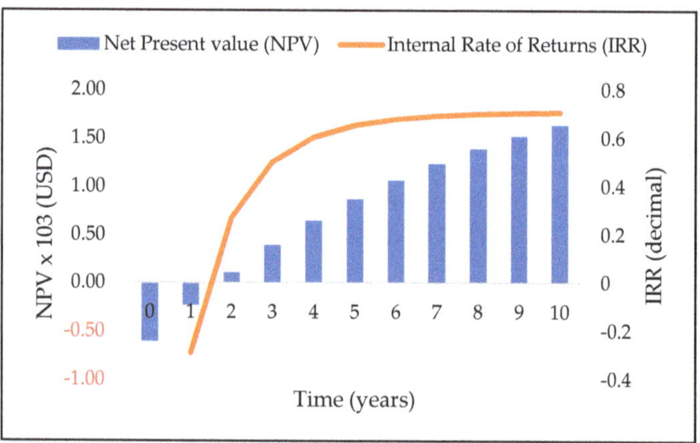


**Figure 10.** Variation of NPV and IRR over the 10-year operation period.

### 3.2.4. Sensitivity Analysis

The effect of price variations for maize drying ($/bag of maize) on the economic outlook of the case scenario is presented in Table 5. The analysis of the results shows that at a constant discount rate of 14%, the NPV, IRR, and BCR values increased considerably at an increased cost of drying. For instance, at 20% increase in drying charge from $0.94 to $1.13, the NPV increased by 33% (from $1633.00 to $2174.00). A similar increasing trend was observed for the other economic indicators as IRR and BCR increased by 24% and 20%, respectively. However, the PBP decreased by 20% when the drying price was increased by 20%. This indicates that an investor will regain his investment in a relatively shorter time as the price charged for drying maize is raised, and more revenue is expected. However, a reduction of the drying price also showed a reverse effect on the economic indicators seen in Table 4. Similar results have been reported by Abbood et al. [15], who worked on the financial analysis of a 1 MW PV plant, and observed that NPV and IRR increased considerably with an increase in the selling price of electricity. The result shows that variations in the cost of maize drying using the drying system can affect the economic potential of the business model.

**Table 5.** Variation of NPV, IRR, PBP, and BCR with drying price charged per bag (130 kg) of maize.

| Drying Charge (USD/Bag of Maize) | NPV ($) | IRR (%) | PBP (years) | BCR |
|---|---|---|---|---|
| 0.75 | 1084 | 53 | 1.86 | 2.26 |
| 0.94 | 1633 | 71 | 1.41 | 2.83 |
| 1.13 | 2174 | 88 | 1.13 | 3.39 |
| 1.32 | 2723 | 106 | 0.95 | 3.96 |

The effect of discount rate on the economic indicators at a constant drying price of $ 0.94 per bag of maize was also investigated, and the result is presented in Table 6. The analysis shows that when the discount rate increases, the economic viability of the business model tends to be affected negatively with respect to the NPV and BCR and vice versa. For instance, with a 50% increase in the discount rate, from 14% to 21%, NPV and BCR decreased by 31% and 17%, respectively. On the other hand, PBP and IRR were not affected by variations in the discount rate. This is attributed to the independence of both economic indicators on the discount rate reported by Abbood et al. [15].

**Table 6.** Variation of NPV, IRR, PBP, and BCR in relation to the discount rate.

| Discount Rate (%) | NPV ($) | IRR (%) | PBP (years) | BCR |
|---|---|---|---|---|
| 7 | 2409 | 71 | 1.41 | 3.47 |
| 14 | 1633 | 71 | 1.41 | 2.83 |
| 21 | 1135 | 71 | 1.41 | 2.34 |
| 28 | 798 | 71 | 1.41 | 1.98 |

The final sensitivity analysis was done in anticipation of manufacturers, investors, and/or distributors who may sell or distribute the drying system. The study considered a situation where an operator tends to buy the column drying system from a manufacturer or an investor at a cost that is 20%, 50%, and 80% more than the actual manufacturing cost (investment cost). The profit margins on the investment cost of the dryer were simulated at a drying charge of $ 0.94/bag and using a discount rate of 14%. The reflection in the economic indicators is presented in Table 7 to see their effect on the economic indicators. NPV, IRR, and BCR tend to decrease as the profit margin on the investment cost increases, although PBP increases. This is justified since a higher investment cost means an extended period to break even on an investment. Although the economic feasibility of the case scenario tends to decline in values, even at a higher profit margin of 80% increase on the investment cost, the economic indicators demonstrate a viable case with a positive NPV and IRR of $1100 and 37%, respectively.

**Table 7.** Variation of NPV, IRR, PBP, and BCR in relation to increasing dryer cost.

| Estimated Investment Cost (USD) | NPV ($) | IRR (%) | PBP (years) | BCR |
|---|---|---|---|---|
| 604 (base cost) | 1633 | 71 | 1.41 | 2.83 |
| 725 (20% more) | 1500 | 58 | 1.70 | 2.54 |
| 906 (50% more) | 1300 | 46 | 2.14 | 2.21 |
| 1089 (80% more) | 1100 | 37 | 2.59 | 1.95 |

## 4. Conclusions

The techno-economic performance of a half-tonne capacity crossflow column dryer with a biomass burner heat source was successfully assessed. Maize at 22.30% was dried to a final moisture content of 13.25% within a period of 5 h. The average drying rate recorded during the study was 1.81%/h with a drying efficiency of 64.65%. The economic viability of a case study was assessed to be viable for a smallholder maize farmer or an investor who operates a unit to provide drying services to maize farmers. The economic analysis over a 10-year lifespan operation of the dryer resulted in an NPV of $1633 and IRR of 71%. At an assumed drying charge of $0.94/bag, which is one-third lower the drying charges of a typical commercial drying facility in Ghana, an investor is expected to recoup his investment in the shortest possible time at a PBP of 1.48 years with a BCR of 2.55. Finally, the positive performance indicators provide confidence for scale-up and adoption by smallholder maize farmers in Ghana. It is recommended that manual mixing of grains should be incorporated in the unloading of maize from the drying system to minimize the difference in moisture content between the grains at the inner and outer sections. Smallholder grain farmers should adopt portable and mobile, low-cost grain drying systems in Ghana and sub-Saharan Africa. In order to facilitate this adoption, smallholder farmers should be brought the knowledge of the technical and economic performance of these systems. In addition to this, smallholder farmers should be equipped with entrepreneurial skills to better utilize such technologies for economic benefits and provide on-time drying services to mitigate the substantial loss of grain in most rural grain-growing communities in sub-Saharan Africa.

**Author Contributions:** The following authors contributed to the work: Conceptualization, D.E.M., J.O.A., A.A.; methodology, G.O.-A., J.O.A. and D.E.M.; software, G.O.-A.; formal analysis, G.O.-A.; investigation, G.O.-A.; resources, D.E.M., J.O.A. and A.A.; writing—original draft preparation, G.O.-A.; writing—review and editing, J.O.A. and D.E.M.; supervision, J.O.A., A.A. and D.E.M.; project administration, D.E.M.; funding acquisition, D.E.M. All authors have read and agreed to the published version of the manuscript.

**Funding:** This study is based on research supported by the Post-Harvest Engineering and Feed Technology Group, Department of Agricultural & Biosystems Engineering, Iowa State University.

**Institutional Review Board Statement:** Not applicable.

**Informed Consent Statement:** Not applicable.

**Data Availability Statement:** Not applicable.

**Conflicts of Interest:** The authors declare no conflict of interest.

## References

1. Fisher, M.; Abate, T.; Lunduka, R.W.; Asnake, W.; Alemayehu, Y.; Madulu, R.B. Drought Tolerant Maize for Farmer Adaptation to Drought in Sub-Saharan Africa: Determinants of Adoption in Eastern and Southern Africa. *Clim. Chang.* **2015**, *133*, 283–299. [CrossRef]
2. Akowuah, J.O.; Maier, D.; Opit, G.; McNeill, S.; Amstrong, P.; Campabadal, C.; Ambrose, K.; Obeng-Akrofi, G. Drying Temperature Effect on Kernel Damage and Viability of Maize Dried in a Solar Biomass Hybrid Dryer. *Open J. Appl. Sci.* **2018**, *8*, 506–517. [CrossRef]
3. Bala, B.K. *Drying and Storage of Cereal Grains*; John Wiley & Sons: Hoboken, NJ, USA, 2016. [CrossRef]

4.  Oppong Akowuah, J.; Mensah, L.D.; Chan, C.; Roskilly, A. African Journal of Microbiology Research Effects of Practices of Maize Farmers and Traders in Ghana on Contamination of Maize by Aflatoxins: Case Study of Ejura-Sekyeredumase Municipality. *Afr. J. Microbiol. Res.* **2015**, *9*, 1658–1666. [CrossRef]

5.  Jokiniemi, H.T.; Ahokas, J.M. Drying Process Optimisation in a Mixed-Flow Batch Grain Dryer. *Biosyst. Eng.* **2014**, *121*, 209–220. [CrossRef]

6.  Mutai, E.B.K.; Tonui, K.S.; Mutuli, D.A.; Too, K.V. Design and Evaluation of Solar Grain Dryer with a Back-up Heater. *Res. J. Appl. Sci. Eng. Technol.* **2014**, *7*, 3036–3043.

7.  Folaranmi, J. Design, Construction and Testing of Simple Solar Maize Dryer. *Leonardo Electron. J. Pract. Technol.* **2008**, *1*, 283–299.

8.  Janjai, S. A Greenhouse Type Solar Dryer for Small-Scale Dried Food Industries: Development and Dissemination. *Int. J. Energy Environ.* **2012**, *3*, 383–398.

9.  El-Sebaii, A.; Shalaby, S.M. Solar Drying of Agricultural Products: A Review. *Renew. Sustain. Energy Rev.* **2012**, *16*, 37–43. [CrossRef]

10. Okoroigwe, E.C.; Eke, M.N.; Ugwu, H.U. Design and Evaluation of Combined Solar and Biomass Dryer for Small and Medium Enterprises for Developing Countries. *Int. J. Phys. Sci. Full Length Res. Pap.* **2013**, *8*, 1341–1349. [CrossRef]

11. Kaaya, A.; Kyamuhangire, W. Drying Maize Using Biomass-Heated Natural Convection Dryer Improves Grain Quality During Storage. *Artic. J. Appl. Sci.* **2010**, *11*, 967–974. [CrossRef]

12. Chua, K.J.; Chou, S.K. Low-Cost Drying Methods for Developing Countries. *Trends Food Sci. Technol.* **2003**, *14*, 519–528. [CrossRef]

13. MoFA SRID. Annual Progress Report. 2014. Available online: http://mofa.gov.gh/site/?page_id=79 (accessed on 22 February 2021).

14. Adams, F.; Amankwah, K.; Wongnaa, C.A.; Honny, E.P.; Peters, D.K.; Asamoah, B.J.; Coffie, B.B.; Yildiz, F. Financial Analysis of Small-Scale Mango Chips Processing in Ghana. *Cogent Food Agric.* **2019**, *5*, 1679701. [CrossRef]

15. Abbood, A.A.; Salih, M.A.; Mohammed, A.Y. Modeling and Simulation of 1mw Grid Connected Photovoltaic System in Karbala City. *Int. J. Energy Environ.* **2018**, *9*, 153–168.

16. Baum, W.C.; Tolbert, S.M. Investing in Development Lessons of World Bank Experience. *Dev. South. Afr.* **1986**, *3*, 199–218. [CrossRef]

17. Gittinger, J.P. *Economic Analysis of Agricultural Projects*; John Hopkins University Press: Baltimore, MD, USA, 1982.

18. Bank of Ghana—Central Bank. Available online: https://www.bog.gov.gh/?option=comwrapper&view=wrapper&Itemid=255 (accessed on 22 February 2021).

19. Saltelli, A.; Ratto, M.; Andres, T.; Campolongo, F.; Cariboni, J.; Gatelli, D.; Saisana, M.; Tarantola, S. Introduction to Sensitivity Analysis. In *Global Sensitivity Analysis. The Primer*; John Wiley & Sons, Ltd.: Hoboken, NJ, USA, 2008; pp. 1–51. [CrossRef]

20. Alam, M.A.; Kumer, S.C.; Alam, M. Field Performance of BAU-STR Dryer in Rural Area of Bangladesh. *Asian J. Poverty Stud.* **2017**, *3*, 170–174.

21. Kumar, V.; Rajak, D.; Kalita, P.; Rausch, K. Performance Evaluation of Modified STR Dryer. *Int. J. Chem. Stud.* **2018**, *6*, 1915–1918.

22. Chakraverty, A.; Singh, R. *Postharvest Technology and Food Process Engineering*; CRC Press: Boca Raton, FL, USA, 2014.

23. Bosomtwe, A.; Danso, J.K.; Osekre, E.A.; Opit, G.P.; Mbata, G.; Armstrong, P.; Arthur, F.H.; Campbell, J.; Manu, N.; McNeill, S.G.; et al. Effectiveness of the Solar Biomass Hybrid Dryer for Drying and Disinfestation of Maize. *J. Stored Prod. Res.* **2019**, *83*, 66–72. [CrossRef]

24. Weiss, W.; Buchinger, J. *Solar Drying*; AEE INTEC Publication: Gleisdorf, Austria, 2012.

25. Etwire, P.M.; Atokple, I.D.K.; Buah, S.S.J.; Abdulai, A.L.; Karikari, A.S.; Asungre, P. Analysis of the Seed System in Ghana. *Int. J. Adv. Agric. Res.* **2013**, *1*, 7–13.

26. PURC. Approved Electricity Tariffs. Available online: http://purc.com.gh/purc/sites/default/files/approved_electricity_tariffs_1719.pdf (accessed on 22 February 2021).

27. Mensah, J. Assessing the Feasibility of Commercial Meat Rabbit Production in the Kumasi Metropolis of Ghana. *Am. J. Exp. Agric.* **2014**, *4*, 183–192. [CrossRef]

*Article*

# Effect of Storage Conditions on Storability and Antioxidant Potential of Pears cv. 'Conference'

**Grzegorz P. Łysiak [1,*], Krzysztof Rutkowski [1] and Dorota Walkowiak-Tomczak [2]**

[1] Department of Ornamental Plants, Dendrology and Pomology, Poznan University of Life Science, ul. Dąbrowskiego 159, 60-594 Poznań, Poland; krzysztof.rutkowski@up.poznan.pl

[2] Department of Food Technology of Plant Origin, Poznan University of Life Sciences, ul. Wojska Polskiego 28, 60-637 Poznań, Poland; dorota.walkowiak@up.poznan.pl

* Correspondence: glysiak@up.poznan.pl; Tel.: +48-61-848-7946

**Abstract:** Late pear cultivars, such as 'Conference', can be stored for a long period if kept in good storage conditions. A three-year study (2011–2013) compared the impact of six-month storage using four technologies—normal atmosphere, normal atmosphere + 1-methylcyclopropene (1-MCP), controlled atmosphere, and controlled atmosphere + 1-MCP—on the quality parameters of 'Conference' pears, such as mass loss, firmness, total soluble solids, acidity, antioxidant capacity, and the incidence of diseases and disorders. Additionally, the study analysed different storage conditions in terms of profitability, based on the market prices for pears in the seasons during which the pears were stored. The storage conditions had a very strong influence on the fruit quality parameters, and were found to affect most visibly the mass loss and the incidence of postharvest diseases and disorders. The storage of 'Conference' pears for 180 days in normal atmosphere is not economically viable, even if the fruit is subjected to 1-MCP treatment; at the same time, it is profitable to store 'Conference' pears in controlled atmosphere for the same period, no matter whether 1-MCP was applied or not.

**Keywords:** rootstock; 1-MCP; cost-effectiveness of technology; controlled atmosphere; cold storage; ORAC; TSS; acidity; firmness

**Citation:** Łysiak, G.P.; Rutkowski, K.; Walkowiak-Tomczak, D. Effect of Storage Conditions on Storability and Antioxidant Potential of Pears cv. 'Conference'. *Agriculture* **2021**, *11*, 545. https://doi.org/10.3390/agriculture11060545

Academic Editor: Dirk E. Maier

Received: 19 May 2021
Accepted: 11 June 2021
Published: 13 June 2021

**Publisher's Note:** MDPI stays neutral with regard to jurisdictional claims in published maps and institutional affiliations.

## 1. Introduction

Pears are the most cultivated pome species in the world after apples [1]. Pear cv. 'Conference' is the most important cultivars in Europe with a yearly production of around 1 million tonnes [2]. It is also one of the most commonly stored pear cultivars. The storability of pear, which is a typical climacteric species [3], depends on various factors [4], most notably, the optimal harvest date [5], the fruit cooling rate after harvest [6], the degree of pollination and rootstock [7], storage conditions [8], fertilization, and health [9]. For example, studies conducted on the influence of pollination on the quality properties of 'Conference' pears showed that the number of seeds was positively correlated with fruit mass and calcium content, but was negatively correlated with total soluble solids and firmness [7], and the initial TSS value and firmness are crucial for storability assessment.

Sometimes, however, as in the case of apples, other factors may contribute, such as the rootstock used, which, by affecting the nutrition of trees, can influence the properties of the stored fruit [10]. Temperature is a factor that crucially influences the rate of any reaction, in particular, of respiration-related reactions [11]. The optimal fruit storage temperature depends on the species, and sometimes on the cultivar [12]. Pears belong to the few fruit species which suffer no damage if stored at a temperature below zero. The optimal temperature is between $-1$ and $0\ {}^\circ C$, with a clear preference towards the negative temperature [13]. As early as 1964, Porrit found out that the storage life of 'd'Anjou' and 'Bartlett' pear was, respectively, 35% and 40% longer at $-1\ {}^\circ C$ than at $0\ {}^\circ C$ [14].

Controlled atmosphere (CA) storage significantly extends the storability of pears compared to normal atmosphere (NA) [4]. CA storage delays ripening and preserves fruit quality [15], but it may cause a decrease in the production of aromatic compounds [16]. The very low oxygen content (ULO) commonly used for long-term storage of apples cannot be used for pears due to their greater susceptibility to damage due to oxygen deficiency [17]. Low oxygen levels trigger anaerobic respiration resulting in the accumulation of alcohol in pears, which is directly responsible for damage to the flesh [18]. This can be prevented by the monitoring of changes in the alcoholic respiration in the atmosphere [19]. However, in controlled atmosphere, it is the internal disorders, such internal browning, which are a major limiting factor [20]. In addition, high carbon dioxide content may cause internal browning in many cultivars, and 'Conference' pears are considered to be sensitive to high levels of this gas [21].

Another method used in recent years to improve storability is to use substances that limit ethylene production in climacteric fruit species [22]. One such substance is 1-methylcyclopropene (1-MCP), which has been used for nearly two decades [23]. In both NA and CA cold storage, 1-MCP can be applied to maintain the quality attributes of fruit, especially firmness. 1-MCP, a gaseous ethylene binding inhibitor, has proven useful in preventing the formation of ethylene in fruit, thus increasing its shelf life after harvest and enabling greater flexibility in distribution and retailing [24]. However, it has been shown that, unlike in apples, low doses of 1-MCP do not completely inhibit ethylene production in pears. 1-MCP interacts with fruit in manifold ways. It can limit the development of fungal diseases. It can also reduce the occurrence of superficial scald [25], but it may also cause an unexpected increase in the incidence of this physiological disorder [26]. However, despite the risk of this negative effect, the benefits of 1-MCP application for other quality parameters and storability are significant and can be observed in all climacteric fruit species [24]. The most important of them include the limitation of the incidence of fungal diseases strictly associated with senescence, and the reduction in transpiration and respiration, which translates into lower storage costs, as the fruit produces less heat at the same temperature. Other important advantages of using 1-MCP include the limitation of: vitamin C losses during storage, the incidence of chilling disorders in tropical fruit species (avocado, mango, papaya) and unwanted changes in flesh structure, such as woolliness (mealiness) and internal breakdown in peaches and nectarines [24]. However, the occurrence of disease is specific to particular fruit species and the conditions in which they are grown and stored, and therefore more detailed studies are still needed in this area [24,27].

Pears have moderate antioxidant activity, but because of relatively high consumption of pears in Europe [1], they are an important source of health-promoting compounds [28]. Pears owe their antioxidant potential to such bioactive compounds as polyphenols, triterpenoids, carotenoids, and chlorophylls, and also have anti-inflammatory and anti-proliferative properties [29].

The antioxidant capacity of food, including fruit and vegetables, depends on the presence of complex bioactive compounds which differ considerably not only in terms of compound class and chemical makeup, but also bioavailability, due to the complex composition of food and the interaction between individual nutrients [30]. The literature provides ample information on the complex nature of the antioxidant activity of food products [31]. In this context, a distinction is often made between extractable and non-extractable antioxidants, the presence (or absence) of which determines how nutritional or healthy a food product is [31]. Residues left after the extraction of bioactive compounds from fruit still show antioxidant activity, so various complex extraction methods are applied to determine more precisely the total antioxidant capacity of food, including that of compounds linked by covalent or hydrogen bonds or forming hydrophobic interactions with other nutritional components such as carbohydrates or proteins. Antioxidant capacity is typically measured in water–alcohol or acidic extracts; however, in complex systems, such as food products, other hydrolysis and extraction methods are also needed to determine the total value of this parameter. A method commonly applied in science to determine the antioxidant capacity of food is to measure the ABTS* cation radical scavenging activity of 70% methanol extracts ($v/v$) from food samples [32].

The objective of this study was to assess (1) the impact of storage conditions on monthly qualitative changes in 'Conference' pears during six months of storage, (2) the effect of the rootstock on the storability of 'Conference' pears, (3) the impact of 1-MCP application on the storability of 'Conference' pears, and how all these factors translate into revenues from the sale of pears after storage.

## 2. Materials and Methods

The experiment was conducted in the experimental orchard and laboratory of the Department of Pomology of the University of Life Sciences in Poznan (52°31′ north latitude and 16°38′ east longitude). 'Conference' pears were collected from trees planted in spring 2002 at spacing of 4 × 1.5 m. Pears were grafted on three different rootstocks: Pyrus caucasica Federov, Pyrodwarf, and Quince S1. There were 64 trees on each rootstock. The pear orchard was maintained according to the standard commercial practice for integrated fruit production.

### 2.1. Sampling

The experiment was carried out from autumn 2011 to spring 2014. The harvest occurred on dates determined as the optimum harvest dates (OHD), using the starch test [33] and the sum of active temperatures (growing degree units) method proposed by Łysiak [34]. The sum of active temperatures determined according to the latter method was 2580 degrees.

After harvest, pears intended for storage were graded to eliminate those not meeting the highest commercial quality standard applicable in OECD countries [35]. According to those standards, pears of superior quality ("extra"), have to be intact, sound, clean, and free of any damage. The experiment was carried out using 20 boxes of 15 kg of graded pears. Each box contained about 75 pears.

### 2.2. Storage Conditions

After harvesting and sorting, the fruit was put in the cold chamber for 48 h to stabilize the fruit temperature. Then half of the fruit (10 boxes) was inserted into a gas-tight chamber where 1-MCP was applied at a dose of 0.05 gm$^{-3}$ for 24 h. After the application, the fruit was placed in four experimental gas-tight chambers with a capacity of 1 m$^3$ each (5 boxes per each chamber). The following storage conditions were applied:

1. Normal atmosphere (NA), temp. −1 °C.
2. Normal atmosphere (NA), + 1-MCP, temp. −1 °C.
3. Controlled atmosphere (CA) 2% $O_2$ + 1% $CO_2$, temp. −1 °C.
4. Controlled atmosphere (CA) 2% $O_2$ + 1% $CO_2$, + 1-MCP, temp. −1 °C.

### 2.3. Quality Measurements

1. Loss of fruit mass was measured in each stored box. Ten pears were numbered and weighed with an accuracy of 0.1 g before and after each month of storage. The mass loss is shown as a percentage of the initial mass.
2. Firmness was measured using a Fruit Tester 327 EFFEGI FT327 penetrometer (Facchini srl, Alfonsine, Italy), mounted on a stand. The maximum penetration force of a probe of 8 mm in length and 11 mm in diameter, applied to a small area with skin removed, on two opposite sides of the fruit, was recorded.
3. Total soluble solids (TSS) were determined using an ATAGO PAL-1 digital refractometer with automatic temperature compensation (Atago, Tokyo, Japan). The results were shown as an average of nine repetitions per sample and expressed as percentage values.
4. Titratable acidity (TA): titration with 1n NaOH to 8.1 pH, mval 100 mL$^{-1}$; the results were expressed as mmol of malic acid per kg of fresh mass.

5. Starch pattern during harvest was determined with Lugol's iodine (measured according to a 10-point scale where 10 means no starch on the pear cross section) [36].
6. The Streif index is a combination of firmness (F), soluble solids content (R) and starch index (S) according the formula:

$$Index = F/RS$$

### 2.4. Measurement of Antioxidant Capacity

Antioxidant capacity of methanol extracts was determined by means of spectrophotometric method using a cationic radical (ABTS+) [2,2′-azinobis-(3-ethylbenzothiazoline-6-sulfonic acid)] [32]. The ABTS+ cation was generated by mixing 7 mM ABTS and 2.45 mM $K_2S_2O_8$ (potassium persulfate or potassium peroxydisulfate) solutions at a ratio of 1:0.5. The ABTS+ cationic radical solution and 70% methanol extract ($v/v$) samples were diluted using Phosphate Buffer Solution (PBS) pH 7.4. Absorbance was measured at a wavelength of 734 nm on samples incubated for 6 min at 30 °C against PBS as a reference assay. Antioxidant capacity was determined based on the percentage reduction in absorbance of the ABTS+ cationic radical solution by the sample compared to the reducing power of Trolox (6-hydroxy-2.5.7.8-tetramethylylchromate-2-carboxylic acid). The measurements were conducted using a Helios Alpha spectrophotometer (Thermo Electron Corporation Waltham, MA, USA) equipped with a water bath for the thermostatting of samples. The results were expressed as an average of nine repetitions per sample and expressed in μmol Trolox/g d·m.

### 2.5. Incidence of Diseases and Disorders

Fruit was harvested in line with commercial quality standards and was free from diseases and disorders. After 6 months of storage, the pears were assessed for the presence of physiological disorders and fungal diseases. Fruit showing symptoms of, respectively, physiological diseases (internal browning and senescent scald) and fungal diseases was counted for each treatment. The identified fungal diseases included: gray mold caused by *Botrytis cinerea* Pers., blue mold caused by *Penicillium* spp., bitter rot caused by *Colletrotrichum* spp., and brown rot caused by *Monilinia* spp. or *Sclerotinia fructigena*. Next, all pears were counted per box, and each box was treated as one repetition. The results were expressed as a percentage share of infected/damaged fruit in the total number of evaluated fruit.

### 2.6. Economic Viability

Prices were obtained from "Fruga", a company trading on the Polish market in fruit produced in Poland and imported from other EU countries. The calculation was based on the average monthly prices received by the grower. Fruit mass losses measured each month were due to transpiration and differences in respiration of fruit undergoing the treatments. In addition, the losses identified after 6 months of storage included fruit mass loss due to fungal diseases and physiological disorders disqualifying the affected fruit as not meeting the highest commercial quality standard applicable in OECD countries [35].

### 2.7. Statistical Analysis

The results were analysed by one-way and multiple-way ANOVA according to the experimental design, using Statistica 13.3 (TIBCO Software Inc., Palo Alto, CA, USA). The assumed sources of variation included the used rootstocks, the storage atmosphere (CA, NA) and the application of 1-MCP. The mean values were compared using Duncan's test at $p \leq 0.05\%$. The correlation coefficient analysis was carried out using Microsoft Office 365 Excel tools.

## 3. Results and Discussion

### 3.1. Rootstock Effect

Fruit growers use different rootstocks to match the tree vigour to the climatic, soil and agricultural conditions [37]. In apples, the rootstock also influences the speed of fruit ripening and thus the date of harvest [38]. However, in this study, the ripening speed and the harvest date were hardly affected by the rootstock type (Table 1). The Streif index, which is considered a good indicator of harvest maturity [39,40], varied little in the discussed experiment. There were no index differences in 2013, and very small differences in the other two years. The starch index, which affects the Streif index most strongly, also showed only slight, although significant, differences. Differences in firmness between the fruit coming from trees grown on different rootstocks were very small as well. No firmness differences were found in two years (2011 and 2012), and a significant difference, although of only 2 N, was identified in 2013.

**Table 1.** The influence of rootstock on the quality parameters of pears at harvest in 2011–2013.

| Rootstock | Streif Index | | Starch Index | | Firmness (N) | | TSS (%) | | TSS/TA | | Total Acidity (% Malic Acid) | |
|---|---|---|---|---|---|---|---|---|---|---|---|---|
| | | | | | | 2011 | | | | | | |
| Q S1 | 0.09 | a [1] | 6.6 | b | 63.7 | a | 12.9 | a | 62.3 | a | 0.21 | b |
| PC | 0.10 | b | 5.4 | a | 64.7 | a | 12.9 | a | 68.6 | a | 0.19 | a |
| PD | 0.10 | b | 5.8 | ab | 63.7 | a | 13.3 | a | 65.8 | a | 0.20 | b |
| | | | | | | 2012 | | | | | | |
| Q S1 | 0.08 | a | 6.5 | b | 60.8 | a | 12.0 | a | 57.9 | ab | 0.21 | a |
| PC | 0.09 | ab | 6.3 | ab | 61.8 | a | 12.4 | b | 61.6 | b | 0.20 | a |
| PD | 0.09 | b | 6.0 | a | 62.8 | a | 12.2 | ab | 57.0 | a | 0.21 | b |
| | | | | | | 2013 | | | | | | |
| Q S1 | 0.06 | a | 9.2 | b | 66.7 | b | 12.4 | b | 69.3 | b | 0.18 | a |
| PC | 0.07 | a | 8.6 | a | 66.7 | b | 11.9 | a | 66.1 | a | 0.18 | a |
| PD | 0.06 | a | 8.8 | a | 64.7 | a | 12.1 | a | 67.1 | a | 0.18 | a |
| | | | | | | Mean | | | | | | |
| Q S1 | 0.07 | a | 7.4 | b | 63.7 | a | 12.1 | a | 62.8 | a | 0.20 | b |
| PC | 0.08 | a | 6.8 | a | 64.7 | a | 12.4 | a | 65.4 | a | 0.19 | a |
| PD | 0.08 | a | 6.5 | a | 63.7 | a | 12.5 | a | 63.3 | a | 0.20 | b |

[1] One-way analyses of variance; data in the same column marked with the same letter, separately for each year of experiment, are not significantly different at $\alpha = 0.05$ (Duncan's test). Q S1—Quince S1 rootstock; PC—Pyrus caucasica Federov rootstock; PD—Pyrodwarf rootstock.

A study assessing the influence of six rootstocks on the TSS of 'Forelle' pears found that the differences in TSS at harvest were very small and did not exceed 0.5° Brix within two years [10]. TSS differences were also small in our study. No differences were detected in the first year of the experiment, and even though they occurred in the two subsequent years, they were rather incidental and random. Total acidity followed a clearer pattern and was the lowest in fruit from trees growing on *Pyrus caucasica*. However, the TSS/TA ratio, which is crucial for the subjective perception of fruit taste [41], did not vary considerably between fruit from trees grown on different rootstocks.

As the present study did not show any significant and long-term influence of the rootstock on the basic quality features of pears, this factor was omitted in our further analyses regarding storage and storability, except for mass loss (Tables S1–S12).

### 3.2. Mass Loss

Fruit mass loss during storage depends on a number of factors occurring both before and after harvest, such as the content of minerals (especially calcium) in fruit, fruit maturity at harvest, incidence of diseases and disorders, and storage conditions [11]. As all fruit analysed in the present study grew under the same conditions and was treated in the same way at the time of harvest, and its selection was entirely random, in line with the experimental design, it can be assumed that the loss of fruit mass was influenced only by the experimental factors. All factors applied in this study affected the loss of fruit mass during storage (Tables 2–4). The rootstock may affect the quality parameters of fruit [42], as well as the speed and time of ripening [43]. In 2011, the effect of the rootstock type on the mass loss of stored fruit was significant from the first measurement until the 120th day of storage, but this factor became less and less significant with time. In the following year, no such effect was found, but it was observed that the interaction between rootstock and 1-MCP application had a very strong influence on transpiration. This interaction did not decrease with time, which means that 1-MCP application was the predominant factor. In 2013, the rootstock clearly affected the loss of fruit mass during the storage period, except the first month after 1-MCP treatment. The interaction between rootstock and 1-MCP application was weaker, such as in the first year, and was significant only in the middle of the storage period.

**Table 2.** Mass loss (%) of non-1-MCP-treated (control) and of 1-MCP-treated 'Conference' pears after storage in normal (NA) and controlled (CA) atmosphere in 2011.

| Rootstock | Storage Atmosphere | 1-MCP Dose | Days of Storage | | | | | | | | | | | |
|---|---|---|---|---|---|---|---|---|---|---|---|---|---|---|
| | | | 30 | | 60 | | 90 | | 120 | | 150 | | 180 | |
| Q S1 | NA | Control | 1.82 [1] | (0.31) | 3.72 | (0.32) | 5.13 | (0.21) | 6.14 | (0.23) | 7.10 | (0.54) | 7.50 | (0.42) |
| | | 1-MCP | 2.15 | (0.29) | 2.65 | (0.44) | 3.23 | (0.59) | 4.39 | (0.97) | 5.56 | (1.32) | 6.06 | (1.47) |
| | CA | Control | 1.49 | (0.51) | 1.97 | (0.51) | 2.52 | (0.55) | 3.00 | (0.61) | 3.48 | (0.70) | 4.11 | (0.85) |
| | | 1-MCP | 0.92 | (0.19) | 1.37 | (0.29) | 1.88 | (0.42) | 2.02 | (0.37) | 2.42 | (0.38) | 3.32 | (0.61) |
| PC | NA | Control | 1.52 | (0.47) | 2.69 | (0.28) | 4.74 | (0.34) | 5.60 | (0.59) | 6.24 | (0.56) | 6.77 | (0.59) |
| | | 1-MCP | 2.31 | (0.79) | 3.01 | (1.00) | 3.89 | (1.43) | 5.38 | (1.80) | 5.51 | (1.58) | 6.22 | (1.86) |
| | CA | Control | 1.01 | (0.06) | 2.02 | (0.30) | 3.05 | (0.44) | 3.46 | (0.60) | 4.32 | (0.62) | 5.18 | (0.64) |
| | | 1-MCP | 0.93 | (0.13) | 1.78 | (0.15) | 2.26 | (0.46) | 2.77 | (0.41) | 3.43 | (0.44) | 3.93 | (0.47) |
| PD | NA | Control | 1.41 | (0.23) | 2.82 | (0.31) | 4.52 | (0.42) | 5.41 | (0.55) | 6.30 | (0.55) | 6.64 | (0.58) |
| | | 1-MCP | 1.83 | (0.25) | 2.34 | (0.28) | 3.41 | (0.51) | 4.59 | (1.01) | 5.23 | (1.12) | 5.74 | (1.29) |
| | CA | Control | 1.07 | (0.17) | 1.98 | (0.49) | 2.56 | (0.50) | 3.08 | (0.60) | 3.59 | (0.74) | 4.44 | (0.75) |
| | | 1-MCP | 0.94 | (0.19) | 1.42 | (0.28) | 1.94 | (0.39) | 2.26 | (0.45) | 2.79 | (0.56) | 3.31 | (0.68) |
| **Main effects [2]** | | | | | | | | | | | | | | |
| Rootstock (A) | | | ** | | ** | | * | | * | | ns | | ns | |
| Storage atmosphere (B) | | | *** | | *** | | *** | | *** | | *** | | *** | |
| 1-MCP dose (C) | | | ns | | *** | | *** | | *** | | *** | | *** | |
| **Interaction** | | | | | | | | | | | | | | |
| A × B | | | ns | | ** | | ns | | ns | | ** | | * | |
| A × C | | | * | | *** | | ns | | * | | ns | | ns | |
| B × C | | | *** | | ns | | ** | | ns | | ns | | ns | |
| A × B × C | | | ns | | * | | ns | | ns | | ns | | ns | |

[1] Numbers in parentheses are the standard deviation of the mean (n = 10). [2] $p$-value of F ratio: ns—not significantly different; * $p < 0.05$; ** $p < 0.01$; *** $p < 0.001$. Q S1—Quince S1 rootstock; PC– Pyrus caucasica Federov rootstock; PD—Pyrodwarf rootstock. NA—normal atmosphere; CA—controlled atmosphere.

The use of 1-MCP reduces autocatalytic ethylene production and, thus, significantly slows down respiration [27]. The strongest impact of 1-MCP application was found in the last year of the research (Table 4), in which the reduction in fruit mass loss as compared to untreated pears had the highest level of significance in each month of testing. In 2011 and 2012, no differences between 1-MCP-treated and non-1-MCP-treated samples were identified after the first month of storage, whereas the differences were highly significant in the subsequent months.

**Table 3.** Mass loss of non-1-MCP-treated (control) and of 1-MCP-treated 'Conference' pears after storage in normal (NA) and controlled (CA) atmosphere in 2012.

| Rootstock | Storage Atmosphere | 1-MCP Dose | Days of Storage | | | | | | | | | | | |
|---|---|---|---|---|---|---|---|---|---|---|---|---|---|---|
| | | | 30 | | 60 | | 90 | | 120 | | 150 | | 180 | |
| Q S1 | NA | Control | 1.96 [1] | (0.60) | 4.00 | (0.90) | 5.47 | (0.78) | 6.32 | (0.66) | 7.17 | (0.53) | 7.96 | (0.50) |
| | | 1-MCP | 1.94 | (0.28) | 2.78 | (0.56) | 3.78 | (0.57) | 4.42 | (0.53) | 5.13 | (0.69) | 5.98 | (0.95) |
| | CA | Control | 1.21 | (0.29) | 1.92 | (0.27) | 2.59 | (0.31) | 2.93 | (0.25) | 3.90 | (0.43) | 4.70 | (0.44) |
| | | 1-MCP | 1.23 | (0.30) | 1.68 | (0.30) | 1.99 | (0.32) | 2.75 | (0.57) | 3.26 | (0.76) | 4.23 | (0.93) |
| PC | NA | Control | 2.54 | (0.62) | 3.90 | (0.70) | 5.15 | (0.62) | 5.92 | (0.49) | 6.71 | (0.56) | 7.88 | (0.65) |
| | | 1-MCP | 1.96 | (0.43) | 2.90 | (0.76) | 4.06 | (0.62) | 4.74 | (0.58) | 5.50 | (0.84) | 6.36 | (0.89) |
| | CA | Control | 1.15 | (0.23) | 1.89 | (0.33) | 2.49 | (0.37) | 3.01 | (0.38) | 3.79 | (0.44) | 4.62 | (0.48) |
| | | 1-MCP | 1.19 | (0.13) | 1.55 | (0.20) | 1.86 | (0.24) | 2.37 | (0.33) | 2.96 | (0.45) | 3.92 | (0.59) |
| PD | NA | Control | 2.18 | (0.25) | 4.02 | (0.37) | 5.39 | (0.65) | 6.05 | (0.59) | 6.70 | (0.53) | 7.53 | (0.47) |
| | | 1-MCP | 1.84 | (0.57) | 2.55 | (0.51) | 3.63 | (0.59) | 4.38 | (0.59) | 5.10 | (0.51) | 6.06 | (0.47) |
| | CA | Control | 1.01 | (0.14) | 1.88 | (0.34) | 2.67 | (0.53) | 3.03 | (0.66) | 3.78 | (0.77) | 4.73 | (0.81) |
| | | 1-MCP | 1.17 | (0.07) | 1.55 | (0.14) | 1.85 | (0.20) | 2.36 | (0.37) | 3.09 | (0.70) | 4.10 | (0.93) |
| **Main effects** [2] | | | | | | | | | | | | | | |
| Rootstock (A) | | | ns | | ns | | ns | | ns | | ns | | ns | |
| Storage atmosphere (B) | | | *** | | *** | | *** | | *** | | *** | | *** | |
| 1-MCP dose (C) | | | ns | | *** | | *** | | *** | | *** | | *** | |
| **Interaction** | | | | | | | | | | | | | | |
| A × B | | | ns | | ns | | ns | | ns | | ns | | ns | |
| A × C | | | ns | | ns | | ns | | ns | | ns | | ns | |
| B × C | | | ** | | *** | | *** | | *** | | *** | | *** | |
| A × B × C | | | ns | | ns | | ns | | * | | ns | | ns | |

[1] Numbers in parentheses are the standard deviation of the mean (n = 10). [2] $p$-value of F ratio: ns—not significantly different; * $p < 0.05$; ** $p < 0.01$; *** $p < 0.001$. Q S1—Quince S1 rootstock; PC—Pyrus caucasica Federov rootstock; PD—Pyrodwarf rootstock. NA—normal atmosphere; CA—controlled atmosphere.

**Table 4.** Mass loss of non-1-MCP-treated (control) and of 1-MCP-treated 'Conference' pears after storage in normal (NA) and controlled (CA) atmosphere in 2013.

| Rootstock | Storage Atmosphere | 1-MCP Dose | Days of Storage | | | | | | | | | | | |
|---|---|---|---|---|---|---|---|---|---|---|---|---|---|---|
| | | | 30 | | 60 | | 90 | | 120 | | 150 | | 180 | |
| Q S1 | NA | Control | 1.86 [1] | (0.31) | 3.41 | (0.40) | 4.48 | (0.43) | 5.09 | (0.52) | 5.93 | (0.50) | 6.63 | (0.45) |
| | | 1-MCP | 1.79 | (0.52) | 2.60 | (0.51) | 3.31 | (0.50) | 4.02 | (0.44) | 4.71 | (0.43) | 5.36 | (0.45) |
| | CA | Control | 1.10 | (0.15) | 1.61 | (0.22) | 2.12 | (0.37) | 3.02 | (0.28) | 3.69 | (0.25) | 4.09 | (0.33) |
| | | 1-MCP | 0.92 | (0.17) | 1.63 | (0.60) | 1.98 | (0.57) | 2.48 | (0.71) | 2.98 | (0.86) | 3.16 | (0.60) |
| PC | NA | Control | 2.15 | (0.27) | 3.66 | (0.68) | 5.01 | (0.90) | 5.86 | (0.95) | 6.28 | (0.62) | 7.13 | (0.68) |
| | | 1-MCP | 1.93 | (0.28) | 3.29 | (0.38) | 3.67 | (0.38) | 4.51 | (0.37) | 5.26 | (0.39) | 6.33 | (0.48) |
| | CA | Control | 1.02 | (0.13) | 1.88 | (0.18) | 2.73 | (0.23) | 3.21 | (0.30) | 4.07 | (0.28) | 4.83 | (0.38) |
| | | 1-MCP | 0.92 | (0.08) | 1.82 | (0.19) | 2.16 | (0.41) | 2.53 | (0.65) | 3.02 | (0.49) | 3.88 | (0.50) |
| PD | NA | Control | 2.27 | (0.27) | 3.79 | (0.43) | 5.27 | (0.77) | 5.83 | (0.85) | 6.13 | (0.74) | 6.96 | (0.44) |
| | | 1-MCP | 1.68 | (0.61) | 2.54 | (0.77) | 3.57 | (0.45) | 4.08 | (0.56) | 4.82 | (0.58) | 5.47 | (0.64) |
| | CA | Control | 0.98 | (0.06) | 1.67 | (0.29) | 2.37 | (0.50) | 3.07 | (0.60) | 3.62 | (0.59) | 4.03 | (0.49) |
| | | 1-MCP | 0.98 | (0.06) | 1.75 | (0.54) | 1.99 | (0.81) | 2.26 | (0.72) | 2.77 | (0.85) | 3.27 | (0.99) |
| **Main effects** [2] | | | | | | | | | | | | | | |
| Rootstock (A) | | | ns | | ** | | ** | | * | | * | | *** | |
| Storage atmosphere (B) | | | *** | | *** | | *** | | *** | | *** | | *** | |
| 1-MCP dose (C) | | | *** | | *** | | *** | | *** | | *** | | *** | |

**Table 4.** *Cont.*

| Rootstock | Storage Atmosphere | 1-MCP Dose | Days of Storage | | | | | |
|---|---|---|---|---|---|---|---|---|
| | | | 30 | 60 | 90 | 120 | 150 | 180 |
| Interaction | | | | | | | | |
| A × B | | | ns | ns | ns | ns | ns | ns |
| A × C | | | ns | ns | ns | ns | ns | ns |
| B × C | | | ns | ns | *** | ** | ns | ns |
| A × B × C | | | * | ns | ns | ns | ns | ns |

[1] Numbers in parentheses are the standard deviation of the mean (n = 10). [2] $p$-value of F ratio: ns—not significantly different; * $p < 0.05$; ** $p < 0.01$; *** $p < 0.001$. Q S1—Quince S1 rootstock; PC—Pyrus caucasica Federov rootstock; PD—Pyrodwarf rootstock. NA—normal atmosphere; CA—controlled atmosphere.

However, it was the gaseous composition of the storage atmosphere which had by far the greatest impact on the mass loss during storage. During respiration, sugar and oxygen are combined to produce carbon dioxide and water, the excess of which is transpired into the environment, thus resulting in fruit mass loss [8,11]. The composition of gases in the storage atmosphere strongly influences the rate of fruit respiration [12]. Reducing the oxygen level and increasing the carbon dioxide level slow down respiration and, along with it, the consumption of respiration substrates. This study, during which the oxygen content was reduced to 2%, and the carbon dioxide content was increased to 1%, confirmed the beneficial effect of CA in each year and after each month of storage. This impact was highly significant in each year of the study. Of all the three main factors, the influence of the gaseous composition of the storage atmosphere on fruit mass loss had the highest level of statistical significance.

Mass loss during storage might also have been caused by parthenocarpy, and 'Conference' pear is known for producing a lot of parthenocarpic fruit in years of adverse weather conditions. It was observed in some earlier studies that parthenocarpy and the number of seeds produced in pears affected both their quality and storability [7].

### 3.3. Changes in Quality Parameters during Storage

Firmness is one of the most important quality criteria for traders and consumers alike [12]. Fruit starts to soften when still on the tree, about 4 weeks prior to the optimum harvest date [44]. After harvest, fruit continues to soften to finally reach edible firmness, and the softening rate depends on such factors as fertilization, harvest date and cultivar, but it is mostly influenced by the length and conditions of storage. The softening rate of late-maturing pear cultivars, such as 'Conference', was found to depend on the synthesis of two enzymes: acetyl-CoA synthase (ACS) and 1-aminocyclopropane carboxylic acid (ACC) [6]. In this study, both the softening rate and the total loss of firmness were clearly dependent on storage conditions (Tables 5–7). The initial firmness measured at harvest depends of many factors, which was visible in our study because it varied little between years, with the lowest value observed in 2012 and the highest in 2011. In addition, there were virtually no differences between years in the final firmness after 180 days of NA storage, and the total loss of firmness ranged from 34 to 40%. No clear regularity was revealed as regards firmness loss per month in individual years, but the firmness showed a tendency to decrease, as the greatest loss of firmness was noted in the last (sixth) month of NA storage in two years and in the next-to-last (fifth) month in one year of the study.

1-MCP application significantly reduces the softening rate of climacteric fruit and vegetables [24]. This study revealed that 1-MCP application visibly slows down the softening rate, particularly in the first months of storage—it reduced the total firmness loss by about 1/3 compared to the untreated fruit after six months of NA storage. Additionally, CA storage with the concentration of $O_2$ of 1.5–3% and $CO_2$ of 0.5–1.0% for 'Conference' slows down the softening rate [45]. Such CA storage conditions were applied in our study, and the CA-stored fruit softened significantly more slowly than the NA-stored fruit, but at an approximately equal rate as the NA-stored + 1-MCP-treated fruit. No differences were observed after 180 days of storage in two years, whereas there was a significant difference in one year of the study, but it did not exceed 10%.

CA-stored + 1-MCP-treated pears had, by far, the highest firmness. Their average firmness loss within the three years of the study was only about 15% of the initial value, and such a difference does not affect consumer preferences [46]. During the first months of storage, the firmness loss was considerably higher in the non-1-MCP-treated fruit. 1-MCP application had a stronger influence on the softening rate than the gaseous composition of storage atmosphere during the first two months of storage, but the reverse was true (storage atmosphere more strongly reduced the softening rate than 1-MCP) at the end of the storage period. This could be explained by the way in which 1-MCP works: it slows down respiration by blocking ethylene receptors, but new non-blocked receptors are formed on the fruit skin with time [24], and the treated fruit becomes similar to untreated fruit.

**Table 5.** Effect of storage technology on the quality parameters of pears in 2011.

| Storage Duration | Firmness (N) | | Monthly Loss (%) | | Total Loss (%) | TSS (%) | | TA (% Malic Acid) | | TSS/TA | | ORAC μmol TE/100 g (d.m.) | |
|---|---|---|---|---|---|---|---|---|---|---|---|---|---|
| **NA** | | | | | | | | | | | | | |
| 0 | 64.2 [1] | n | 0 | | | 12.6 | a | 0.21 | l | 60.0 | a | 2858.8 | d |
| 30 | 62.2 | mn | 3.0 | a | 3.0 | 12.9 | a–c | 0.18 | i | 73.2 | b | | |
| 60 | 55.9 | g–i | 10.1 | c | 12.8 | 13.3 | b–d | 0.15 | ef | 89.9 | c–e | | |
| 90 | 52.2 | f | 6.8 | b | 18.7 | 14.6 | g–i | 0.11 | c | 134.5 | g | | |
| 120 | 50.9 | ef | 2.4 | a | 20.7 | 14.9 | i | 0.11 | c | 135.5 | g | | |
| 150 | 43.0 | ab | 15.5 | d | 32.9 | 14.6 | g–i | 0.10 | b | 154.0 | h | | |
| 180 | 40.6 | a | 5.7 | b | 36.8 | 13.9 | ef | 0.06 | a | 231.7 | i | 2562.5 | a |
| **NA + 1-MCP** | | | | | | | | | | | | | |
| 0 | 64.2 | n | 0 | | | 12.6 | a | 0.21 | l | 60.0 | a | 2858.8 | d |
| 30 | 63.4 | n | 1.1 | a | 1.1 | 12.8 | ab | 0.19 | i-l | 67.8 | ab | | |
| 60 | 58.1 | i–k | 8.4 | c | 9.5 | 12.9 | a–c | 0.19 | i-k | 69.3 | ab | | |
| 90 | 53.5 | fg | 7.9 | b | 16.6 | 13.4 | c–e | 0.16 | f-h | 84.8 | cd | | |
| 120 | 47.6 | cd | 11.0 | e | 25.8 | 14.3 | f–i | 0.16 | e-h | 92.1 | c–e | | |
| 150 | 43.4 | b | 8.9 | d | 32.4 | 14.5 | f–i | 0.15 | e | 99.5 | e | | |
| 180 | 42.9 | ab | 1.1 | a | 33.2 | 14.7 | hi | 0.10 | bc | 147.8 | h | 2782.5 | b |
| **CA** | | | | | | | | | | | | | |
| 0 | 64.2 | n | 0 | | | 12.6 | a | 0.21 | l | 60.0 | a | 2858.8 | d |
| 30 | 64.6 | n | −0.7 | a | −0.7 | 12.5 | a | 0.18 | ij | 69.0 | ab | | |
| 60 | 62.3 | mn | 3.5 | b | 2.9 | 12.7 | a | 0.19 | i-l | 66.7 | ab | | |
| 90 | 57.8 | h–j | 7.3 | e | 10.0 | 13.1 | a–d | 0.18 | i | 74.5 | b | | |
| 120 | 54.8 | g | 5.1 | c | 14.6 | 14.1 | fg | 0.15 | e-g | 93.9 | c–e | | |
| 150 | 48.6 | de | 11.3 | f | 24.2 | 14.5 | g-i | 0.15 | e-h | 95.1 | de | | |
| 180 | 45.3 | bc | 6.7 | d | 29.3 | 14.8 | i | 0.12 | d | 123.7 | f | 2817.5 | c |
| **CA + 1-MCP** | | | | | | | | | | | | | |
| 0 | 64.2 | n | 0 | | | 12.6 | a | 0.21 | l | 60.0 | a | 2858.8 | d |
| 30 | 63.6 | n | 0.8 | b | 0.8 | 12.8 | ab | 0.21 | m | 60.6 | a | | |
| 60 | 59.4 | kl | 6.7 | f | 7.4 | 12.6 | a | 0.20 | kl | 64.1 | ab | | |
| 90 | 60.7 | lm | −2.2 | a | 5.4 | 13.1 | a–d | 0.19 | ij | 71.0 | ab | | |
| 120 | 59.1 | jk | 2.7 | c | 7.9 | 13.5 | de | 0.16 | gh | 84.0 | c | | |
| 150 | 55.3 | gh | 6.4 | e | 13.8 | 13.9 | ef | 0.17 | h | 84.7 | cd | | |
| 180 | 53.3 | fg | 3.5 | d | 16.9 | 14.3 | f–i | 0.16 | e-h | 89.4 | c–e | 2962.5 | e |

[1] One-way analyses of variance; data in the same column marked with the same letter are not significantly different within a year at $\alpha = 0.05$ (Duncan's test). NA—normal atmosphere; NA + 1-MCP—normal atmosphere + 1-methylcyclopropene; CA—controlled atmosphere; CA + 1-MCP—controlled atmosphere + 1-methylcyclopropene.

**Table 6.** Effect of storage technology on the quality parameters of pears in 2012.

| Storage Duration | Firmness (N) | | Monthly Loss (%) | | Total Loss (%) | TSS (%) | | TA (% Malic Acid) | | TSS/TA (N) | | ORAC μmol TE/100 g (d.m.) | |
|---|---|---|---|---|---|---|---|---|---|---|---|---|---|
| NA | | | | | | | | | | | | | |
| 0 | 60.6[1] | l | 0 | | | 12.2 | a | 0.23 | m | 53.0 | a | 2635.0 | c |
| 30 | 56.4 | hi | 6.9 | c | 6.9 | 12.6 | a–d | 0.20 | j–l | 64.7 | a–d | | |
| 60 | 53.5 | fg | 5.1 | a | 11.6 | 13.1 | e–h | 0.19 | h–j | 70.4 | c–f | | |
| 90 | 50.5 | de | 5.6 | b | 16.6 | 13.5 | f–i | 0.17 | f | 79.4 | gh | | |
| 120 | 46.9 | c | 7.1 | d | 22.6 | 14.0 | j–l | 0.15 | e | 93.3 | i–k | | |
| 150 | 43.7 | b | 6.9 | c | 27.9 | 14.2 | l | 0.11 | b–e | 137.4 | l | | |
| 180 | 39.8 | a | 8.9 | e | 34.3 | 13.6 | k | 0.07 | a | 194.3 | m | 2435.0 | a |
| NA + 1-MCP | | | | | | | | | | | | | |
| 0 | 60.6 | l | 0 | | | 12.2 | a | 0.23 | m | 53.0 | a | 2635.0 | c |
| 30 | 58.9 | jl | 2.8 | b | 2.8 | 12.4 | ab | 0.19 | i–k | 66.5 | b–e | | |
| 60 | 58.3 | i–l | 1.1 | a | 3.8 | 12.5 | a–c | 0.19 | h–j | 67.4 | b–f | | |
| 90 | 56.3 | hi | 3.4 | c | 7.1 | 13.0 | d–g | 0.18 | g–i | 71.9 | d–g | | |
| 120 | 52.0 | ef | 7.6 | d | 14.2 | 13.4 | f–i | 0.17 | f | 79.4 | gh | | |
| 150 | 47.9 | c | 7.8 | e | 20.9 | 13.9 | j–l | 0.12 | c | 122.8 | k | | |
| 180 | 44.2 | b | 7.8 | e | 27.1 | 14.3 | l | 0.11 | bc | 130.9 | l | 2537.5 | b |
| CA | | | | | | | | | | | | | |
| 0 | 64.2 | m | 0 | | | 12.2 | a | 0.23 | m | 53.0 | a | 2635.0 | c |
| 30 | 60.4 | l | 5.9 | d | 5.9 | 12.2 | a | 0.20 | lm | 60.9 | a–c | | |
| 60 | 58.6 | i–l | 2.9 | b | 8.7 | 12.5 | a–c | 0.19 | j–l | 65.3 | a–d | | |
| 90 | 57.4 | h–k | 2.1 | a | 10.6 | 12.8 | b–e | 0.18 | g–i | 70.7 | c–f | | |
| 120 | 55.7 | h | 2.8 | b | 13.1 | 13.2 | f–i | 0.18 | fg | 75.4 | f–h | | |
| 150 | 53.1 | fg | 4.7 | c | 17.2 | 13.6 | i–k | 0.15 | de | 92.6 | i–k | | |
| 180 | 48.9 | cd | 8.1 | e | 23.9 | 14.2 | l | 0.14 | d–g | 100.4 | j–l | 2627.5 | c |
| CA + 1-MCP | | | | | | | | | | | | | |
| 0 | 60.6 | l | 0 | | | 12.2 | a | 0.23 | m | 53.0 | a | 2635.0 | c |
| 30 | 60.6 | l | 0.0 | b | 0.0 | 12.2 | a | 0.21 | m | 58.8 | ab | | |
| 60 | 59.5 | kl | 1.8 | c | 1.8 | 12.3 | ab | 0.20 | l | 62.5 | a–c | | |
| 90 | 58.3 | i–l | 2.0 | d | 3.8 | 12.7 | b–e | 0.20 | j–l | 65.5 | a–d | | |
| 120 | 58.5 | i–l | −0.4 | a | 3.3 | 12.9 | c–f | 0.18 | f-h | 72.7 | d–g | | |
| 150 | 56.9 | h–j | 2.8 | e | 6.0 | 13.5 | h–j | 0.18 | g–i | 73.9 | e–h | | |
| 180 | 55.2 | gh | 3.0 | f | 8.8 | 14.0 | kl | 0.17 | fg | 81.0 | h–j | 2777.5 | d |

[1] One-way analyses of variance; data in the same column marked with the same letter are not significantly different within a year at α = 0.05 (Duncan's test). NA—normal atmosphere; NA + 1-MCP—normal atmosphere + 1-methylcyclopropene; CA—controlled atmosphere; CA + 1-MCP—controlled atmosphere + 1-methylcyclopropene.

**Table 7.** Effect of storage technology on the quality parameters of pears in 2013.

| Storage Duration | Firmness (N) | | Monthly Loss (%) | | Total Loss (%) | TSS (%) | | TA (% Malic Acid) | | TSS/TA (N) | | ORAC μmol TE/100 g (d.m.) | |
|---|---|---|---|---|---|---|---|---|---|---|---|---|---|
| NA | | | | | | | | | | | | | |
| 0 | 65.9[1] | k | 0 | | | 12.1 | a | 0.18 | m | 67.5 | a | 2975.0 | c |
| 30 | 64.4 | i–k | 2.2 | a | 2.2 | 12.5 | a–c | 0.18 | m | 70.7 | ab | | |
| 60 | 62.8 | hi | 2.5 | b | 4.7 | 12.7 | b–d | 0.15 | h | 87.6 | cd | | |
| 90 | 56.1 | f | 10.6 | d | 14.8 | 13.6 | f | 0.13 | fg | 101.7 | ef | | |
| 120 | 49.7 | cd | 11.4 | e | 24.6 | 14.3 | gh | 0.11 | c | 130.0 | h | | |
| 150 | 45.4 | b | 8.6 | c | 31.0 | 14.8 | jk | 0.07 | b | 208.3 | i | | |
| 180 | 39.9 | a | 12.3 | f | 39.5 | 13.7 | f | 0.06 | a | 228.3 | j | 2750.0 | a |

**Table 7.** *Cont.*

| Storage Duration | Firmness (N) | | Firmness Monthly Loss (%) | | Total Loss (%) | TSS (%) | | TA (% Malic Acid) | | TSS/TA (N) | | ORAC μmol TE/100 g (d.m.) | |
|---|---|---|---|---|---|---|---|---|---|---|---|---|---|
| | | | | | NA + 1-MCP | | | | | | | | |
| 0 | 65.9 | k | 0 | | | 12.1 | a | 0.18 | m | 67.5 | a | 2975.0 | c |
| 30 | 64.4 | i–k | 2.2 | b | 2.2 | 12.4 | ab | 0.18 | m | 70.7 | ab | | |
| 60 | 64.1 | i–k | 0.5 | a | 2.7 | 12.4 | a–c | 0.16 | j–l | 76.8 | ab | | |
| 90 | 61.1 | h | 4.6 | c | 7.2 | 12.8 | c–e | 0.15 | h | 88.3 | cd | | |
| 120 | 55.8 | f | 8.8 | f | 15.4 | 13.6 | f | 0.12 | de | 107.7 | f | | |
| 150 | 51.4 | de | 7.9 | e | 22.0 | 14.6 | ij | 0.12 | d | 121.7 | g | | |
| 180 | 47.6 | bc | 7.3 | d | 27.7 | 14.6 | ij | 0.11 | c | 132.7 | h | 2890.8 | b |
| | | | | | CA | | | | | | | | |
| 0 | 65.9 | k | 0 | | | 12.1 | a | 0.18 | m | 67.5 | a | 2975.0 | c |
| 30 | 66.2 | k | −0.5 | a | −0.5 | 12.3 | ab | 0.18 | m | 69.1 | a | | |
| 60 | 63.3 | h–j | 4.4 | c | 4.0 | 12.4 | ab | 0.18 | m | 69.0 | a | | |
| 90 | 62.0 | hi | 2.1 | b | 6.0 | 12.6 | a–c | 0.16 | jk | 80.4 | bc | | |
| 120 | 57.7 | fg | 6.9 | e | 12.4 | 13.8 | fg | 0.15 | h | 93.4 | de | | |
| 150 | 52.9 | e | 8.4 | f | 19.8 | 14.7 | ij | 0.14 | g | 107.9 | f | | |
| 180 | 49.9 | cd | 5.6 | d | 24.2 | 15.1 | k | 0.13 | ef | 119.0 | g | 2990.0 | c |
| | | | | | CA + 1-MCP | | | | | | | | |
| 0 | 65.9 | k | 0 | | | 12.1 | a | 0.18 | m | 67.5 | a | 2975.0 | c |
| 30 | 66.3 | k | −0.6 | a | −0.6 | 12.2 | a | 0.18 | m | 68.9 | a | | |
| 60 | 64.3 | i–k | 3.0 | c | 2.4 | 12.4 | ab | 0.17 | l | 73.9 | ab | | |
| 90 | 63.3 | h–k | 1.5 | b | 3.9 | 12.5 | a–c | 0.16 | kl | 75.9 | ab | | |
| 120 | 61.4 | h | 3.1 | c | 6.9 | 13.1 | de | 0.16 | j–l | 80.6 | bc | | |
| 150 | 58.6 | g | 4.5 | d | 11.1 | 13.7 | fg | 0.16 | ij | 88.5 | cd | | |
| 180 | 55.8 | f | 4.7 | e | 15.3 | 14.1 | hi | 0.15 | hi | 96.7 | de | 3107.5 | d |

[1] One-way analyses of variance; data in the same column marked with the same letter are not significantly different within a year at α = 0.05 (Duncan's test). NA—normal atmosphere; NA + 1-MCP—normal atmosphere + 1-methylcyclopropene; CA—controlled atmosphere; CA + 1-MCP—controlled atmosphere + 1-methylcyclopropene.

### 3.3.1. Total Soluble Solids, Total Acidity and TSS/TA Ratio

Two very important qualitative criterions are: TSS, which is the content of solids, notably sugars, in a liquid, and TA, which is the content of acids and is assessed as the sum of acids converted into malic acid [7,12]. The TSS value usually grows in the initial storage period, which is caused by the degradation of polysaccharides into monosaccharides, but may decrease later as the fruit uses stored energy for respiration [47]. The speed of changes depends on the storage time and storage conditions [11]. The TSS value at harvest was similar and ranged between 12.1 and 12.6% in all years of the study. The TSS value changed most rapidly in the NA-stored fruit: it increased relatively quickly to reach the maximum after four or five months. After six months of storage, the TSS content dropped but was still significantly higher than at harvest. In NA-stored 1-MCP-treated pears, the TSS content grew more slowly and reached its peak in the last 2–3 months of storage. CA storage considerably reduced the ripening speed expressed by TSS, regardless of whether 1-MCP was applied or not. 1-MCP application may have an inconsistent effect, both in CA and NA storage [48]. In this study, TSS in the CA-stored pears gradually grew and rose to the maximum value after the entire storage period. This suggests that after 180 days of CA storage the 'Conference' pears did not start yet the excessive consumption of sugars in the respiration process.

Total acidity in the stored fruit changed according to a clearer pattern than TSS. Monthly measurements showed a steady decrease in TA at a rate dependent on the storage conditions and 1-MCP treatment. This corresponds with the findings by Hedges et al. [23], who additionally pointed to the harvest date as a factor affecting TA as a harvest delay resulted in a visible decrease in the initial TA value. What deserves mentioning as regards our study is the weather conditions prevailing during the growing season because the initial TA values measured at the optimal harvest date varied between years. The differences were not large, but were consistent with the findings from a study on apples harvested in the same year in three European countries characterized by different weather conditions [49]. In this study, TA declined during storage irrespective of storage conditions, but it reached the lowest values in the NA-stored fruit and amounted to about 1/3 of the initial value after storage in each year. The 1-MCP-treated pears had a significantly higher TA, which never fell below 50% of the initial value. CA storage slowed down the degradation of acids still further and the combination of CA and 1-MCP prevented TA from dropping below 20% of the initial value after six months. Such conditions were shown to be highly effective not only for 'Conference' pears, but also for another popular European pear cultivar, 'Alexander Lucas' [23].

Consumers' perception of fruit sweetness and acidity depends not only on the absolute TSS or TA values, but also on the TSS/TA ratio [50]. In our study, the TSS/TA ratio changed very rapidly in each of the three years and was strongly influenced by storage conditions. For example, whereas the TSS of NA-stored pears did not vary by more than 20% in any of the years, their TSS/TA ratio varied by at least 300%. This was the only quality parameter that was different in virtually every month, regardless of storage conditions and 1-MCP application.

### 3.3.2. Oxygen Radical Absorbance Capacity (ORAC)

Free radical scavenging activity is highly dependent on the species, cultivar, climatic conditions and harvest date [11]. During storage, it can remain unchanged, as was the case for 'Rocha' pears in Portugal [28], or it can grow, as shown by [51] for 'Golden Smoothee' apples. In our research, ORAC varied depending on storage conditions. Antioxidant capacity dropped considerably after NA storage, whereas the decrease was significantly smaller in the 1-MCP-treated sample. No changes were observed after CA storage, the ORAC value even rose after 1-MCP application. These differences can be explained by the differences in the ripening processes induced by increased respiration. Larrigaudiere et al. [52] found out that ripening may involve a noticeable decline in the content of ascorbic acid, which is one of the substances making up the antioxidant potential of fruit.

### 3.4. Revenue Differences Related to Differences in Storage Technology

Pome fruits, the most popular of which are apples and pears, are characterized by high storability [12]. As apples and pears can endure long-term storage, they can be supplied on the market all year round. The price of stored fruit is often similar to, and in some years even higher than, that of freshly harvested fruit [53]. In addition to market dependencies, including mainly the supply of fruit over a given period, the most important factor affecting the price is fruit quality [54]. For a grower who prepares fruit for sale it is important that it meets the parameters allowing its classification as top-class fruit for fresh consumption [55]. Fruit discarded after grading generates either no or very low revenue. This study compared the values of fruit stored in different conditions, taking into account the losses which arose during storage. Two types of loss were considered that could be measured based on the study results. Fruit mass loss caused by transpiration depends on relative humidity and respiration [12]. Assuming that the storage humidity was equal for all fruit, fruit mass loss resulted mainly from the rate of respiration. It has been shown in Section 3.2 that CA storage and 1-MCP application have a significant impact on respiration. The figures in Table 8 demonstrate that this translated directly in to the grower's revenue.

**Table 8.** Differences in revenues from the sale of the average pear yield in the EU at market prices in each year of the study, depending on storage conditions.

| Storage Days | NA | NA 1-MCP | CA | CA 1-MCP | Price kg in (EUR) | NA | NA 1-MCP | CA | CA 1-MCP |
|---|---|---|---|---|---|---|---|---|---|
| | | | | | **2011/2012 (Av. Yield = 14.03 t·h$^{-1}$) [1]** | | | | |
| | **Transpiration (%)** | | | | | **Value after Storage (EUR)** | | | |
| 0 | 0.0 | 0.0 | 0.0 | 0.0 | 0.55 [2] | 7723 [3] | 7723 | 7723 | 7723 |
| 30 | 1.6 | 2.1 | 1.2 | 0.9 | 0.55 | 7600 | 7561 | 7631 | 7651 |
| 60 | 3.1 | 2.7 | 2.0 | 1.5 | 0.55 | 7485 | 7517 | 7569 | 7605 |
| 90 | 4.8 | 3.5 | 2.7 | 2.0 | 0.57 | 7659 | 7762 | 7827 | 7882 |
| 120 | 5.7 | 4.8 | 3.2 | 2.3 | 0.57 | 7585 | 7659 | 7789 | 7856 |
| 150 | 6.5 | 5.4 | 3.8 | 2.9 | 0.62 | 8119 | 8216 | 8358 | 8438 |
| 180 | 7.0 | 6.0 | 4.6 | 3.5 | 0.64 | 8382 | 8469 | 8598 | 8693 |
| | Storage diseases (%) | | | | | | | | |
| | 16.0 ± 3.9 [4] | 11.6 ± 3.5 | 9.1 ± 3.3 | 7.4 ± 3.5 | | 6944 | 7424 | 7782 | 8026 |
| | Value difference after storage | | | | | −779 | −299 | 59 | 303 |
| | | | | | **2012/2013 (Av. Yield = 11.78 t·h$^{-1}$)** | | | | |
| | **Transpiration (%)** | | | | | **Value after storage (EUR)** | | | |
| 0 | 0.0 | 0.0 | 0.0 | 0.0 | 0.58 | 6846 | 6846 | 6846 | 6846 |
| 30 | 2.2 | 1.9 | 1.1 | 1.2 | 0.58 | 6693 | 6714 | 6769 | 6763 |
| 60 | 4.0 | 2.7 | 1.9 | 1.6 | 0.65 | 7395 | 7490 | 7555 | 7579 |
| 90 | 5.3 | 3.8 | 2.6 | 1.9 | 0.73 | 8100 | 8230 | 8336 | 8395 |
| 120 | 6.1 | 4.5 | 3.0 | 2.5 | 0.82 | 9106 | 9260 | 9408 | 9456 |
| 150 | 6.9 | 5.2 | 3.8 | 3.1 | 0.99 | 10,892 | 11,082 | 11,247 | 11,332 |
| 180 | 7.8 | 6.1 | 4.7 | 4.1 | 1.14 | 12,361 | 12,583 | 12,778 | 12,858 |
| | Storage diseases (%) | | | | | | | | |
| | 25.0 ± 3.7 | 19.9 ± 3.6 | 14.0 ± 2.7 | 9.9 ± 2.6 | | 9004 | 9916 | 10,899 | 11,531 |
| | Value difference after storage | | | | | 3070 | 4053 | 4686 | 3070 |
| | | | | | **2013/2014 (Av. Yield = 18.85 t·h$^{-1}$)** | | | | |
| | **Transpiration (%)** | | | | | **Value after storage (EUR)** | | | |
| 0 | 0.0 | 0.0 | 0.0 | 0.0 | 0.43 | 8122 | 8122 | 8122 | 8122 |
| 30 | 2.1 | 1.8 | 1.0 | 0.9 | 0.53 | 9719 | 9748 | 9824 | 9834 |
| 60 | 3.6 | 2.8 | 1.7 | 1.7 | 0.53 | 9567 | 9648 | 9756 | 9754 |
| 90 | 4.9 | 3.5 | 2.4 | 2.0 | 0.57 | 10,296 | 10,448 | 10,568 | 10,608 |
| 120 | 5.6 | 4.2 | 3.1 | 2.4 | 0.60 | 10,649 | 10,806 | 10,930 | 11,007 |
| 150 | 6.1 | 4.9 | 3.8 | 2.9 | 0.60 | 10,590 | 10,724 | 10,852 | 10,950 |
| 180 | 6.9 | 5.7 | 4.3 | 3.4 | 0.60 | 10,501 | 10,635 | 10,793 | 10,892 |
| | Storage diseases (%) | | | | | | | | |
| | 29.2 ± 4.4 | 23.3 ± 3.1 | 17.9 ± 3.9 | 11.9 ± 1.8 | | 7211 | 8006 | 8776 | 9550 |
| | Value difference after storage | | | | | −911 | −115 | 654 | 1428 |

[1] Average pear yield per 1 hectare in the EU countries where production exceeds 1000 ha. [2] Price of pears obtained by growers in the period. [3] Average pear yield per ha in the EU countries in which pear production exceeds 1000 ha x average pear price for growers. [4] Standard deviation. NA—normal atmosphere; NA + 1-MCP—normal atmosphere + 1-methylcyclopropene; CA—controlled atmosphere; CA + 1-MCP—controlled atmosphere + 1-methylcyclopropene.

The difference between the highest and the lowest value of the yield that could be obtained per 1 ha did not exceed EUR 100 after the first month of storage (NA, NA + 1MCP, CA, CA + 1MCP), but it grew with each month. Even in the 2011/2012 season, in which the market prices were the most stable, this difference amounted to over EUR 350 after six-month storage. In the following season, during which the market prices rose much faster and the average yield was 2.5 t h$^{-1}$ lower, the difference was as much as about EUR 500. With the previous season's yield, the potential difference between revenues from the yield subjected to the simplest treatment (NA) and the most advanced treatment (CA + 1MCP) would have exceeded EUR 1500. In the last season, the difference decreased, but only to EUR 400, due to a high average yield.

The least advanced storage technologies give low protection against increased transpiration and the incidence of physiological disorders and fungal diseases [12,53]. The total losses caused by fungal diseases varied among individual years, because many of them originate in the orchard. The biggest difference was observed in the last year of the study—the revenues from the sale of the average pear yield in the EU after CA + 1-MCP treatment were over EUR 2300 higher than those calculated for the same yield after NA storage. CA + 1MCP-treated pears will be certainly easier to sell because also their other parameters are superior to those of the NA-stored pears. Since the costs of building a controlled atmosphere storage room, if properly designed, can be only about 5% higher than the costs of building a cold storage room [56], the financial advantages that can be achieved each year will surely more than make up for higher capital expenditures.

### 3.5. Incidence of Diseases and Disorders

Postharvest diseases of pome fruit result in substantial economic losses during storage worldwide every year [9]. In this study, the occurrence of fungal diseases, physiological disorders and visible physical damage of fruit flesh changed every year but always depended on treatment (Figures 1 and 2). Fungal diseases occurred more often than physiological disorders—this tendency is stronger in pears than in apples [53]. The biggest total losses— of about 30%—were observed after six-month storage in 2013 (Figure 1C). The 1-MCP treatment of NA-stored pears reduced the total losses and visibly curbed the incidence of fungal diseases despite an increased incidence of internal browning. CA storage reduced the total losses by about half and the CA+ 1-MCP combination cut the incidence of diseases and disorders significantly. The smallest losses caused by diseases and disorders were noted in the first year of the study (Figure 1A), but 1-MCP application reduced the incidence of diseases and disorders only in the NA-stored fruit. In the second year, model results were obtained—starting from NA storage, each further treatment caused a drop in the total losses caused by diseases and disorder after six-month storage (Figure 1B). Other studies report different outcomes, though, which certainly depended on weather conditions in the orchard [23,57–59].

Fungal pathogens are the main source of losses during the storage and sale of pears [53]. This finding is confirmed by the results of this study except the CA-stored + 1MCP-treated pears. In 2011–2012, the share of fruit with physiological disorders was similar to that of those with fungal diseases (Figure 2). It seems that it was because the treatments very strongly reduced the incidence of fungal diseases. The most important fungal pathogens causing losses due to rotting are *Penicillium expansum*, *Botrytis cinerea*, and *Mucor piriformis* [58]. Other etiological factors include *Phialophora malorum*, *Alternaria* spp., *Cladosporium herbarum*, and *Neofabrea* spp. [60]. Their spores are ubiquitous in the orchard and infect also other tree parts. Fungal diseases were found to be the main loss factor in each year of the study. In 2013, the year of increased incidence of fungal diseases during storage, the percentage of fruit infected with fungal diseases in NA storage was about 10 times higher than the percentage of fruit showing symptoms of physiological disorders. Pears suffer from physiological disorders more rarely than apples [12], which was also clearly apparent in the two other years of the study. In 2011 and 2013, no differences in the incidence of fungal diseases were found between the NA-stored + 1-MCP-treated pears and the CA-stored pears, but the combination of CA and 1-MCP treatment allowed a reduction in the incidence of fungal diseases by half (2013) or by one-third (2011). In 2012, the incidence of fungal diseases decreased gradually and significantly between samples from the level observed after NA storage to the level noted after CA +1-MCP treatment.

Even though physiological disorders occur more rarely compared to fungal diseases, they are more difficult to contain by changing the composition of the atmosphere or applying 1-MCP [61–63]. The share of fruit affected by physiological disorders varied considerably between treatments, but it did not exceed 7.5% in any of the samples, which shows that physiological disorders are a minor cause of losses during storage.

The share of individual diseases and disorders causing losses during storage varied between years (Figure 3). Blue mold led to the biggest losses in all years of the study. It is caused by *Penicillium epansum*, a fungus commonly found in orchards [64]. The study clearly shows that the best way to control the disease is to improve the storage conditions. Every method to enhance the storage conditions significantly reduced the infection rate so that the incidence of blue mold in the CA-stored + 1-MCP-treated fruit constituted 20–30% of that found in the NA-stored fruit. Blue mold control has not only economic relevance for growers, but it also makes it possible to limit the development and spread of strains that produce patulin, a mycotoxin that affects humans [64].

Grey mold, which is caused by *Botrytis cinerea*, was another fungal disease that was observed to infect the fruit in each year of the study. The lowest grey mold incidence rate was noted in 2011 and it is probably due to the low number of infected pears that the differences between the treatments were small and generally insignificant. In the two subsequent years, the differences were bigger and CA storage had a noticeable limiting effect on the disease. A controlled atmosphere is recommended for the storage of grapes that are very susceptible to grey mold for fresh consumption [65].

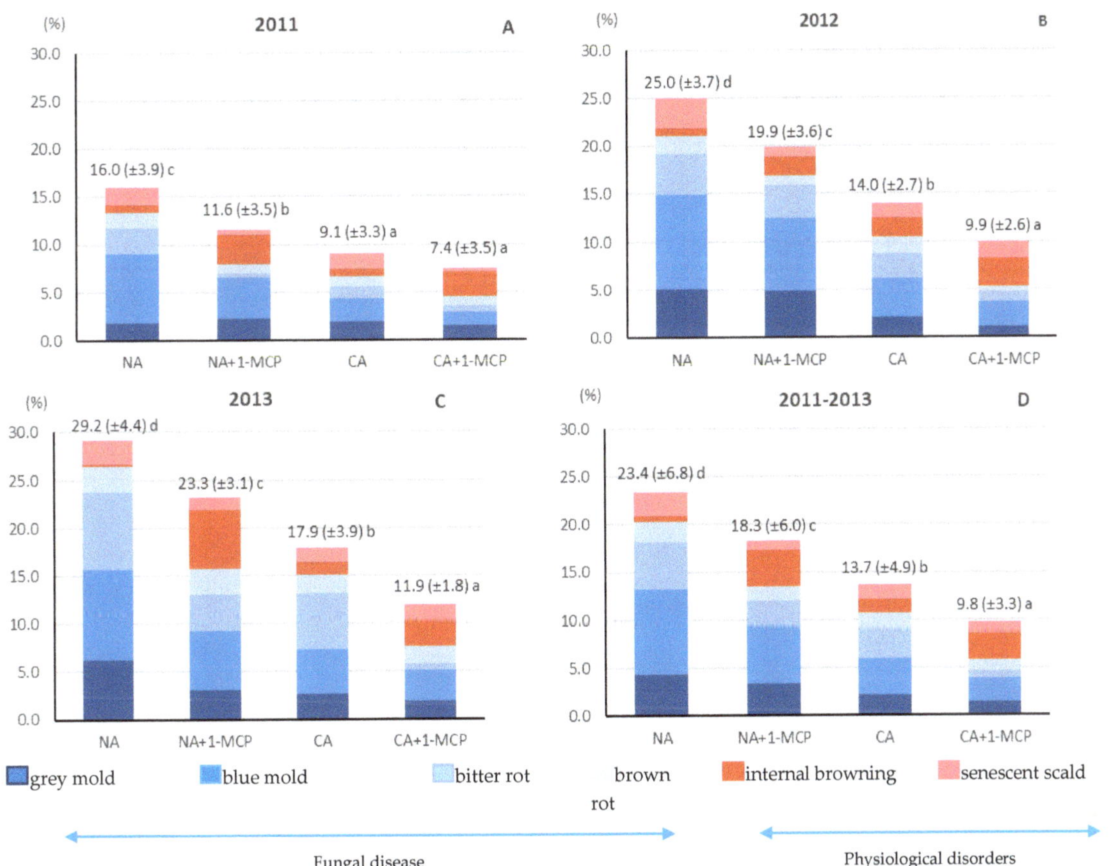


**Figure 1.** Total fruit losses caused by fungal diseases and physiological disorders. Numbers with different letters were significantly different at *p* = 0.05 according to Duncan's test. The data are expressed as mean ± SD (n = 10). NA—normal atmosphere; NA + 1-MCP—normal atmosphere + 1-methylcyclopropene; CA—controlled atmosphere; CA + 1-MCP—controlled atmosphere + 1-methylcyclopropene.

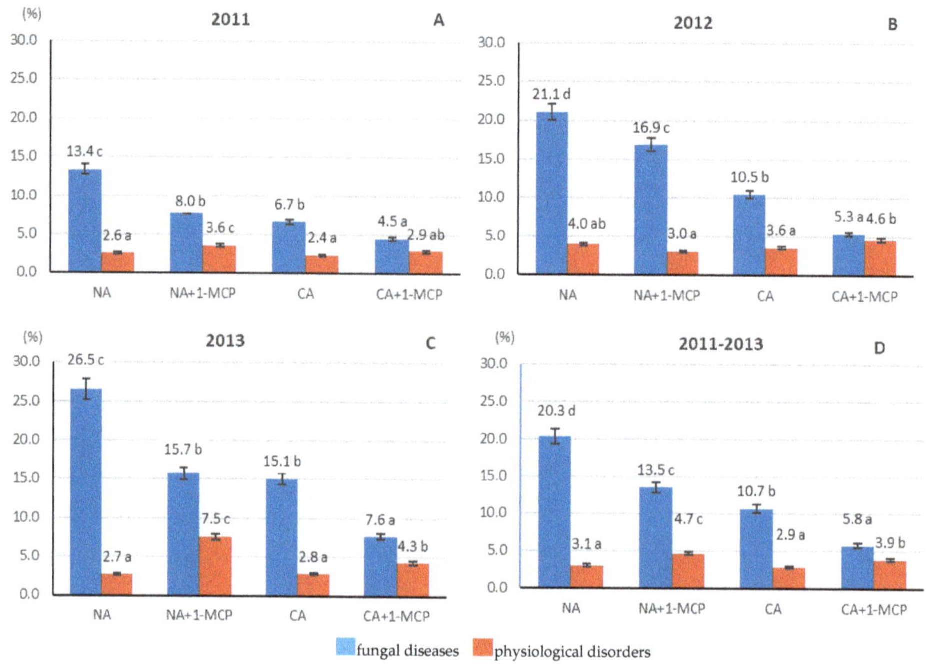



**Figure 2.** Fungal diseases and physiological disorders after storage in different conditions. Numbers with different letters were significantly different at $p = 0.05$ according to Duncan's test. The data are expressed as mean $\pm$ SD (n = 10). NA—normal atmosphere; NA + 1-MCP—normal atmosphere + 1-methylcyclopropene; CA—controlled atmosphere; CA + 1-MCP—controlled atmosphere + 1-methylcyclopropene.

Postharvest pathogens with economic importance for stored fruit also include bitter rot caused by *Colletotrichum* spp. [9]. In this study, the application of 1-MCP was much more effective in controlling this disease than the modification of storage atmosphere. Such a positive outcome of 1-MCP treatment was observed during the storage of apples [66]. Brown rot occurred in less fruit compared to the other fungal diseases. The differences in the incidence of bitter rot between treatments were small, although significant in some cases. In 2011, the highest number of infected pears was found in the samples stored in NA for six months. The other samples showed no differences. In 2012, 1-MCP had a beneficial effect on both NA- and CA-stored fruit, whereas, in 2013, a difference was noted only between the NA- and CA-stored samples, regardless of 1-MCP treatment.

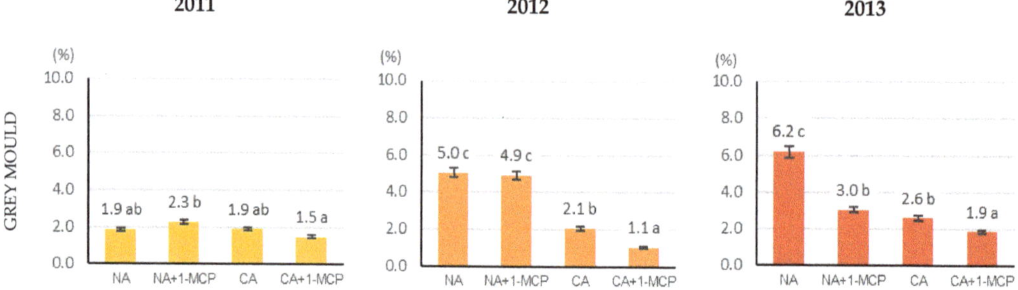



**Figure 3.** *Cont.*

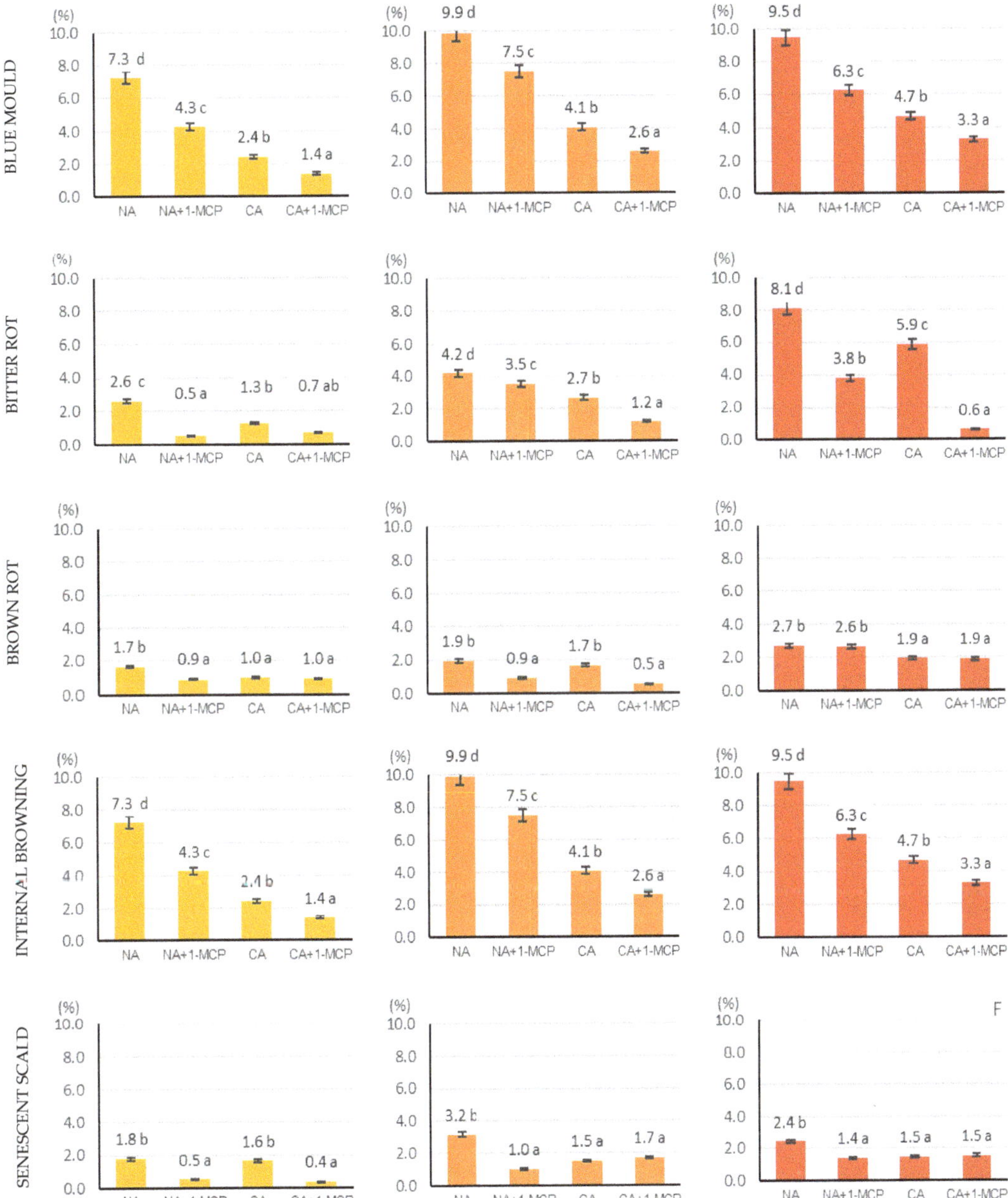

**Figure 3.** Impact of storage technology on the incidence of postharvest diseases and disorders in 2011–2013. Numbers with different letters were significantly different at $p = 0.05$ according to Duncan's test. The data are expressed as mean ± SD (n = 10). NA—normal atmosphere; NA + 1-MCP—normal atmosphere + 1-methylcyclopropene; CA—controlled atmosphere; CA + 1-MCP—controlled atmosphere + 1-methylcyclopropene.

Physiological disorders caused much smaller losses than fungal diseases, which confirms the findings usually reported after the storage of pears [7,11]. The differences between individual treatments appeared to be random and there was even a growth in the number of fruit showing the symptoms of physiological diseases after the application of 1-MCP. 'Conference' pears are prone to internal browning, the incidence of which may increase during CA storage [67,68]. This study only partially supported those findings and the 1-MCP treatment seemed to more strongly promote the development of internal browning, which agreed with the observations already made by Hendges et al. [63] during and after the storage of 'Alexander Lucas' pears. The possible reason for this is the loss of the antioxidant capacity and/or energy deficit caused by a reduction in the respiratory activity of fruit stored under CA. It has also been suggested that the inhibition of ethylene production may induce a stress response and thereby cause cell damage [69]. In our study, the significant increase in the incidence of internal browning was noted in 2011 and 2013 in both NA- and CA-stored fruit.

Senescent scald was a second physiological disorder observed in our study. It manifests itself with skin decolourization [12]. 1-MCP effectively reduces senescent scald in apples, but it is not clear how successfully it helps to control this disorder in pears [70]. Our study did not yield a clear result either; however, much more often than not, 1-MCP limited the occurrence of senescent scald. In 2011, the positive impact of 1-MCP was found after both NA and CA storage. In the following years, the difference was significant for NA-stored fruit. This suggests that the period of 6 months is too long for the storage of 'Conference' pears under NA conditions. The same conclusions were presented in an Italian study assessing the influence of 1-MCP on 'Abbé Fétel' stored in normal atmosphere [57].

## 4. Conclusions

A three-year study showed that the rootstock type, storage atmosphere, and 1-MCP application affected the storability of 'Conference' pears. This is the first study that presents a simultaneous assessment of the influence of the above factors on the quality parameters, the losses caused by diseases and disorders, the antioxidant capacity, and of the economic profitability of long-term storage of an important European pear cultivar.

Rootstock had the weakest influence on storability, and its effects were identified only when determining the fruit mass loss caused by transpiration and respiration.

Antioxidant capacity, just like various other quality parameters, was strongly affected by storage conditions. It grew during six-month CA storage after applying 1-MCP, whereas it stayed at the same level or declined in other storage conditions. This is an important fact that may enable the promotion of the consumption of 'Conference' pear long after harvest.

Most of the results obtained in the study on how six-month storage affects fruit quality and proceeds from its sale show that 'Conference' pears should not be stored in NA for so long. The high incidence of fungal diseases and physiological disorders after such a long storage period and the resulting losses cannot be compensated by the benefits of long-term storage. The economic analysis has revealed that it pays off much more to sell the fruit directly after harvest than after six months of NA storage. The application of 1-MCP alleviates the above-mentioned drawbacks, but does not fully make up for the expenditures. The best solution is to keep the fruit under CA and to additionally apply 1-MCP. This technology is recommended as it allows the preservation of firmness, an appropriate proportion between sugars and acids, and a high content of antioxidant substances.

**Supplementary Materials:** The following are available online at https://www.mdpi.com/article/10.3390/agriculture11060545/s1, Table S1–S3: Firmness (N) of untreated (control) and 1-MCP treated 'Conference' pears analysis after storage in normal (NA) and controlled (CA) atmospheres at 2011–2013 year, Table S4–S6: Soluble solid content (%) of untreated (control) and 1-MCP treated 'Conference' pears analysis after storage in normal (NA) and controlled (CA) atmospheres at 2011–2013 year, Table S7–S9: Acidity of untreated (control) and 1-MCP treated 'Conference' pears analysis after storage in normal (NA) and controlled (CA) atmospheres at 2011–2013 year. Table S10–S12: Impact of storage technology on the incidence of postharvest diseases and disorders in 2011–2013.

**Author Contributions:** Conceptualization, G.P.Ł.; methodology, G.P.Ł.; software, K.R.; validation, G.P.Ł., K.R. and D.W.-T.; formal analysis, K.R.; investigation, G.P.Ł. and D.W.-T.; resources, G.P.Ł., K.R. and D.W.-T.; data curation, K.R.; writing—original draft preparation, G.P.Ł. and K.R.; writing—review and editing, G.P.Ł.; visualization, K.R.; supervision, G.P.Ł.; project administration, G.P.Ł.; funding acquisition, K.R. All authors have read and agreed to the published version of the manuscript.

**Funding:** The publication was co-financed within the framework of the "Regional Initiative Excellence" programme implemented at the initiative of the Polish Ministry of Science and Higher Education in 2019–2022 (No. 005/RID/2018/19)', financing amount: PLN 12,000,000.

**Institutional Review Board Statement:** Not applicable.

**Informed Consent Statement:** Not applicable.

**Data Availability Statement:** Data available on request.

**Conflicts of Interest:** The authors declare no conflict of interest.

# References


1. FAO. Food and Agriculture Organization of the United Nations. 2020. Available online: www.fao.org (accessed on 20 December 2020).
2. European Commission Directorate-General For Agriculture And Rural Development 2019. Brussels, DDG3.G2/JG/Rr (2021) 3000924. Available online: https://ec.europa.eu/info/sites/default/files/food-farming-fisheries/key_policies/documents/cdg-horticulture-olives-spirits-2019-03-18-minutes_en.pdf (accessed on 5 May 2021).
3. Jackson, J.E. *The Biology of Apples and Pears*; The Biology of Horticultural Crops; Cambridge University Press: Cambridge, UK, 2009; ISBN 978-0-521-38018-8.
4. Saquet, A.A. Storage of Pears. *Sci. Hortic.* **2019**, *246*, 1009–1016. [CrossRef]
5. Verlinden, B.E.; de Jager, A.; Lammertyn, J.; Schotsmans, W.; Nicolai, B.M. PH—Postharvest Technology: Effect of Harvest and Delaying Controlled Atmosphere Storage Conditions on Core Breakdown Incidence in 'Conference' Pears. *Biosyst. Eng.* **2002**, *83*, 339–347. [CrossRef]
6. Chiriboga, M.-A.; Saladié, M.; Giné Bordonaba, J.; Recasens, I.; Garcia-Mas, J.; Larrigaudière, C. Effect of Cold Storage and 1-MCP Treatment on Ethylene Perception, Signalling and Synthesis: Influence on the Development of the Evergreen Behaviour in 'Conference' Pears. *Postharvest Biol. Technol.* **2013**, *86*, 212–220. [CrossRef]
7. Łysiak, G.P.; Antkowiak, W. Quality Features of Parthenocarpic Pears Collected from Trees Grown on Different Rootstocks. *Acta Sci. Pol.-Hortorum Cultus* **2015**, *14*, 69–82.
8. Bertolini, P.; Bottardi, S.; Folchi, A.; Dalla Rosa, M. Effect of Controlled Atmosphere Storage on the Physiological Disorders and Quality of Conference Pears. *Ital. J. Food Sci. Italy* **1997**, *9*, 303–312.
9. Wenneker, M.; Thomma, B.P.H.J. Latent Postharvest Pathogens of Pome Fruit and Their Management: From Single Measures to a Systems Intervention Approach. *Eur. J. Plant Pathol.* **2020**, *156*, 663–681. [CrossRef]
10. North, M.S.; de Kock, K.; Booyse, M. Effect of Rootstock, Harvest Date and Storage Time on 'Forelle' Pear Fruit Quality after Cold Storage. *Acta Hortic.* **2012**, 491–497. [CrossRef]
11. Valero, D.; Serrano, M. *Postharvest Biology and Technology for Preserving Fruit Quality*; CRC Press: Boca Raton, FL, USA, 2010; ISBN 1-4398-0267-X.
12. Kader, A.A. *Postharvest Technology of Horticultural Crops*; University of California: Berkeley, CA, USA, 1992.
13. Villalobos-Acuña, M.; Mitcham, E.J. Ripening of European Pears: The Chilling Dilemma. *Postharvest Biol. Technol.* **2008**, *49*, 187–200. [CrossRef]
14. Porritt, S.W. The Effect of Temperature on Postharvest Physiology and Storage Life of Pears. *Can. J. Plant Sci.* **1964**. [CrossRef]
15. Moya-León, M.A.; Vergara, M.; Bravo, C.; Montes, M.E.; Moggia, C. 1-MCP Treatment Preserves Aroma Quality of 'Packham's Triumph' Pears during Long-Term Storage. *Postharvest Biol. Technol.* **2006**, *42*, 185–197. [CrossRef]
16. Lara, I.; Miró, R.M.; Fuentes, T.; Sayez, G.; Graell, J.; López, M.L. Biosynthesis of Volatile Aroma Compounds in Pear Fruit Stored under Long-Term Controlled-Atmosphere Conditions. *Postharvest Biol. Technol.* **2003**, *29*, 29–39. [CrossRef]
17. Wright, A.H.; Delong, J.M.; Arul, J.; Prange, R.K. The Trend toward Lower Oxygen Levels during Apple (Malus × Domestica Borkh) Storage. *J. Hortic. Sci. Biotechnol.* **2015**, *90*, 1–13. [CrossRef]
18. Weber, A.; Brackmann, A.; Both, V.; Pavanello, E.P.; de Oliveira Anese, R.; Thewes, F.R.; Weber, A.; Brackmann, A.; Both, V.; Pavanello, E.P.; et al. Respiratory Quotient: Innovative Method for Monitoring 'Royal Gala' Apple Storage in a Dynamic Controlled Atmosphere. *Sci. Agric.* **2015**, *72*, 28–33. [CrossRef]
19. Veltman, R.H.; Verschoor, J.A.; van Dugteren, J.H.R. Dynamic Control System (DCS) for Apples (Malus Domestica Borkh. Cv 'Elstar'): Optimal Quality through Storage Based on Product Response. *Postharvest Biol. Technol.* **2003**, *27*, 79–86. [CrossRef]
20. Lum, G.B.; Shelp, B.J.; DeEll, J.R.; Bozzo, G.G. Oxidative Metabolism Is Associated with Physiological Disorders in Fruits Stored under Multiple Environmental Stresses. *Plant Sci.* **2016**, *245*, 143–152. [CrossRef] [PubMed]
21. Streif, J.; Saquet, A.A.; Xuan, H. CA-Related Disorders of Apples and Pears. *Acta Hortic.* **2003**, 223–230. [CrossRef]
22. Ekman, J.H.; Clayton, M.; Biasi, W.V.; Mitcham, E.J. Interactions between 1-MCP Concentration, Treatment Interval and Storage Time for 'Bartlett' Pears. *Postharvest Biol. Technol.* **2004**, *31*, 127–136. [CrossRef]

23. Hendges, M.V.; Neuwald, D.A.; Steffens, C.A.; Vidrih, R.; Zlatić, E.; do Amarante, C.V.T. 1-MCP and Storage Conditions on the Ripening and Production of Aromatic Compounds in Conference and Alexander Lucas Pears Harvested at Different Maturity Stages. *Postharvest Biol. Technol.* **2018**, *146*, 18–25. [CrossRef]
24. Watkins, C.B. The Use of 1-Methylcyclopropene (1-MCP) on Fruits and Vegetables. *Biotechnol. Adv.* **2006**, *24*, 389–409. [CrossRef] [PubMed]
25. Spotts, R.A.; Sholberg, P.L.; Randall, P.; Serdani, M.; Chen, P.M. Effects of 1-MCP and Hexanal on Decay of d'Anjou Pear Fruit in Long-Term Cold Storage. *Postharvest Biol. Technol.* **2007**, *44*, 101–106. [CrossRef]
26. Chen, P.M.; Spotts, R.A. Changes in Ripening Behaviors of 1-MCP-Treated 'd'Anjou' Pears After Storage. *Int. J. Fruit Sci.* **2005**, *5*, 3–18. [CrossRef]
27. Golding, J.B.; Singh, S.P. Use of 1-MCP in the Storage Life Extension of Fruit. In *Reference Module in Food Science*; Elsevier: Amsterdam, The Netherlands, 2017; ISBN 978-0-08-100596-5.
28. Silva, F.J.P.; Gomes, M.H.; Fidalgo, F.; Rodrigues, J.A.; Almeida, D.P.F. Antioxidant Properties and Fruit Quality During Long-Term Storage of 'Rocha' Pear: Effects of Maturity and Storage Conditions. *J. Food Qual.* **2010**, *33*, 1–20. [CrossRef]
29. Kolniak-Ostek, J.; Kłopotowska, D.; Rutkowski, K.P.; Skorupińska, A.; Kruczyńska, D.E. Bioactive Compounds and Health-Promoting Properties of Pear (*Pyrus Communis* L.) Fruits. *Molecules* **2020**, *25*, 4444. [CrossRef] [PubMed]
30. Rein, M.J.; Renouf, M.; Cruz-Hernandez, C.; Actis-Goretta, L.; Thakkar, S.K.; da Silva Pinto, M. Bioavailability of Bioactive Food Compounds: A Challenging Journey to Bioefficacy. *Br. J. Clin. Pharmacol.* **2013**, *75*, 588–602. [CrossRef] [PubMed]
31. Durazzo, A. Study Approach of Antioxidant Properties in Foods: Update and Considerations. *Foods* **2017**, *6*, 17. [CrossRef]
32. Re, R.; Pellegrini, N.; Proteggente, A.; Pannala, A.; Yang, M.; Rice-Evans, C. Antioxidant Activity Applying an Improved ABTS Radical Cation Decolorization Assay. *Free Radic. Biol. Med.* **1999**, *26*, 1231–1237. [CrossRef]
33. Ctifl (Centre Technique Interprofessionnel Des Fruits et Légumes) Code Amidon Pomme (Starch Conversion Chart for Apples). Available online: http://www.ctifl.fr/Pages/Kiosque/DetailsOuvrage.aspx?idouvrage=819 (accessed on 20 December 2020).
34. Łysiak, G. The Sum of Active Temperatures as a Method of Determining the Optimum Harvest Date of 'Sampion' and 'Ligol' Apple Cultivars. *Acta Sci. Pol.-Hortorum Cultus* **2012**, *11*, 3–13.
35. OECD Organisation for Economic Co-Operation and Development. International Standards for Fruits and Vegetables, Pears. Available online: https://www.oecd-ilibrary.org/docserver/87244e04-en-fr.pdf?expires=1616519112&id=id&accname=guest&checksum=32A0D385371504C147657DAE88E6A245 (accessed on 5 May 2021).
36. Brookfield, P.; Murphy, P.; Harker, R.; MacRae, E. Starch Degradation and Starch Pattern Indices; Interpretation and Relationship to Maturity. *Postharvest Biol. Technol.* **1997**, *11*, 23–30. [CrossRef]
37. Westwood, M.N. *Temperate-Zone Pomology: Physiology and Culture*, 3rd ed.; Timber Press. Inc.: Portland, OR, USA, 1993; ISBN 0-88192-253-6.
38. Łysiak, G.P.; Kurlus, R. Rootstock Effect on Optimum Harvest Date and Storability of Two Apple Cultivars. In Proceedings of the Proc. Int. Conf. Fruit Production and Fruit Breeding, Tartu, Estonia, 12–13 September 2000; pp. 12–13.
39. Streif, J. Optimum Picking Date for Cox Orange Apples Grown in Bodensee Region. *Workshop Optim. Harvest Date* **1992**, *1*, 2–5.
40. Łysiak, G. The Determination of Harvest Index of Šampion Apples Intended for Long Storage. *Acta Sci. Pol. Hortorum Cultus* **2011**, *10*, 273–282.
41. Mendes da Silva, T.; Torello Marinoni, D.; Peano, C.; Roberta Giuggioli, N. A New Sensory Approach Combined with a Text-Mining Tool to Create a Sensory Lexicon and Profile of Monovarietal Apple Juices. *Foods* **2019**, *8*, 608. [CrossRef] [PubMed]
42. Kullaj, E. Chapter 8—Rootstocks for Improved Postharvest Quality of Fruits: Recent Advances. In *Preharvest Modulation of Postharvest Fruit and Vegetable Quality*; Siddiqui, M.W., Ed.; Academic Press: Cambridge, MA, USA, 2018; pp. 189–207, ISBN 978-0-12-809807-3.
43. Stern, R.A.; Doron, I. Performance of 'Coscia' Pear (*Pyrus Communis*) on Nine Rootstocks in the North of Israel. *Sci. Hortic.* **2009**, *119*, 252–256. [CrossRef]
44. Murayama, H.; Takahashi, T.; Honda, R.; Fukushima, T. Cell Wall Changes in Pear Fruit Softening on and off the Tree. *Postharvest Biol. Technol.* **1998**, *14*, 143–149. [CrossRef]
45. Saquet, A.A. Storability of 'Conference' Pear Under Various Controlled Atmospheres. *Erwerbs-Obstbau* **2018**, *60*, 275–280. [CrossRef]
46. Torregrosa, L.; Echeverria, G.; Illa, J.; Giné-Bordonaba, J. Ripening Behaviour and Consumer Acceptance of 'Conference' Pears during Shelf Life after Long Term DCA-Storage. *Postharvest Biol. Technol.* **2019**, *155*, 94–101. [CrossRef]
47. Wawrzyńczak, A.; Rutkowski, K.P.; Kruczyńska, D.E. Changes in Fruit Quality in Pears during CA Storage. *J. Fruit Ornam. Plant Res.* **2006**, *14*, 77–84.
48. Watkins, C.B.; Nock, J.F.; Whitaker, B.D. Responses of Early, Mid and Late Season Apple Cultivars to Postharvest Application of 1-Methylcyclopropene (1-MCP) under Air and Controlled Atmosphere Storage Conditions. *Postharvest Biol. Technol.* **2000**, *19*, 17–32. [CrossRef]
49. Łysiak, G.P.; Michalska-Ciechanowska, A.; Wojdyło, A. Postharvest Changes in Phenolic Compounds and Antioxidant Capacity of Apples Cv. Jonagold Growing in Different Locations in Europe. *Food Chem.* **2020**, *310*, 125912. [CrossRef]
50. Charles, M.; Aprea, E.; Gasperi, F. Factors Influencing Sweet Taste in Apple. In *Sweeteners*; Merillon, J.-M., Ramawat, K.G., Eds.; Reference Series in Phytochemistry; Springer International Publishing: Cham, Switzerland, 2018; pp. 1–22, ISBN 978-3-319-26478-3.

51.   Vilaplana, R.; Valentines, M.C.; Toivonen, P.; Larrigaudière, C. Antioxidant Potential and Peroxidative State of 'Golden Smoothee' Apples Treated with 1-Methylcyclopropene. *J. Am. Soc. Hortic. Sci.* **2006**, *131*, 104–109. [CrossRef]

52.   Larrigaudiere, C.; Pintó, E.; Lentheric, I. Oxidative Behaviour of Conference Pears Stored in Air and Controlled-Atmosphere Storage. *Acta Hortic.* **2003**, 355–360. [CrossRef]

53.   Florkowski, W.; Shewfelt, R.L.; Brueckner, B.; Prussia, S.E. *Postharvest Handling. A Systems Approach*, 3rd ed.; Academic Press, Elsevier: Amsterdam, The Netherlands, 2014; ISBN 978-0-12-408137-6.

54.   Nótári, M.; Ferencz, Á. The Harvest and Post-Harvest of Traditional Pear Varieties in Hungary. *APCBEE Procedia* **2014**, *8*, 305–309. [CrossRef]

55.   Florkowski, W.J.; Łysiak, G. Quality Attribute-Price Relationship: Modernization of the Sweet Cherry Sector in Poland. *Sci. J. Wars. Univ. Life Sci. SGGW Probl. World Agric.* **2015**, *15*, 1–15. [CrossRef]

56.   Waelti, H.; Bartsch, J.A. Controlled atmosphere storage facilities. In *Food Preservation by Modified Atmospheres*; CRC Press: Boca Raton, FL, USA, 1990; pp. 373–389, ISBN 978-0-8493-6569-0.

57.   Rizzolo, A.; Grassi, M.; Vanoli, M. Influence of Storage (Time, Temperature, Atmosphere) on Ripening, Ethylene Production and Texture of 1-MCP Treated 'Abbé Fétel' Pears. *Postharvest Biol. Technol.* **2015**, *109*, 20–29. [CrossRef]

58.   Sardella, D.; Muscat, A.; Brincat, J.-P.; Gatt, R.; Decelis, S.; Valdramidis, V. A Comprehensive Review of the Pear Fungal Diseases. *Int. J. Fruit Sci.* **2016**, *16*, 351–377. [CrossRef]

59.   Saquet, A.; Almeida, D. Internal Disorders of 'Rocha' Pear Affected by Oxygen Partial Pressure and Inhibition of Ethylene Action. *Postharvest Biol. Technol.* **2017**, *128*, 54–62. [CrossRef]

60.   Sutton, T.B.; Aldwinckle, H.S.; Agnello, A.M.; Walgenbach, J.F. (Eds.) *Compendium of Apple and Pear Diseases and Pests*, 2nd ed.; The American Phytopathological Society: St. Paul, MN, USA, 2014; ISBN 978-0-89054-433-4.

61.   Xie, X.; Song, J.; Wang, Y.; Sugar, D. Ethylene Synthesis, Ripening Capacity, and Superficial Scald Inhibition in 1-MCP Treated 'd'Anjou' Pears Are Affected by Storage Temperature. *Postharvest Biol. Technol.* **2014**, *97*, 1–10. [CrossRef]

62.   Dong, Y.; Wang, Y.; Einhorn, T.C. Postharvest Physiology, Storage Quality and Physiological Disorders of 'Gem' Pear (*Pyrus Communis* L.) Treated with 1-Methylcyclopropene. *Sci. Hortic.* **2018**, *240*, 631–637. [CrossRef]

63.   Hendges, M.V.; Steffens, C.A.; Espindola, B.P.; Amarante, C.V.T.; Neuwald, D.A.; Kittemann, D. 1-MCP Treatment Increases Internal Browning Disorders in 'Alexander Lucas' Pears Stored under Controlled Atmosphere. *Acta Hortic.* **2015**, 511–517. [CrossRef]

64.   Bautista-Baños, S. *Postharvest Decay: Control Strategies*; Elsevier: Amsterdam, The Netherlands, 2014; ISBN 978-0-12-411568-2.

65.   Domingues, A.R.; Roberto, S.R.; Ahmed, S.; Shahab, M.; José Chaves Junior, O.; Sumida, C.H.; De Souza, R.T. Postharvest Techniques to Prevent the Incidence of Botrytis Mold of 'BRS Vitoria' Seedless Grape under Cold Storage. *Horticulturae* **2018**, *4*, 17. [CrossRef]

66.   Tomala, K.; Grzęda, M.; Guzek, D.; Głąbska, D.; Gutkowska, K. The Effects of Preharvest 1-Methylcyclopropene (1-MCP) Treatment on the Fruit Quality Parameters of Cold-Stored 'Szampion' Cultivar Apples. *Agriculture* **2020**, *10*, 80. [CrossRef]

67.   Saquet, A.A.; Streif, J.; Bangerth, F. Changes in ATP, ADP and Pyridine Nucleotide Levels Related to the Incidence of Physiological Disorders in 'Conference' Pears and 'Jonagold' Apples during Controlled Atmosphere Storage. *J. Hortic. Sci. Biotechnol.* **2000**, *75*, 243–249. [CrossRef]

68.   Veltman, R.H.; Kho, R.M.; van Schaik, A.C.R.; Sanders, M.G.; Oosterhaven, J. Ascorbic Acid and Tissue Browning in Pears (*Pyrus Communis* L. Cvs Rocha and Conference) under Controlled Atmosphere Conditions. *Postharvest Biol. Technol.* **2000**, *19*, 129–137. [CrossRef]

69.   Jung, S.-K.; Watkins, C.B. Involvement of Ethylene in Browning Development of Controlled Atmosphere-Stored 'Empire' Apple Fruit. *Postharvest Biol. Technol.* **2011**, *59*, 219–226. [CrossRef]

70.   Lurie, S.; Watkins, C.B. Superficial Scald, Its Etiology and Control. *Postharvest Biol. Technol.* **2012**, *65*, 44–60. [CrossRef]

 *agriculture*

**MDPI**

*Article*

# Application of Dynamic Controlled Atmosphere Technologies to Reduce Incidence of Physiological Disorders and Maintain Quality of 'Granny Smith' Apples

Tatenda Gift Kawhena [1,2], Olaniyi Amos Fawole [2,3] and Umezuruike Linus Opara [2,4,*]

1   Department of Horticultural Science, Faculty of AgriSciences, Stellenbosch University, Stellenbosch 7600, South Africa; 19547129@sun.ac.za
2   SARChI Postharvest Technology Research Laboratory, Africa Institute for Postharvest Technology, Faculty of AgriSciences, Stellenbosch University, Stellenbosch 7600, South Africa; olaniyi@sun.ac.za
3   Postharvest Research Laboratory, Department of Botany and Plant Biotechnology, University of Johannesburg, Johannesburg 2006, South Africa
4   UNESCO International Centre for Biotechnology, Nsukka 410001, Enugu State, Nigeria
*   Correspondence: opara@sun.ac.za; Tel.: +27-21-808-4068

**Abstract:** The efficacy of dynamic controlled atmosphere technologies; repeated low oxygen stress (RLOS) and dynamic controlled atmosphere-chlorophyll fluorescence (DCA-CF) to control superficial scald development on 'Granny Smith' apples during long-term storage was studied. Fruit were stored for 2, 4, 6, 8, and 10 months at 0 °C in DCA-CF (0.6% $O_2$ and 0.8% $CO_2$), regular atmosphere (RA)($\approx$21% $O_2$ and 90–95% RH), and RLOS treatments: (1) 0.5% $O_2$ for 10 d followed by ultra-low oxygen (ULO) (0.9% $O_2$ and 0.8% $CO_2$) for 21 d and 0.5% $O_2$ for 7 d or (2) 0.5% $O_2$ for 10 d followed by controlled atmosphere (CA) (1.5% $O_2$ and 1% $CO_2$) for 21 d and 0.5% $O_2$ for 7 d. Development of superficial scald was inhibited for up to 10 months and 7 d shelf life (20 °C) under RLOS + ULO and DCA-CF treatments. Apples stored in RLOS + ULO, RLOS + CA, and DCA-CF had significantly ($p < 0.05$) higher flesh firmness and total soluble solids. The RLOS phases applied with CA or ULO and DCA-CF storage reduced the development of superficial scald by possibly suppressing the oxidation of volatiles implicated in superficial scald development.

**Keywords:** chlorophyll; fluorescence; storage atmosphere; superficial scald

**Citation:** Kawhena, T.G.; Fawole, O.A.; Opara, U.L. Application of Dynamic Controlled Atmosphere Technologies to Reduce Incidence of Physiological Disorders and Maintain Quality of 'Granny Smith' Apples. *Agriculture* **2021**, *11*, 491. https://doi.org/10.3390/agriculture11060491

Academic Editor: Dirk E. Maier

Received: 24 April 2021
Accepted: 18 May 2021
Published: 26 May 2021

**Publisher's Note:** MDPI stays neutral with regard to jurisdictional claims in published maps and institutional affiliations.



## 1. Introduction

'Granny Smith' apples (*Malus × domestica* Borkh.) are susceptible to superficial scald, a serious postharvest physiological disorder adversely affecting fruit quality and marketability [1–3]. Superficial scald is observed as black or brown patches on fruit skin during cold storage and is associated with cell death or necrosis in hypodermal cortical tissue [4]. Although internal quality is usually not affected, development of superficial scald renders fruit unmarketable because of reduced appearance quality [5]. The autooxidation of naturally occurring sesquiterpene α-farnesene volatile to conjugated trienols and 6-methyl-5-hepten-2-one (MHO) is probably the main reaction resulting in the manifestation of symptoms of superficial scald [2,6]. However, the loss of natural antioxidant metabolites (tocopherol and phenolic compounds) and enzymes, which prevent cell damage by reactive oxygen species, contributes to the development of scald symptoms [5,7]. In addition, low phenolic content in apple peel has been correlated with high superficial scald incidence [8]. Both lipophilic and hydrophilic antioxidants may be involved in superficial scald prevention; however, no specific antioxidant has been consistently linked to α-farnesene, MHO, or superficial scald [4,7,9]. Additionally, pre-harvest factors such as cultivar, maturity, rootstock, and seasonal differences determine susceptibility to scald development [4].

In an attempt to control superficial scald, several methods have been applied to fruit that were either chemical or non-chemical in nature. Of note was diphenylamine, a synthetic antioxidant that successfully inhibited superficial scald until it was banned in Europe because of consumer safety concerns [10]. Low oxygen stress (LOS) storage of apples has been used for many decades as a non-chemical storage technology alternative for apples [3,4]. For instance, initial low oxygen stress (ILOS), anaerobic treatment for 9–14 d, before controlled atmosphere (CA), ultra-low oxygen (ULO), or regular atmosphere (RA) has been reported to effectively prevent the development of superficial scald in 'Granny Smith', Starkimson', 'Delicious', and 'Royal Gala' apples [11–14]. The mechanism of action is not fully understood. However, hypoxic conditions during storage lead to stimulated and rapid ethanol production in the fruit pulp, which presumably limits the oxidation of $\alpha$-farnesene in the peel [4,15]. Ethanol vapors have shown inhibitory effects against superficial scald when exposed to various apple cultivars, further supporting this hypothesis [16,17].

In recent years, CA technologies have gained a lot of attention, particularly dynamic controlled atmosphere (DCA) [3]. The storage technology regularly adjusts gas composition during storage using biosensors, namely chlorophyll fluorescence (DCA-CF), respiration quotient (DCA-RQ), and ethanol (DCA-ET) [2,3]. Research studies by Mditshwa et al. [2] demonstrated the efficacy of repeated application of DCA-CF to control superficial scald (2%) on 'Granny Smith' apples when stored for 16 w in DCA-CF with a 14 d of interruption with regular atmosphere (RA) at −0.5 °C and 95% relative humidity (RH). Similarly, research work has also shown that DCA-CF storage maintains fruit firmness, inhibits the development of decay, and preserves Gala's internal quality [18] and 'Granny Smith' apples [19].

Despite the demonstrated efficacy of DCA-CF, more studies are still relevant to developing cultivar-specific storage protocols and validating existing results. For example, no evidence of a substantial difference between the sensory parameters of 'Greenstar' apples stored in DCA-CF at two oxygen concentration regimes (0.4 and 0.7%) (1.2 ± 0.2 °C) was observed after storage for 10 months, which suggests the need for further optimization studies [20]. Additionally, DCA-CF is subject to errors in determining the low oxygen limit of fruit because chlorophyll fluorescence depends on the metabolic activity of fruit [21,22]. For example, Feng et al. [23] reported variations in metabolic activity of fruit depending on canopy position in a tree. For the three apple cultivars 'McIntosh', 'Gala', and 'Mutsu', the sun-exposed side exhibited elevated rates of metabolism (higher soluble sugars, sugar alcohols, ascorbic acid, and succinic acids in the peel) compared to the shaded side.

Recent studies, according to Bessemans et al. [24], showed better fruit quality in 'Granny Smith' apples subjected to DCA-RQ (0.25–0.4 kPa $O_2$) compared to standard CA storage at low ethanol concentration (<0.028 g $L^{-1}$) in the fruit pulp. The quality of the fruit resembled that of 1-Methylcyclopropene (1-MCP) treated (preceded with CA) apples after 7 d at 18 °C. In a study on 'Royal Gala' apples, Weber et al. [22] showed that fruit stored in DCA-RQ had superior quality (less flesh breakdown) compared to static CA after 8 months of cold storage (1 °C). Despite the notable benefits in adopting DCA-RQ, the use of RQ is usually feasible under strict and gas tight conditions, unattainable in most CA rooms due to leakage [25]. In addition, this storage technique is feasible in controlled laboratory conditions with sensitive instruments that can accurately measure oxygen consumption rates and $CO_2$ production [22,24].

Application of DCA-ET, also known as repeated low oxygen stress (RLOS), is based on determination of low oxygen limit (LOL) through either the destructive measurements of ethanol content from fruit pulp (estimated to be 1 ppm) or headspace analysis with sensors, notably DCS™ (Storex, Gravendeel) [3]. Few studies have reported the commercial application of DCA-ET beyond the ongoing research work in the Netherlands on different apple cultivars ('Elstar' and 'Jonagold') [21]. The study by Veltman et al. [26] investigated the effects of DCA-ET using Chrompack gas chromatography to regulate ethanol levels in the fruit pulp of 'Elstar' apples during cold storage (1 °C). The results showed that, in addition to less skin spot development, the fruit had better color and firmness retention than standard CA (1.2% $O_2$ and 2.5% $CO_2$). However, there is limited information available on the application of the RLOS technology in other important apple cultivars such as 'Granny Smith', and the mechanism of action of RLOS is not clearly understood.

This study will further evaluate RLOS phases' effects on the incidence of physiological disorders and internal quality of the 'Granny Smith' apples during long-term storage. The study also assessed changes in radical scavenging activity, total phenolic content, and selected volatiles of 'Granny Smith' apples subjected to DCA-CF and RLOS storage technologies.

## 2. Materials and Methods

### 2.1. Fruit Supply, Treatments and Storage

'Granny Smith' apples were harvested at 172 days after full bloom (DAFB) with an average starch breakdown = 36.3% and firmness = 79 N from Grabouw (34°12'12'' S, 19°02'35'' E), Western Cape, South Africa. Fruit were transported to the research laboratory at the Agricultural Research Council (ARC), Stellenbosch, South Africa. 'Granny Smith' apples were sorted for external damages, packaged in crates (90–120 fruit per crate), and stored at 0 °C. Each treatment had 3 replications (1 crate = 1 replication = 90–120 fruit). Subsequently, 'Granny Smith' apples were subjected to the following repeated low oxygen stress (RLOS) treatments: (1) low oxygen stress at 0.5% $O_2$ for 10 d followed by ultra-low oxygen (ULO) (0.9% $O_2$ and 0.8% $CO_2$) for 21 d and low oxygen stress at 0.5% for 7 d, (2) low oxygen stress at 0.5% $O_2$ for 10 d followed by controlled atmosphere (CA) (1.5% $O_2$ and 1% $CO_2$) for 21 d, and low oxygen stress at 0.5% $O_2$ for 7 d.

Existing CA cold rooms were installed with the HarvestWatch® (SAtlantic Inc., Halifax, NS, Canada) chlorophyll fluorescence non-destructive monitoring system. The interactive response monitor sensor (FIRM) was used to detect the low oxygen limit (LOL) of the fruit and monitor the physiological response of the fruit to low oxygen levels hourly. Following the storage protocol reported by Mditshwa et al. [2], 'Granny Smith' apples were loaded into cold rooms, and a core temperature of −0.5 °C was achieved within 48–96 h after harvest. Low oxygen limit of the fruit was detected within 48 h after harvest, using compressed air, carbon dioxide, and nitrogen from a membrane generator (Isosep, Isolcell, Italy). The LOL was set at 0.3% $O_2$ for both harvest seasons (2015 and 2016), after which it was set to 0.6% $O_2$ for the entire storage period. In this study, gas composition needed to be adjusted to prevent a shift to the anaerobic respiration of fruit, so the storage room chamber was analyzed at 90 min intervals. Fruit were also stored in RA at 0 °C (≈21% $O_2$ and 90–95% RH). The study was repeated over two consecutive seasons (2015 and 2016). In summary, 'Granny Smith' apples were stored using the following treatments (1–4):

1. RLOS + ULO: Cycles of RLOS (0.5% $O_2$ for 10 d) followed ULO (0.9% $O_2$ and 0.8% $CO_2$ for 21 d and 0.5% $O_2$ for 7 d);
2. RLOS + CA: Cycles of RLOS (0.5% $O_2$ for 10 d) followed by CA (1.5% $O_2$ and 1% $CO_2$ for 21 d and 0.5% $O_2$ for 7 d);
3. DCA-CF: Storage at 0.6% $O_2$ and 0.8% $CO_2$;
4. RA: Storage at ≈21% $O_2$ and 90–95% RH.

### 2.2. Assessment of Quality and Analysis

#### 2.2.1. Physiological Disorders

Superficial Scald

Fruit with superficial scald were counted and expressed as a percentage of the total number of fruit in a single replication (90–120 fruit) [5]. Observations of the symptoms were done after a 6 w simulated shipping and handling period at −0.5 °C and 7 d shelf life (20 °C and 65% RH).

Coreflush

A sample of 10 fruit was taken of each replicate (90–120 fruit) from each treatment, each fruit being cut open for a rating of coreflush incidence. Fruit with coreflush were counted and expressed as a percentage of the total number of fruit per sample [27]. This was done after a 6 w simulated shipping and handling period at −0.5 °C and 7 d shelf life (20 °C and 65% RH).

### 2.2.2. Physicochemical Properties
Firmness

A Fruit Texture Analyzer (FTA 20, Güss, South Africa) with 11.1 mm compression probe was used to measure flesh firmness [28]. To measure flesh firmness for each replicate (10 fruit), each fruit was peeled equatorially on opposite sides, the plunger was pressed into the peeled flesh, and the firmness reading recorded. The operating conditions of the instrument were: pre-test speed 1.5 mm s$^{-1}$, 0.5 mm s$^{-1}$ test speed, 10.0 mm s$^{-1}$ post-test speed, and 0.20 N trigger force. The average reading from both sides was used.

Total Soluble Solids and Titratable Acidity

To measure total soluble solids (TSS), fruit segments ($\approx$20 g of each fruit) were cut transversely from 10 randomly selected fruit per replicate. The fruit segments were processed for juice using a domestic juicer (Mellerware Liquafresh juice extractor III). The juice was homogenized by mixing and stabilizing for five minutes, and 5 mL was sampled using a syringe for TSS measurements. Total soluble solids content was obtained using a calibrated refractometer (Pocket refractometer PAL$^{-1}$, ATAGO Co. LTD, Tokyo, Japan) [29]. Standardization was done using distilled water (refractive index of 0). The refractometer was rinsed between readings to maintain accurate measurements of TSS. To measure titratable acidity (TA), 20 g of fruit segments were cut from each of the 10 fruit (per replicate) and a 46 mL juice sample was blended and titrated against 0.333 N of sodium hydroxide (NaOH) to a pH of 8.2 using a Crison Titromatic 1S/2B (Crison Instruments, Barcelona, Spain). Titratable acidity was expressed as g of malic acid per 100 g [29].

Background Color

A Chroma Meter (CR 400/410 Konica Minolta Sensing Inc., Japan) was used to obtain background color according to the Commission Internationale De I'Eclairage (CIELAB)(L*, a* and b*) system from two opposite positions along the equatorial region of the fruit [30]. The color coordinate L* = 0 to 100 (describing black to white), a* = red (+)/green (−), and b* = yellow (+)/blue (−). Hue angle (h°) was used to measure background color and determined according to the following Equation (1):

$$h° = \arctan (b^*/a^*) \tag{1}$$

where a* represents redness and greenness and b* represents yellowness and blueness.

### 2.2.3. Headspace Volatile Analysis

Fruit were sampled at harvest, after each storage period, and after a 6 w simulated shipping and handling period followed by 7 d shelf life. Apple peel was carefully obtained from four regions of each fruit using a stainless steel peeler (Sigma–Aldrich, Johannesburg, South Africa). A sample of 10 fruit was obtained from each replicate, peeled, cut into smaller pieces, and 5 g was weighed into 20 mL solid phase microextraction (SMPE) glass vials. Ten microlitres of 3-octanol or anisole/methyoxybenzene solution were added as an internal standard to the vials, after which they were sealed. Three replicates were prepared for each treatment (1 glass vials = 1 replicate = 5 g apple peel). The solid phase microextraction (SPME) method was used for headspace volatile analysis [9,31]. Equilibration of vials was done for 10 min in an autosampler incubator (CTC Analytics AG, Zwingen, Switzerland) set at 50 °C. Volatile compounds in the headspace were trapped on a 50/30 μm divinylbenzenecarboxen-polydimethylsiloxane coated fiber after exposure for 20 min at 50 °C. Volatile compounds on the fiber coating were then desorbed for 2 min in the injection port of the gas chromatograph, operated in splitless mode at a temperature of 250 °C. Preconditioning of the fiber was done for 2 min at a temperature of 50 °C and 250 rpm followed by volatile compound chromatographic separation using a polar capillary column (60 m 0.25 mm i.d., 0.5 μm film thickness) (Agilent Technologies DB-FFAP, model J & W 122–3263). The oven was set at a temperature of 40 °C, held for 5 min, and then rapidly increased to 230 °C for 6 min. The total run time for the method was 30 min. Helium was used as a carrier gas at a constant flow rate of 1 mL/min. The ion source and quadropole were kept at 240 °C and 150 °C, respectively. The transfer line temperature was kept at 280 °C.

### 2.2.4. Biochemical Analysis
#### Total Phenolic Content

Measurement of total phenolic content of apple peel was done using the Folin–Ciocalteu method according to Mditshwa et al. [9], with slight modifications. In triplicates, 50 μL of 4-fold diluted crude extract was added to 450 μL of 50% methanol, followed by 500 μL Folin–C reagent and then sodium carbonate (2%) solution after 2 min. The mixture was vortexed incubated for 40 min in the dark at room temperature, and absorbance was measured at 725 nm using a UV–visible spectrophotometer (Thermo Scientific Technologies, Madison, Wisconsin). Total phenolic content in the extract was extrapolated using a gallic acid calibration curve. Results were expressed as mean (milligrams) of gallic acid equivalents per unit dry matter (mg GAE/g DM) of peel in triplicate samples.

#### Radical Scavenging Activity

Radical scavenging activity (RSA) from apple peel extract was determined according to Mditshwa et al. [9]. In triplicates, 15 μL of 3-fold diluted crude extract was added to 735 μL of 50% methanol followed by 750 μL 2,2-diphenyl-1-picrylhydrazyl (DPPH) (0.1 mM) solution. The mixture was incubated for 30 min in the dark at room temperature before measuring the absorbance at 517 nm using a UV–visible spectrophotometer (Thermo Scientific Technologies, Madison, WI, USA). Ascorbic acid concentration was used to generate a calibration curve, and RSA in the extract was extrapolated from the calibration curve. The RSA was expressed as the mean (millimolar) of ascorbic acid equivalent per milligram of dry matter (mM AAE/mg DM).

### 2.2.5. Statistical Analysis

The experimental design was a completely randomized design with three factors (storage treatment, storage duration, and shelf life). Analysis of variance (ANOVA) was done using SAS software (SAS Enterprise Guide 7.1) and means were separated by least significant difference (LSD; $p = 0.05$), according to Bonferroni (Dunn) $t$-test. Relationship among the measured parameters was determined by subjecting data to the Pearson correlation test in XLSTAT software version 2012.04.1 (Addinsoft, Paris, France). GraphPad Prism software version 8.4.3 (GraphPad Software, Inc., San Diego, CA, USA) was used for graphical presentations.

## 3. Results
### 3.1. Physiological Disorders
#### 3.1.1. Superficial Scald

In the 2015 season (Table 1), superficial scald incidence was significantly ($p = 0.0003$) influenced by three-way interaction amongst the main effects (storage treatments, storage duration, and shelf life). The treatment contribution to the three-way interaction could be attributed to significantly higher superficial scald development on RA stored 'Granny Smith' apples than RLOS (ULO and CA) and DCA-CF stored fruit. No incidence of superficial scald was observed on fruit subjected to RLOS + ULO and DCA-CF treatments at every sampling interval. In the 2016 seasons, superficial scald developed on RLOS (ULO and CA) and DCA-CF stored fruit from 6 months until the end of storage. There was no significant difference in superficial scald in the 2015 season for RLOS + ULO and RLOS + CA, whereas in 2016, there was a significant difference at 6 months and 7 days shelf life (RLOS + ULO = 32.38%, RLOS + CA = 0%, and DCA-CF = 3.36%). Overall, the results showed minimal risk of superficial scald development for DCA-CF stored 'Granny Smith' apples over the two seasons.

**Table 1.** Superficial scald incidence (%) on 'Granny Smith' apples harvested at commercial maturity (with no superficial scald at harvest) and stored for up to 10 months at 0 °C in various storage conditions, evaluated every two months followed by a 6 week simulated shipment and handling period, and at 7 days on the shelf (20 °C and 65% RH).

| Season | Storage Duration (Months) | Shelf-Life (Days) | Superficial Scald (%) | | | |
|---|---|---|---|---|---|---|
| | | | RLOS + ULO | RLOS + CA | DCA-CF | RA |
| 2015 | 2 | 0 | 0.00 ± 0.00 [d] | 0.00 ± 0.00 [d] | 0.00 ± 0.00 [d] | 0.00 ± 0.00 [d] |
| | | 7 | 0.00 ± 0.00 [d] | 0.00 ± 0.00 [d] | 0.00 ± 0.00 [d] | 0.00 ± 0.00 [d] |
| | 4 | 0 | 0.00 ± 0.00 [d] | 0.00 ± 0.00 [d] | 0.00 ± 0.00 [d] | 90.86 ± 8.45 [b] |
| | | 7 | 0.00 ± 0.00 [d] | 0.00 ± 0.00 [d] | 0.00 ± 0.00 [d] | 100.00 ± 0.00 [a] |
| | 6 | 0 | 0.00 ± 0.00 [d] | 0.00 ± 0.00 [d] | 0.00 ± 0.00 [d] | 100.00 ± 0.00 [a] |
| | | 7 | 0.00 ± 0.00 [d] | 0.00 ± 0.00 [d] | 0.00 ± 0.00 [d] | 100.00 ± 0.00 [a] |
| | 8 | 0 | 0.00 ± 0.00 [d] | 0.00 ± 0.00 [d] | 0.00 ± 0.00 [d] | 100.00 ± 0.00 [a] |
| | | 7 | 0.00 ± 0.00 [d] | 0.00 ± 0.00 [d] | 0.00 ± 0.00 [d] | 100.00 ± 0.00 [a] |
| | 10 | 0 | 0.00 ± 0.00 [d] | 0.00 ± 0.00 [d] | 0.00 ± 0.00 [d] | 100.00 ± 0.00 [a] |
| | | 7 | 0.00 ± 0.00 [d] | 7.06 ± 7.96 [d] | 0.00 ± 0.00 [d] | 100.00 ± 0.00 [a] |
| 2016 | 2 | 0 | 0.00 ± 0.00 [c] | 0.00 ± 0.00 [c] | 0.00 ± 0.00 [c] | 0.00 ± 0.00 [c] |
| | | 7 | 0.00 ± 0.00 [c] | 0.00 ± 0.00 [c] | 0.00 ± 0.00 [c] | 84.92 ± 6.80 [a] |
| | 4 | 0 | 0.00 ± 0.00 [c] | 0.00 ± 0.00 [c] | 0.00 ± 0.00 [c] | 29.24 ± 10.63 [b] |
| | | 7 | 0.00 ± 0.00 [c] | 0.00 ± 0.00 [c] | 0.00 ± 0.00 [c] | 100.00 ± 0.00 [a] |
| | 6 | 0 | 0.00 ± 0.00 [c] | 0.00 ± 0.00 [c] | 0.00 ± 0.00 [c] | 97.48 ± 2.22 [a] |
| | | 7 | 32.38 ± 24.12 [b] | 0.00 ± 0.00 [c] | 3.36 ± 1.07 [c] | 100.00 ± 0.00 [a] |
| | 8 | 0 | 0.00 ± 0.00 [c] | 0.00 ± 0.00 [c] | 0.00 ±0.00 [c] | 100.00 ± 0.00 [a] |
| | | 7 | 0.00 ± 0.00 [c] | 5.17 ± 5.45 [c] | 1.31 ± 2.26 [c] | 100.00 ± 0.00 [a] |
| | 10 | 0 | 2.78 ± 2.64 [c] | 4.88 ±2.32 [c] | 0.00 ± 0.00 [c] | 100.00 ± 0.00 [a] |
| | | 7 | 32.04 ± 13.60 [b] | 32.15 ± 7.79 [b] | 2.41 ± 1.63 [def] | 100.00 ± 0.00 [a] |

| | Pr > F | |
|---|---|---|
| Season | 2015 | 2016 |
| Treatment (A) | <0.0001 | <0.0001 |
| Storage duration (B) | <0.0001 | <0.0001 |
| Shelf life (C) | 0.0186 | <0.0001 |
| A × B | <0.0001 | <0.0001 |
| A × C | 0.0729 | <0.0001 |
| B × C | 0.0729 | <0.0001 |
| A × B × C | 0.0003 | <0.0001 |

Mean ± standard deviation in the same column followed by different letter(s) are significantly different ($p < 0.05$) according to least significant difference (LSD) *t*-test. RLOS—repeated low oxygen stress; ULO—ultra-low oxygen; CA—controlled atmosphere; DCA-CF—dynamic controlled atmosphere-chlorophyll fluorescence; RA—regular atmosphere.

### 3.1.2. Coreflush

In the 2015 season (Table 2), there was an onset of coreflush after 6 months for apples stored in RLOS (ULO = 77% and CA = 87%) and DCA-CF (80%). However, fruit subjected to RA storage had an onset of coreflush incidence (100%) after 8 months. This observation suggests that the maximum storage duration before the risk of coreflush development for 'Granny Smith' apples was 6 months. Contrary to the first season, in the 2016 season, coreflush developed at certain storage intervals without forming a trend. Fruit subjected to RLOS + ULO phases developed coreflush after 2, 6, and 10 months of storage. Moreover, fruit subjected to RLOS + CA treatment recorded 100% coreflush after 2 months of storage. Overall, the results highlighted the inefficacy of low oxygen technologies to prevent coreflush incidence over the entire 10 months of storage and shelf life.

**Table 2.** Coreflush incidence (%) on 'Granny Smith' apples harvested at commercial maturity (with no coreflush incidence at harvest) and stored for up to 10 months at 0 °C in various storage conditions, evaluated every two months followed by a 6 week simulated shipment and handling period, and at 7 days on the shelf (20 °C and 65% RH).

| Season | Storage Duration (Months) | Shelf-Life (Days) | Coreflush (%) | | | |
|---|---|---|---|---|---|---|
| | | | RLOS + ULO | RLOS + CA | DCA-CF | RA |
| 2015 | 2 | 0 | $0.00 \pm 0.00$ [g] | $0.00 \pm 0.00$ [g] | $0.00 \pm 0.00$ [g] | $0.00 \pm 0.00$ [g] |
| | | 7 | $0.00 \pm 0.00$ [g] | $0.00 \pm 0.00$ [g] | $0.00 \pm 0.00$ [g] | $0.00 \pm 0.00$ [g] |
| | 4 | 0 | $0.00 \pm 0.00$ [g] | $0.00 \pm 0.00$ [g] | $0.00 \pm 0.00$ [g] | $0.00 \pm 0.00$ [g] |
| | | 7 | $0.00 \pm 0.00$ [g] | $0.00 \pm 0.00$ [g] | $0.00 \pm 0.00$ [g] | $0.00 \pm 0.00$ [g] |
| | 6 | 0 | $0.00 \pm 0.00$ [g] | $66.66 \pm 5.77$ [f] | $80.00 \pm 0.00$ [d] | $0.00 \pm 0.00$ [g] |
| | | 7 | $76.67 \pm 5.77$ [de] | $86.67 \pm 11.55$ [c] | $100.00 \pm 0.00$ [a] | $0.00 \pm 0.00$ [g] |
| | 8 | 0 | $100.00 \pm 0.00$ [a] | $100.00 \pm 0.00$ [a] | $100.00 \pm 0.00$ [a] | $100.00 \pm 0.00$ [a] |
| | | 7 | $100.00 \pm 0.00$ [a] | $100.00 \pm 0.00$ [a] | $100.00 \pm 0.00$ [a] | $100.00 \pm 0.00$ [a] |
| | 10 | 0 | $80.00 \pm 0.00$ [d] | $93.33 \pm 5.77$ [b] | $100.00 \pm 0.00$ [a] | $73.33 \pm 11.54$ [e] |
| | | 7 | $100.00 \pm 0.00$ [a] | $100.00 \pm 0.00$ [a] | $100.00 \pm 0.00$ [a] | $100.00 \pm 0.00$ [a] |
| 2016 | 2 | 0 | $26.67 \pm 20.81$ [def] | $100.00 \pm 0.00$ [a] | $100.00 \pm 0.00$ [a] | $0.00 \pm 0.00$ [g] |
| | | 7 | $0.00 \pm 0.00$ [g] | $0.00 \pm 0.00$ [g] | $0.00 \pm 0.00$ [g] | $0.00 \pm 0.00$ [g] |
| | 4 | 0 | $0.00 \pm 0.00$ [g] | $0.00 \pm 0.00$ [g] | $0.00 \pm 0.00$ [g] | $0.00 \pm 0.00$ [g] |
| | | 7 | $0.00 \pm 0.00$ [g] | $0.00 \pm 0.00$ [g] | $0.00 \pm 0.00$ [g] | $0.00 \pm 0.00$ [g] |
| | 6 | 0 | $0.00 \pm 0.00$ [g] | $0.00 \pm 0.00$ [g] | $43.33 \pm 5.77$ [bcd] | $0.00 \pm 0.00$ [g] |
| | | 7 | $0.00 \pm 0.00$ [g] | $0.00 \pm 0.00$ [g] | $0.00 \pm 0.00$ [g] | $0.00 \pm 0.00$ [g] |
| | 8 | 0 | $20.00 \pm 0.00$ [ef] | $30.00 \pm 10.00$ [cde] | $100.00 \pm 0.00$ [a] | $0.00 \pm 0.00$ [g] |
| | | 7 | $0.00 \pm 0.00$ [g] | $0.00 \pm 0.00$ [g] | $0.00 \pm 0.00$ [g] | $0.00 \pm 0.00$ [g] |
| | 10 | 0 | $60.00 \pm 10.00$ [b] | $53.33 \pm 11.55$ [b] | $100.00 \pm 0.00$ [a] | $46.67 \pm 5.77$ [bc] |
| | | 7 | $0.00 \pm 0.00$ [g] | $10.00 \pm 17.32$ [fg] | $0.00 \pm 0.00$ [g] | $0.00 \pm 0.00$ [g] |
| | | Pr > F | | | | |
| Season | | 2015 | | 2016 | | |
| Treatment (A) | | <0.0001 | | <0.0001 | | |
| Storage duration (B) | | <0.0001 | | <0.0001 | | |
| Shelf life (C) | | <0.0001 | | <0.0001 | | |
| A × B | | <0.0001 | | <0.0001 | | |
| A × C | | <0.0001 | | <0.0001 | | |
| B × C | | <0.0001 | | <0.0001 | | |
| A × B × C | | <0.0001 | | <0.0001 | | |

Mean $\pm$ standard deviation in the same column followed by different letter(s) are significantly different ($p < 0.05$) according to least significant difference (LSD) *t*-test. RLOS—repeated low oxygen stress; ULO—ultra-low oxygen; CA—controlled atmosphere; DCA-CF—dynamic controlled atmosphere-chlorophyll fluorescence; RA—regular atmosphere.

### 3.2. Physicochemical Properties

### 3.2.1. Flesh Firmness

In the 2015 season, compared to RLOS and DCA-CF treatments, lower flesh firmness was observed in RA treated apples between 6 and 10 months of storage (Figure 1a). However, at 0 and 7 d shelf life, there was no significant difference in flesh firmness for fruit stored in RLOS (ULO and CA) and DCA-CF treatments. Similarly, in the 2016 season, better firmness retention was associated with fruit stored in low oxygen storage technologies (DCA-CF and RLOS) (Figure 1b) compared to RA.

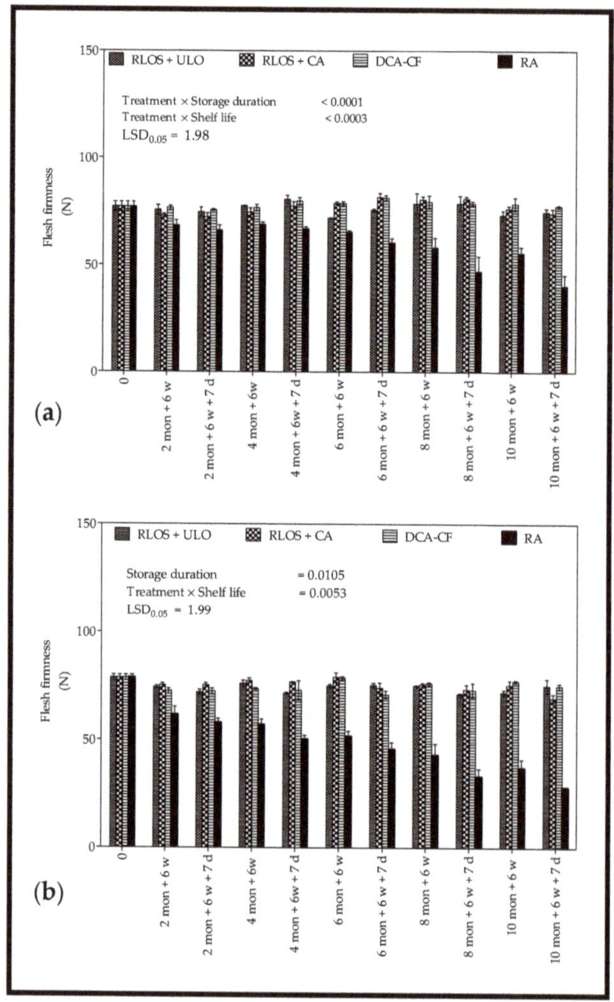

**Figure 1.** Firmness (N) of 'Granny Smith' apples harvested in the 2015 (**a**) and 2016 (**b**) seasons at commercial maturity and stored for up to 10 months at 0 °C in various storage conditions, evaluated every two months followed by a 6 week simulated shipment and handling period, and at 7 days on the shelf (20 °C and 65% RH). Vertical bars represent the standard error (SE) of mean values of 3 replicates (1 replicate = 10 fruit). $LSD_{0.05}$ represent least significant difference ($p < 0.05$). For storage duration, mon—months, w—weeks, and d—days. RLOS—repeated low oxygen stress; ULO—ultra-low oxygen; CA—controlled atmosphere; DCA-CF—dynamic controlled atmosphere-chlorophyll fluorescence; RA—regular atmosphere.

### 3.2.2. Background Color

In the 2015 season, L* values fluctuated across all treatments, from day 0 to 10 months of storage. The pattern of change for L* was not clear. However, between 2 and 8 months, L* values were mostly higher than values at day 0 (Figure 2a). In the period between 2 and 10 months, apples stored using RA recorded lesser negative a* values (less green) compared with RLOS (ULO and CA) and DCA-CF treatments (Figure 2b).

As observed in Figure 2c between 2 and 4 months, apples stored using DCA-CF recorded higher b* values (more yellow) than RLOS (ULO and CA) and RA treatments.

However, from 4 to 10 months, the pattern of change of b* fluctuated, with apples stored using RA recording the lowest b* values (6, 8, and 10 months). There was a marked decrease in h° of 'Granny Smith' apples from 6 to 10 months of storage in RA (Figure 2d). However, RLOS (ULO and CA) and DCA-CF maintained h° and influence background color retention at 0 and 7 d shelf life.

In the 2016 season, the pattern of change in L* was not clear (Figure 3a). However, across all storage treatments, L* values recorded at day 0 were lower than other storage periods. The L* values fluctuated, and at certain points (2, 4, and 8 months), apples stored using RLOS + ULO recorded the highest L* values compared with RLOS + CA, DCA-CF, and RA. Similar to the 2015 season, apples stored using RA mostly recorded less negative a* values (less green) compared with RLOS (ULO and CA) and DCA-CF (Figure 3b). The values for color attribute b* mostly fluctuated during storage, and the differences between storage treatments were mostly not significant (Figure 3c). A similar trend was observed with smaller differences in h° between low oxygen technologies (RLOS + ULO, RLOS + CA, and DCA-CF) and RA storage (Figure 3d). Overall, RLOS (ULO and CA) and DCA-CF appeared to reduce loss of h°, especially in the 2015 season.

### 3.2.3. Total Soluble Solids and Titratable Acidity

In the first harvest season (2015), the results showed no statistical differences in TSS content between RLOS (ULO and CA) and DCA-CF treated apples (Figure 4a). Across all treatments, TSS content fluctuated during storage. However, in the second harvest season (2016) (Figure 4b), the results showed that low oxygen treatments retained significantly higher TSS content than RA treated fruit. Nevertheless, based on the two seasons, RLOS (ULO and CA) and DCA-CF treatments maintained TSS content significantly higher with RA during long term storage.

In the 2015 harvest season, TA measured from RLOS + CA stored apples decreased from 4 to 10 months storage (Figure 5a). There was a general decrease in TA content during storage of RA treated fruit, with significant changes occurring between 4 to 10 months. The TA content was lower in RA treated fruit than in DCA-CF and RLOS (ULO and CA) treatments, especially from 6 to 10 months. Moreover, there was a decrease in TA from 0 to 7 d shelf life for RLOS + ULO, RA, and DCA-CF treated fruit. As with the 2015 harvest season, RA treated fruit generally had lower TA content than fruit subjected to RLOS (ULO and CA) and DCA-CF treatments throughout storage (Figure 5b). The rate of decrease in TA was higher for RA stored apples compared RLOS (ULO and CA) and DCA-CF treated apples.

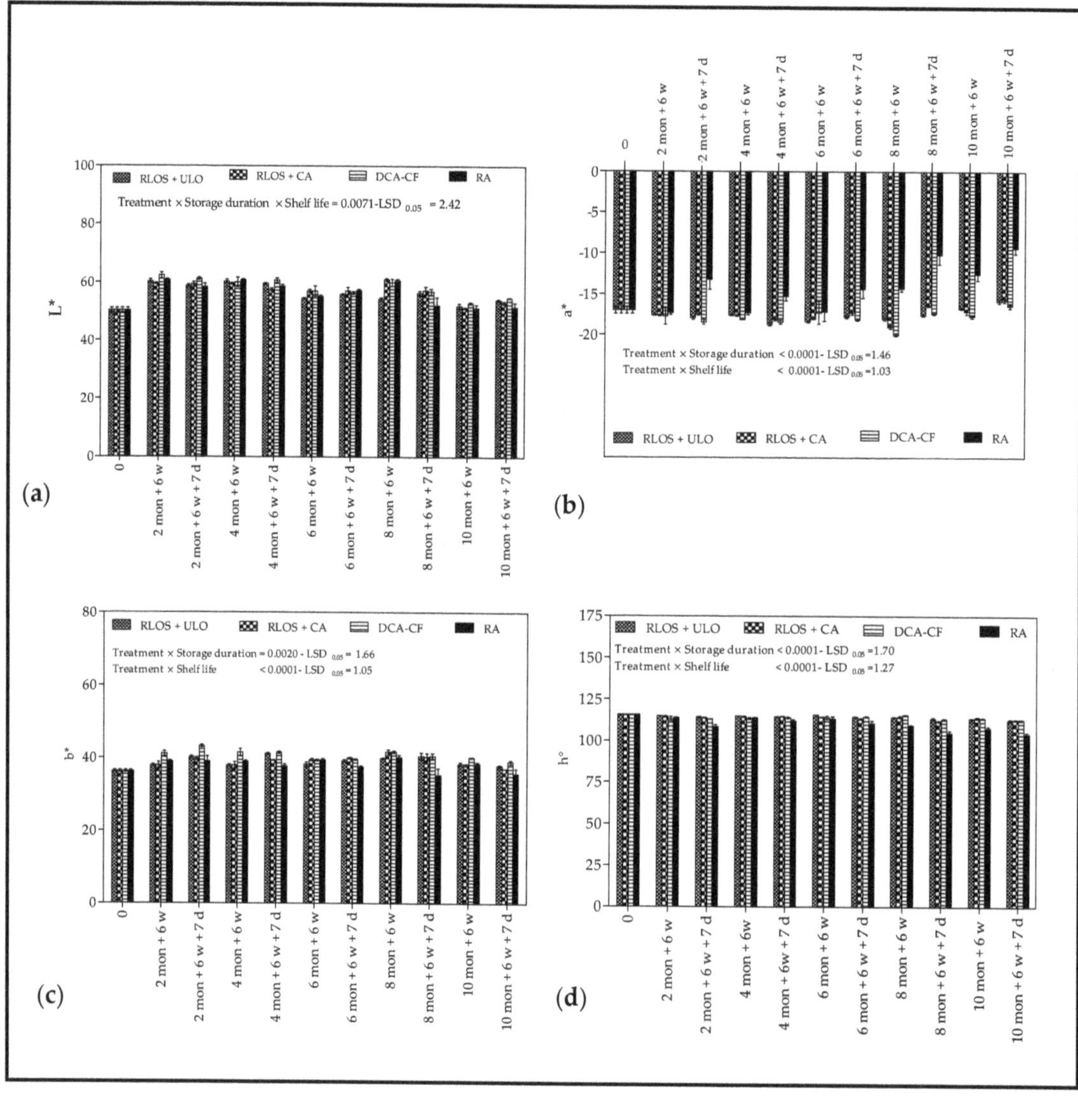

**Figure 2.** Background color (**a**) lightness (L*), (**b**) redness/greenness (a*), (**c**) yellowness/blueness (b*) and (**d**) hue angle (h°) of 'Granny Smith' apples harvested in the 2015 season at commercial maturity and stored for up to 10 months at 0 °C in various storage conditions, evaluated every two months followed by a 6 week simulated shipment and handling period, and at 7 days on the shelf (20 °C and 65% RH). Vertical bars represent the standard error (SE) of mean values of 3 replicates (1 replicate = 10 fruit). $LSD_{0.05}$ represents least significant difference ($p < 0.05$). For storage duration, mon—months; w—weeks; and d—days. RLOS—repeated low oxygen stress; ULO—ultra-low oxygen; CA—controlled atmosphere; DCA-CF—dynamic controlled atmosphere-chlorophyll fluorescence; RA—regular atmosphere.

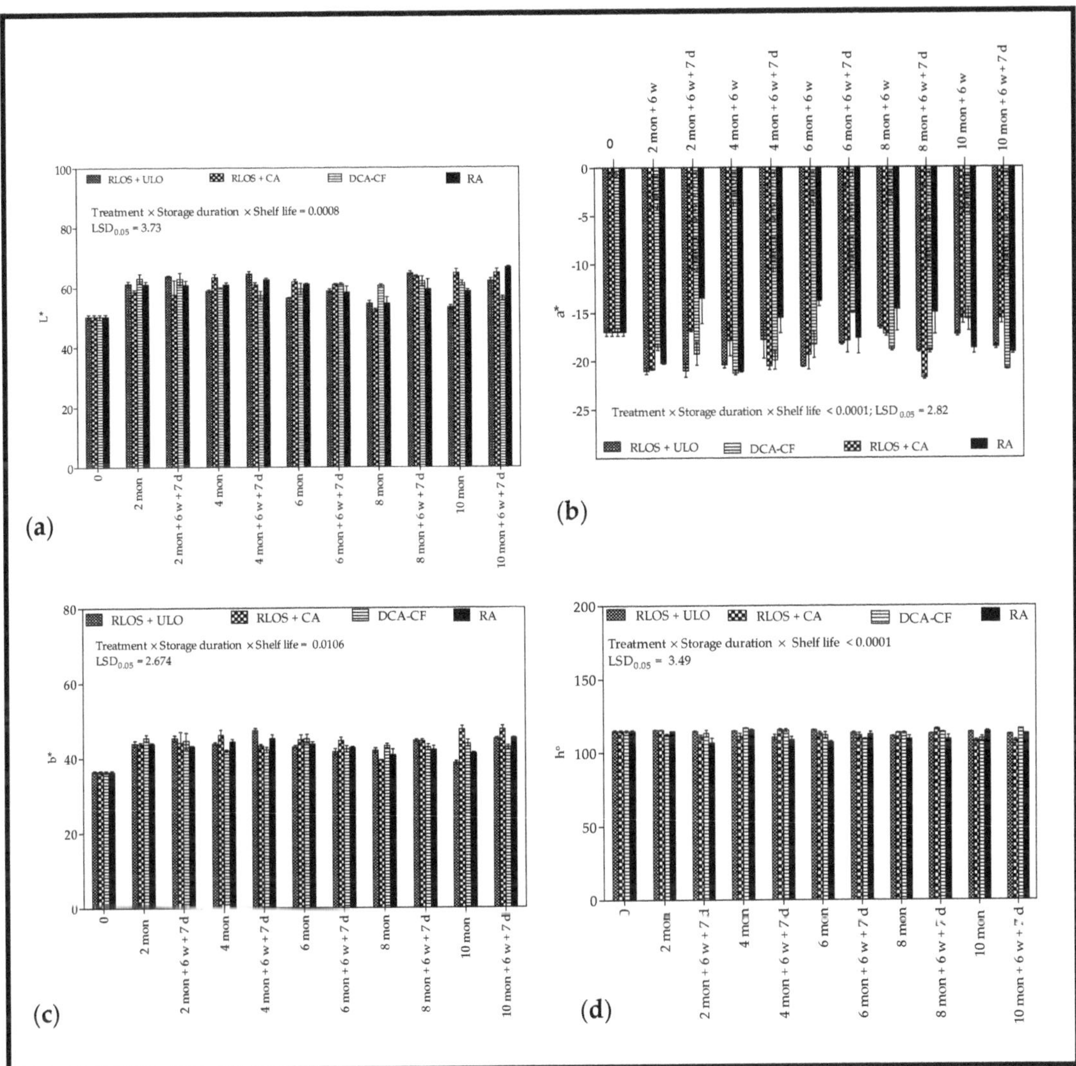

**Figure 3.** Background color (**a**) lightness (L*), (**b**) redness/greenness (a*), (**c**) yellowness/blueness (b*) and (**d**) hue angle (h°) of 'Granny Smith' apples harvested in the 2016 season at commercial maturity and stored for up to 10 months at 0 °C in various storage conditions, evaluated every two months followed by a 6 week simulated shipment and handling period, and at 7 days on the shelf (20 °C and 65% RH). Vertical bars represent the standard error (SE) of mean values of 3 replicates (1 replicate = 10 fruit). $LSD_{0.05}$ represents least significant difference ($p < 0.05$). For storage duration, mon—months; w—weeks; and d—days. RLOS—repeated low oxygen stress; ULO—ultra-low oxygen; CA—controlled atmosphere; DCA-CF—dynamic controlled atmosphere-chlorophyll fluorescence; RA—regular atmosphere.

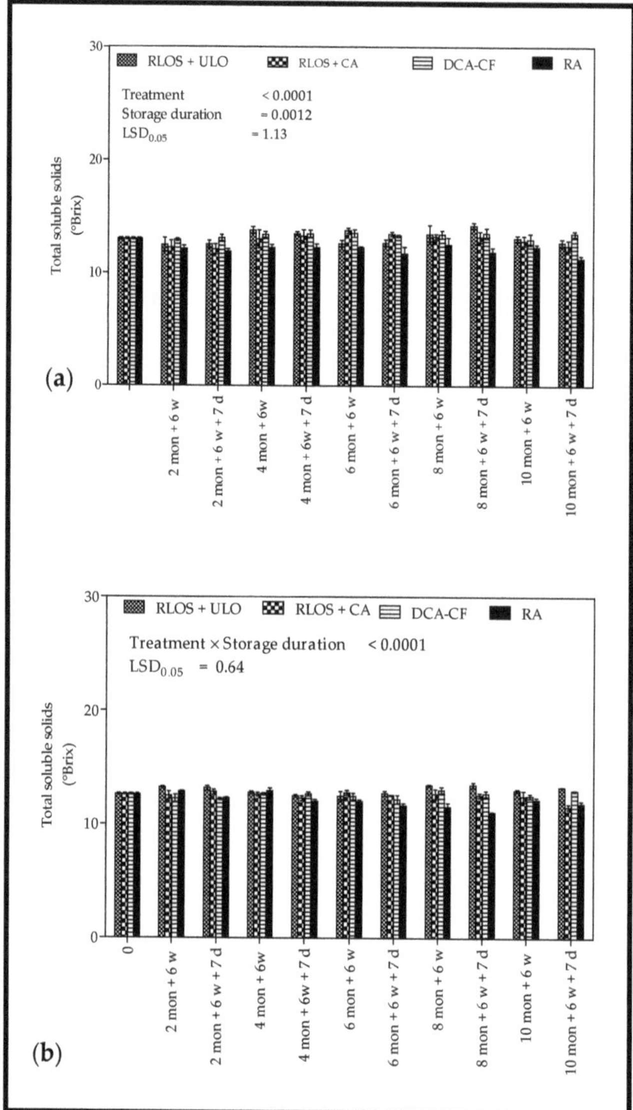

**Figure 4.** Total soluble solids (°Brix) of 'Granny Smith' apples harvested in the 2015 (**a**) and 2016 (**b**) seasons at commercial maturity and stored for up to 10 months at 0 °C in various storage conditions, evaluated every two months followed by a 6 week simulated shipment and handling period, and at 7 days on the shelf (20 °C and 65% RH). Vertical bars represent the standard error (SE) of mean values of 3 replicates (1 replicate = 10 fruit). $LSD_{0.05}$ represents least significant difference ($p < 0.05$). For storage duration, mon—months; w—weeks; and d—days. RLOS—repeated low oxygen stress; ULO—ultra-low oxygen; CA—controlled atmosphere; DCA-CF—dynamic controlled atmosphere-chlorophyll fluorescence; RA—regular atmosphere.

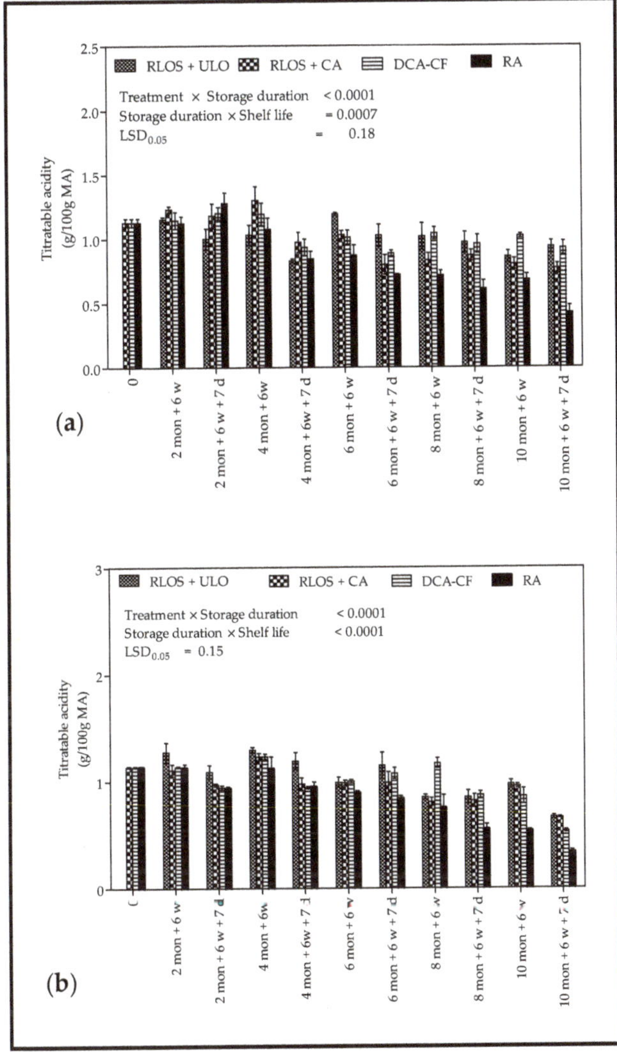

**Figure 5.** Titratable acidity (g/100 g MA) of 'Granny Smith' apples harvested in the 2015 (**a**) and 2016 (**b**) seasons at commercial maturity and stored for up to 10 months at 0 °C in various storage conditions, evaluated every two months followed by a 6 week simulated shipment and handling period, and at 7 days on the shelf (20 °C and 65% RH). Vertical bars represent the standard error (SE) of mean values of 3 replicates (1 replicate = 10 fruit). $LSD_{0.05}$ represents least significant difference ($p < 0.05$). For storage duration, mon—months; w—weeks; and d—days. RLOS—repeated low oxygen stress; ULO—ultra-low oxygen; CA—controlled atmosphere; DCA-CF—dynamic controlled atmosphere-chlorophyll fluorescence; RA—regular atmosphere.

### 3.3. Headspace Volatile Analysis

Changes in MHO and $\alpha$-farnesene contents during storage for both harvest seasons (2015 and 2016) are presented in Figures 6 and 7, respectively. In the first season (2015), MHO accumulation was markedly higher in RA treated peels from 4 to 6 months (7 d shelf life) and from 8 to 10 months (0 and 7 d shelf life), compared to RLOS (ULO and CA) and DCA-CF sampled peels (Figure 6a). Similarly, in the second season (2016), the accumulation of MHO was significantly higher after 6 and 8 months (7 d shelf life), and 10 months (0 d shelf life) for RA treated peels compared to RLOS (ULO and CA) and DCA-CF stored peels (Figure 6b). For both seasons (2015 and 2016), α-farnesene content was significantly higher in RA treated peels after 6 months compared to RLOS (ULO and CA) and DCA-CF treatments (Figure 7a,b). During storage, α-farnesene content was generally stable in peels sampled from RLOS (ULO and CA) and DCA-CF stored apple. On the contrary, in RA treated peels, α-farnesene steadily increased, reaching its peak at around 6 months of storage and subsequently declining.

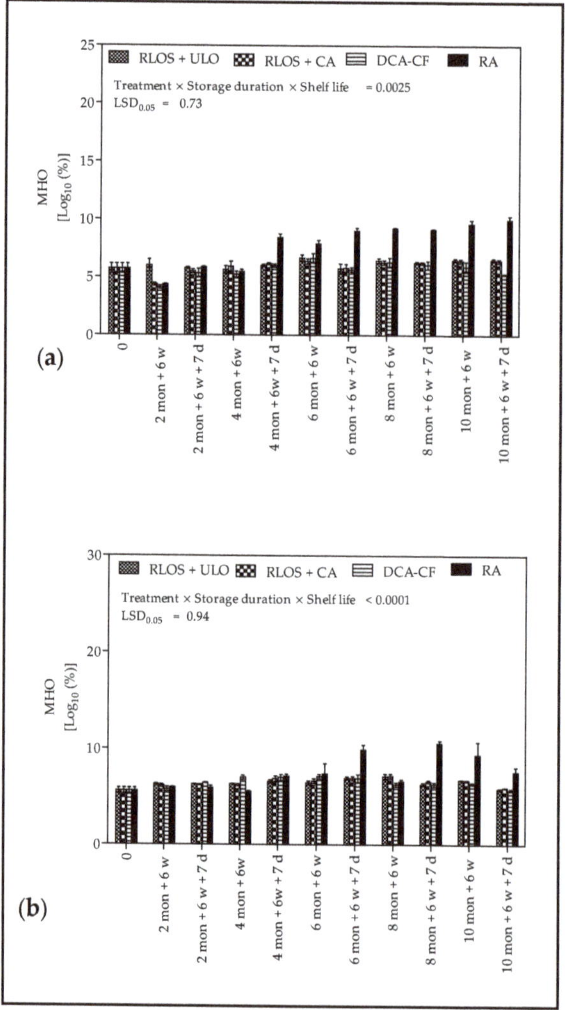


**Figure 6.** Accumulation of MHO (Log (%)) of 'Granny Smith' apples harvested in the 2015 (**a**) and 2016 (**b**) seasons at commercial maturity and stored for up to 10 months at 0 °C in various storage conditions, evaluated every two months followed by a 6 week simulated shipment and handling period, and at 7 days on the shelf (20 °C and 65% RH). Vertical bars represent the standard error (SE) of mean values of 3 replicates (1 replicate = 10 fruit). LSD$_{0.05}$ represents least significant difference ($p < 0.05$). For storage duration, mon—months; w—weeks; and d—days. RLOS—repeated low oxygen stress; ULO—ultra-low oxygen; CA—controlled atmosphere; DCA-CF—dynamic controlled atmosphere-chlorophyll fluorescence; RA—regular atmosphere.

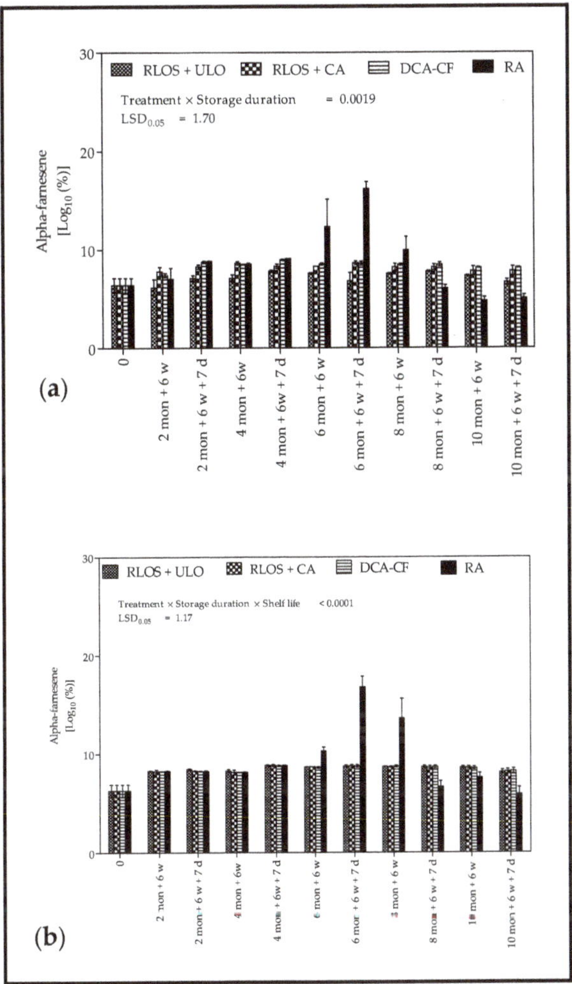

**Figure 7.** Alpha-farnesene (Log% peak area) of 'Granny Smith' apples harvested in the 2015 (**a**) and 2016 (**b**) seasons at commercial maturity and stored for up to 10 months at 0 °C in various storage conditions, evaluated every two months followed by a 6 week simulated shipment and handling period, and at 7 days on the shelf (20 °C and 65% RH). Vertical bars represent the standard error (SE) of mean values of 3 replicates (1 replicate = 10 fruit). $LSD_{0.05}$ represents least significant difference ($p < 0.05$). $LSD_{0.05}$ represent least significant difference ($p < 0.05$). For storage duration, mon—months; w—week; and d—days. RLOS—repeated low oxygen stress; ULO—ultra-low oxygen; CA—controlled atmosphere; DCA-CF—dynamic controlled atmosphere-chlorophyll fluorescence; RA—regular atmosphere.

### 3.4. Biochemical Analysis

#### 3.4.1. Total Phenolic Content

The changes in TPC are shown in Figure 8. There were minimal changes in TPC as storage duration was extended, with no clear difference between fruit stored using low oxygen technologies and RA (Figure 8a). However, TPC reached a maximum of $40.64 \pm 0.82$ mg GAE/g DM for apples stored in RLOS + CA after 6 months. The lowest TPC ($38.75 \pm 0.06$ mg GAE/g DM) was recorded after 8 months for RA stored fruit. In the 2016 season (Figure 8b), at 6 months, TPC appeared to significantly decrease for RA stored apples compared to low oxygen technologies. Following that, all low oxygen technologies, except DCA-CF, maintained TPC higher than RA treatment.

### 3.4.2. Radical Scavenging Activity

In the 2015 season, significantly lower antioxidant activity in RA than in RLOS (ULO and CA) and DCA-CF sampled peels was recorded after 4, 8, and 10 months (Figure 9a). The RSA generally fluctuated during RLOS (ULO and CA) storage and DCA-CF treated peels, with no significant change. For RLOS (ULO and CA) and DCA-CF treated peels, there was no substantial difference in RSA between 0 and 7 d shelf life. As previously observed in the first harvest season (2015), RSA was stable for RLOS (ULO and CA) and DCA-CF treated peels in the 2016 season (Figure 9b). However, between 4 and 8 months of storage, a significant decrease in RSA was observed after 7 d shelf life for RA treated peels.

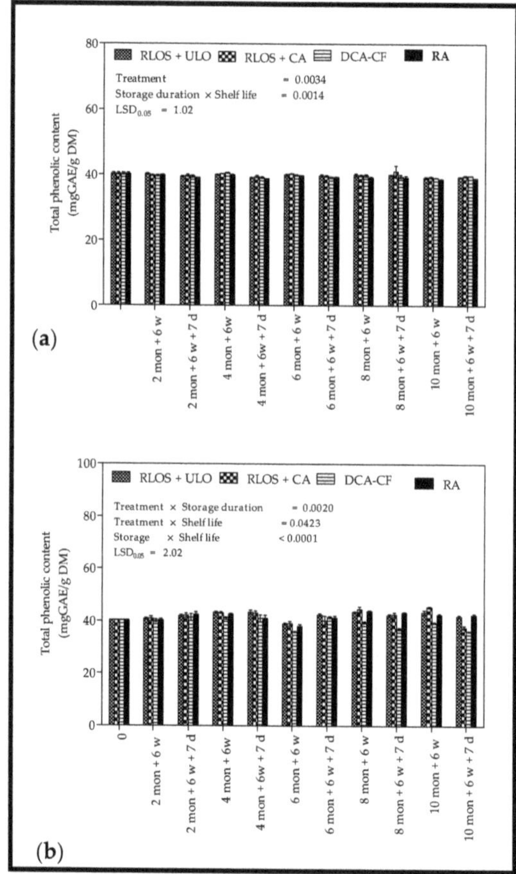

**Figure 8.** Total phenolic content (mgGAE/g DM) of 'Granny Smith' apples harvested in the 2015 (**a**) and 2016 (**b**) seasons at commercial maturity and stored for up to 10 months at 0 °C in various storage conditions, evaluated every two months followed by a 6 week simulated shipment and handling period, and at 7 days on the shelf (20 °C and 65% RH). Vertical bars represent the standard error (SE) of mean values of 3 replicates (1 replicate = 10 fruit). $LSD_{0.05}$ represents least significant difference ($p < 0.05$). $LSD_{0.05}$ represent least significant difference ($p < 0.05$). For storage duration, mon—months; w—weeks; and d—days. RLOS—repeated low oxygen stress; ULO—ultra-low oxygen; CA—controlled atmosphere; DCA-CF—dynamic controlled atmosphere-chlorophyll fluorescence; RA—regular atmosphere.

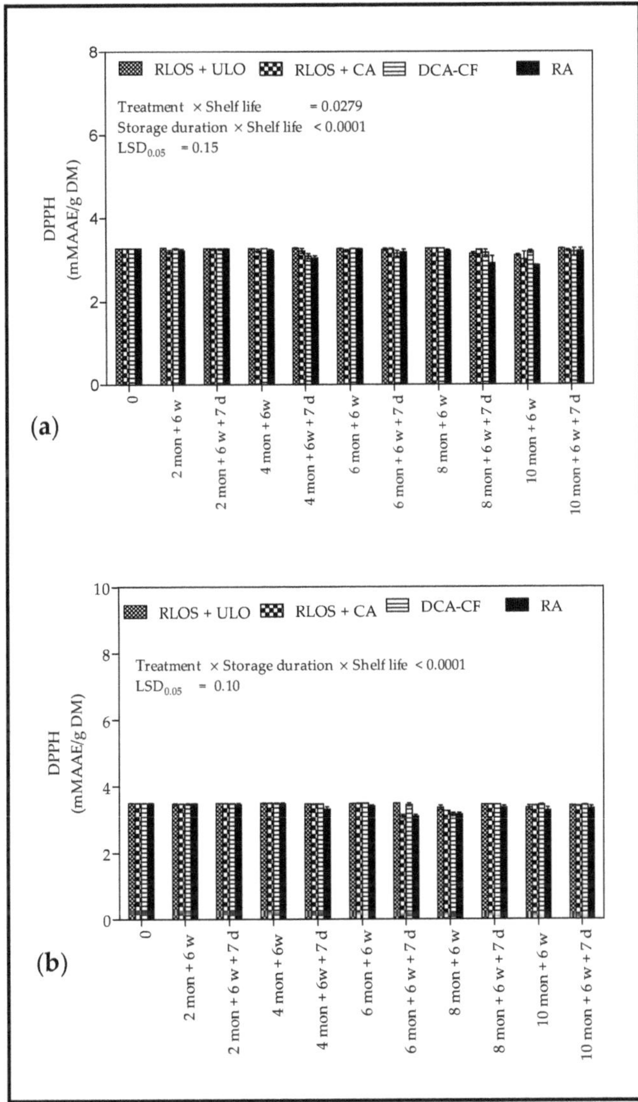

**Figure 9.** Radical scavenging activity (mgGAE/g DM) of 'Granny Smith' apples harvested in the 2015 (**a**) and 2016 (**b**) seasons at commercial maturity and stored for up to 10 months at 0 °C in various storage conditions, evaluated every two months followed by a 6 week simulated shipment and handling period, and at 7 days on the shelf (20 °C and 65% RH). Vertical bars represents the standard error (SE) of mean values of 3 replicates (1 replicate = 10 fruit). $LSD_{0.05}$ represent least significant difference ($p < 0.05$). For storage duration, mon—months; w—weeks; and d—days. RLOS—repeated low oxygen stress; ULO—ultra-low oxygen; CA—controlled atmosphere; DCA-CF—dynamic controlled atmosphere-chlorophyll fluorescence; RA—regular atmosphere.

### 3.5. Correlation Analysis

There was a positive correlation for superficial scald with MHO of r = 0.66103 ($p < 0.0001$). A negative correlation was detected between superficial scald and RSA and between superficial scald and TPC (r = −0.30748 and −0.3527, respectively) (Table 3).

**Table 3.** Correlation of superficial scald and volatiles ($\alpha$-farnesene and MHO), total phenolic content, and total antioxidant capacity.

| | Pearson's Correlation | | | | | | | | | |
| | Superficial Scald | | MHO | | $\alpha$-Farnesene | | Total Phenolic Content | | Radical Scavenging Activity | |
| Parameter | r | p | r | p | r | p | r | p | r | p |
|---|---|---|---|---|---|---|---|---|---|---|
| 6 methyl-5-hepten-2-one | 0.6610 | <0.0001 | - | - | 0.2701 | <0.0001 | −0.2061 | <0.0001 | −0.2547 | <0.0001 |
| $\alpha$-farnesene | 0.3187 | <0.0001 | 0.2701 | <0.0001 | - | - | - | NS * | - | NS |
| Total phenolic content | −0.3527 | <0.0001 | −0.2061 | <0.0001 | - | NS | - | - | 0.4191 | <0.0001 |
| Radical scavenging activity | −0.3048 | <0.0001 | −0.2545 | <0.0001 | - | NS | 0.4191 | <0.0001 | - | - |

*: NS—no significant correlation between two parameters ($p \leq 0.05$); r—correlation coefficient; $p$—$p$ value at 95% confidence interval.

## 4. Discussion

### 4.1. Physiological Disorders

#### 4.1.1. Superficial Scald

In this study, the storage of 'Granny Smith' apples using RLOS + ULO, RLOS + CA, and DCA-CF treatments reduced the development of superficial scald compared with RA. Similarly, several studies have reported the reduction of superficial scald development on 'Granny Smith' apples by low oxygen storage technologies [2,3,9]. For instance, Mditshwa et al. [2] demonstrated that DCA-CF ($O_2$ = 0.3–0.5% and $CO_2$ = 1%) reduced superficial scald to 2% after 16 weeks of cold storage. Likewise, Poirier et al. [32] reported that, over two consecutive seasons, ULO ($\leq$1.0 kPa $O_2$) phase applied before CA (0.5–1 °C and 0.5–0.8 kPa $O_2$; 0.5–0.6 kPa $CO_2$) storage controlled the development superficial scald on 'Granny Smith' apples. Studies have suggested that low oxygen technologies minimize the accumulation of ethylene, conjugated trienols, 6 methyl-5-hepten-2-one, and $\alpha$-farnesene in apple peel to reduce superficial incidence [1,2,33].

The development of superficial scald symptoms is believed to be primarily driven by the oxidation processes of implicated volatiles such as $\alpha$-farnesene, 6-methyl-5-hepten-2-one, and others [4]. Therefore, storing 'Granny Smith' apples in low oxygen technologies such as RLOS + ULO, RLOS + CA, and DCA-CF is expected to reduce superficial scald development. The ULO (0.9% $O_2$ and 0.8% $CO_2$ for 21 d and 0.5% $O_2$ for 7 d) phase exposes apples to lower oxygen concentration than the CA (1.5% $O_2$ and 1% $CO_2$ for 21 d and 0.5% $O_2$ for 7 d); that is, 0.9% $O_2$ for ULO and 1.5% $O_2$ for CA. The assumption would be that the risk of superficial scald development would be higher in the CA phase than in the ULO phase because of the lower oxygen availability. However, in a season with a high risk of superficial scald development, other factors besides oxidation of implicated volatiles become more important, especially factors related to seasonal variations.

Critical factors that vary with harvest season, such as air temperature and light intensity, affect superficial scald susceptibility [4]. In 'Granny Smith', 'Cox's Orange', and 'Pacific Queen' apples, low mean air temperature and high light intensity during a growing season was associated with increased total phenolics and ascorbic acid content, which reduced the risk of superficial scald development [34]. A closer look at the results of apples stored using the RA treatment shows that the 2016 season appeared to have a higher risk of superficial scald than the 2015 season. In the 2016 season, superficial scald was detected earlier at 2 months + 7 days, whereas in the 2015 season, it was observed after 4 months of cold storage. In the 2016 season, the detection of superficial scald in fruit stored using RLOS + ULO and DCA-CF suggests that other factors besides oxygen concentration may have been involved in the development of the physiological disorder. We speculate that these factors are linked to seasonal variation. The seasonal variation in the incidence and severity of superficial scald between the 2015 and 2016 seasons corroborates with previous studies conducted on 'Granny Smith' apples during cold storage [2,14].

#### 4.1.2. Coreflush

Coreflush is a physiological disorder often described as a form of chilling injury affecting the quality of pome fruit [35,36]. The physiological disorder is often observed when apples are cut open as diffuse browning of cortex tissue adjacent to the carpels [37]. The mechanism of development of coreflush is not clearly understood. However, studies suggest that it is more prevalent in late harvested fruit and is induced by low temperature storage [35,38]. Prevention of coreflush incidence is achieved by slow cooling to 0 °C, applying antioxidants before cold storage, and low ethylene storage [38]. In this study, fruit were not step-wise cooled before storage, which could have contributed to the development of coreflush. The results showed that low oxygen technologies were not effective in minimizing the development of coreflush in either harvest seasons.

### 4.2. Physicochemical Properties

#### 4.2.1. Flesh Firmness

Loss of firmness is often associated with the ripening of apples during cold storage [24]. During storage, the tissue strength of apple peel decreases due to ripening associated processes such as enzyme mediated increase in soluble pectin, volume, and internal cell spaces and net loss of non-cellulosic sugars, galactose, and arabinose [39]. Based on the results, RLOS (ULO and CA) and DCA-CF treatments retarded the loss of flesh firmness during storage. Studies have found apples stored in low oxygen technologies to be more firm compared with fruit stored at RA [40,41]. Low oxygen technologies slow down ethylene production, respiratory metabolism, and tissue breakdown during the ripening processes, which reduces firmness loss [3,42].

#### 4.2.2. Background Color

The pattern of change for color attribute a* in the 2015 and 2016 seasons suggested that RLOS (ULO and CA) and DCA-CF treatments minimized the decrease in the green color of 'Granny Smith' apples. The results corroborate with Zanella [35], who observed minimal change in the green color of 'Granny Smith' apples stored using low oxygen (0.4 kPa $O_2$) and CA (1.5 kPa $O_2$ and 1.3 kPa $CO_2$) compared with RA (21 kPa $O_2$ and 0.03 kPa $CO_2$). Likewise, studies by Mditshwa et al. [2] showed that repeated application of DCA-CF (0.3–0.5% $O_2$ and 1% $CO_2$) maintained the green color of 'Granny Smith' apples during cold storage (0 °C and 95% RH).

Overall, RLOS (ULO and CA) and DCA-CF reduced h° loss of 'Granny Smith' apples compared to RA, especially in the first harvest season (2015 season) when the risk of scald development was low. This indicates that low oxygen technologies resulted in fresher green color, while lower hue values denoted a more yellowish background color of apples subjected to RA storage. Similar results were reported for 'Granny Smith' apples and 'd'Anjou' pears subjected to low oxygen technologies such as ULO and DCA-CF during cold storage [13,19,35], which limited the decrease in h° by possibly reducing chlorophyll breakdown due to leakage of organic acid from the vacuole, oxidative stress, and chlorophyllases [24].

#### 4.2.3. Total Soluble Solids and Titratable Acidity

Overall, RLOS (ULO and CA) and DCA-CF maintained TSS content significantly higher than RA during long term storage. The results were in accordance with Rebeaud and Gasser [40], who reported higher TSS content in DCA-CF stored 'Golden Delicious' apples compared to RA stored fruit after 16 and 36 w of cold storage. Titratable acidity content of 'Granny Smith' apples usually decreases as storage duration is extended [43]. The use of fruit organic acids in respiration during storage has been suggested as the cause for decreases in TA content [44–46]. The RLOS (ULO and CA) and DCA-CF treatments reduced the decline in TA as storage duration was extended whilst there was a general decrease in TA content during storage of RA treated fruit. These results agree with Mditshwa et al. [9], who reported that 'Granny Smith' apples stored in repeated DCA-CF had higher TA levels compared to RA stored fruit. Similarly, Lafer [47] reported that DCA-CF storage maintained firmness and TA content at higher levels for 'Uta' pears compared to standard CA.

### 4.3. Headspace Volatile Analysis

The greater accumulation of MHO observed for RA treated peels compared to RLOS (ULO and CA) and DCA-CF stored apples was consistent with results reported by Mditshwa et al. [2] for 'Granny Smith' apple peels subjected to repeated DCA-CF storage for up to 6 w at −0.5 °C and 95% relative humidity (RH). Additionally, Ramokonyane [14] outlined that optimally harvested 'Granny Smith' apples stored in RA at 0 °C for up to 7 months had markedly higher MHO concentration compared to DCA-CF and CA, preceded by initial low oxygen stress (ILOS + CA) stored fruit. Studies have reported that superficial scald is related to the extent of $\alpha$-farnesene oxidation in susceptible apple cultivars [1,2]. Therefore, the accumulation $\alpha$-farnesene in apple peels is expected to vary with storage duration as oxidative products increase.

Alpha-farnesene was detected at harvest, which confirms that it is a naturally occurring volatile in 'Granny Smith' apples [4]. Accumulation of $\alpha$-farnesene in apple peel is reduced by low oxygen storage by applying anaerobic treatments, CA, or hypobaric storage [48,49]. Moreover, Sabban-Amin et al. [50] reported reduced $\alpha$-farnesene production for 'Granny Smith' apples subjected to <0.5% $O_2$ followed by RA at 0 °C for 24 w. Likewise, Mditshwa et al. [2] observed substantially low $\alpha$-farnesene content for 'Granny Smith' apples subjected to repeated DCA-CF treatment with an intermittent RA period in comparison to RA stored fruit for 24–70 d storage. It is generally accepted that the concentration of $\alpha$-farnesene increases when apples are transferred into cold storage, particularly in RA, reaching a maximum between 8 and 12 w and subsequently declining [51–53]. However, dynamic changes in $\alpha$-farnesene content in apple peel varies with storage temperature and atmosphere. For instance, Mditshwa [54] reported that $\alpha$-farnesene concentration in 'Granny Smith' apples was highest after 12 weeks of storage in RA at 0 °C. This is contrary to Ramokonyane [14], who only observed significant differences in $\alpha$-farnesene concentration at shelf life and not during storage for 'Granny Smith' apples subjected to DCA-CF at 0 °C. In this study, the delay in $\alpha$-farnesene accumulation could suggest suppression of oxidative processes in peels, particularly for RLOS (ULO and CA) and DCA-CF treatments. Subsequent decreases in $\alpha$-farnesene may also indicate the accumulation of MHO, its oxidative product [4,55].

### 4.4. Biochemical Analysis

#### 4.4.1. Total Phenolic Content

Slight changes in TPC observed under all treatments during storage could be associated with the respiratory and ethylene climacteric in apple fruit, which initiates ripening, particularly for fruit subjected to RA storage regimes [4,9]. Research work has suggested that the production of phenolics in apple peel during storage is an ethylene-independent process, and phenolics exhibit an ethylene-dependent regulation when ethylene biosynthesis is suppressed [56]. The behavior of RLOS (ULO and CA) and DCA-CF treated peels resembles 1-MCP treated 'Granny Smith' apples, as reported by Shaham et al. [57], in which no significant decline in phenol concentration were observed when fruit was stored at 0 °C. However, Leja et al. [58] observed a significant increase in total phenolics for 'Jonagold' and 'S'ampion' during cold storage at 1 °C. Studies have demonstrated various patterns of change exhibited by phenolics during storage in several apple cultivars. For instance, Tarozzi et al. [59] reported that, during storage of 'Golden Delicious' apples at 0 °C, total phenolic content decreased in fruit peels after 3 months of cold storage. However, Shaham et al. [57] did not observe a distinct pattern of change in phenolics during storage of optimally harvested 'Granny Smith' apples stored at 0 °C for up to 6 w after pre-treatment with 1-MCP or heat. Golding et al. [8] reported that phenol concentration was generally stable, though simple phenols decreased in optimally harvested 'Granny Smith', 'Lady Williams', and 'Crofton' apples stored in air at 0 °C for 9 months.

### 4.4.2. Radical Scavenging Activity

The effectiveness of low oxygen technologies to limit the decrease in total antioxidant capacity has been previously reported. For instance, Mditshwa et al. [9] observed higher RSA for optimally harvested 'Granny Smith' apples stored at 0 °C in DCA-CF compared to RA. The total antioxidant capacity of RLOS (ULO and CA) treated peels resembled the behavior of 1-MCP treated optimally harvested 'Granny Smith' apples according to Shaham et al. [57], who reported higher RSA in 1-MCP treated 'Granny Smith' apples compared to regular stored fruit. The trend in RSA change during storage in RA treated peels corroborates the findings of Barden and Bramlage [7], who observed that water soluble antioxidants decreased with an increase in storage duration of 'Cortland' and 'Delicious' apples stored at 0 °C. In addition, Mditshwa et al. [9] reported inferior quality of optimally harvested 'Granny Smith' apples after storage in RA, which correlated with a decrease in RSA during storage. In contrast, RSA increased regardless of the storage conditions for optimally harvested 'Jonagold' and 'S'ampion' apples stored in CA (2% $CO_2$ and 2% $O_2$) and regular atmosphere for 120 days at 1 °C [58]. Based on the results, it can be argued that both RLOS (ULO and CA) and DCA-CF treatments inhibited the loss of RSA during storage and after 7 d shelf life in 'Granny Smith' apples.

### 4.5. Correlation Analysis

Negative correlations were found between superficial scald and both RSA and TPC, which may indicate minimal involvement of phenolic compounds in superficial scald induction. These results agree with Shaham et al. [57], who observed a fluctuation and little change in phenolic compound constitution and attributed the development of superficial scald to antioxidant enzyme activity. The accumulation of oxidative compounds of $\alpha$-farnesene and MHO has been associated with superficial scald induction [6,7]. However, in this study, correlation analysis suggested that other volatiles, such as conjugated trienols (CTols) and antioxidants (ascorbic acid and tocopherol), not quantified in this study, may be responsible for the development of superficial scald in 'Granny Smith' apples [8,9,60].

## 5. Conclusions

This study showed that storing fruit under low oxygen controlled atmosphere technologies (RLOS and DCA-CF) at 0 °C can inhibit the development of superficial scald on 'Granny Smith' apples in a season of low superficial scald potential. Applying RLOS and DCA-CF maintained some internal quality parameters for up to 10 months of storage at 0 °C and after a simulated 6 w of shipment and handling period plus 7 d shelf life (20 °C). This study demonstrated that both RLOS and DCA-CF inhibited superficial scald in 'Granny Smith' apples, possibly suppressing $\alpha$-farnesene oxidation. The results from this study confirmed the hypothesis that MHO causes superficial scald; however, other underlying mechanisms may have substantial contributions to the induction of superficial scald. This study also showed that RLOS and DCA-CF storage technologies maintain the antioxidant status of 'Granny Smith' apples, which is important in quality preservation. This study also highlighted that, while phenolic compounds possibly contribute to the inhibition of superficial incidence, their role varies significantly with harvest season. Further studies that focus on the emission of other volatiles such as conjugated trienes during storage, lipid peroxidation, and the relationship of superficial scald with other metabolites can elaborate more on the mechanism of action of RLOS technology. Additionally, the possibility of RLOS technology being used in combination with 1-MCP, currently applied on 'Granny Smith' apples, presents an innovative technology that could be investigated on 'Granny Smith' apples and other cultivars.

**Author Contributions:** Conceptualization, O.A.F. and U.L.O.; methodology, T.G.K., O.A.F., and U.L.O.; software, T.G.K. and O.A.F.; formal analysis, T.G.K. and O.A.F.; investigation, T.G.K.; resources, O.A.F. and U.L.O.; writing—original draft preparation, T.G.K., writing—review and editing, O.A.F. and U.L.O.; supervision, O.A.F. and U.L.O.; project administration, U.L.O. and O.A.F.; funding acquisition, U.L.O. and O.A.F. All authors have read and agreed to the published version of the manuscript.

**Funding:** The authors are grateful to the Agricultural Research Council of South Africa, Postharvest Innovation Programme (PHI) and Hortgro Science for financial support.

**Institutional Review Board Statement:** Not applicable.

**Informed Consent Statement:** Not applicable.

**Data Availability Statement:** Restrictions apply to the availability of these data. Data are available from the authors with the permission of funder.

**Acknowledgments:** This work is based on the research supported wholly or in part by the National Research Foundation of South Africa (Grant Numbers: 64813). Research reported in this publication was supported in part by the Foundation for Food and Agriculture Research under award number— Grant ID: DFs-18-0000000008.

**Conflicts of Interest:** The opinions, findings and conclusions or recommendations expressed are those of the author(s) alone, and the NRF accepts no liability whatsoever in this regard.

## References

1. Bordonaba, J.G.; Matthieu-Hurtiger, V.; Westercamp, P.; Coureau, C.; Dupille, E.; Larrigaudière, C. Dynamic changes in conjugated trienols during storage may be employed to predict superficial scald in 'Granny Smith' apples. *LWT Food Sci. Technol.* **2013**, *54*, 535–541. [CrossRef]
2. Mditshwa, A.; Fawole, O.A.; Vries, F.; van der Merwe, K.; Crouch, E.; Opara, U.L. Repeated application of dynamic controlled atmospheres reduced superficial scald incidence in 'Granny Smith' apples. *Sci. Hortic.* **2017**, *220*, 168–175. [CrossRef]
3. Mditshwa, A.; Fawole, O.A.; Opara, U.L. Recent developments on dynamic controlled atmosphere storage of apples—A review. *Food Packag. Shelf Life* **2018**, *16*, 59–68. [CrossRef]
4. Lurie, S.; Watkins, C.B. Superficial scald, its etiology and control. *Postharvest Biol. Technol.* **2012**, *65*, 44–60. [CrossRef]
5. Mditshwa, A.; Fawole, O.A.; Vries, F.; van der Merwe, K.; Crouch, E.; Opara, U.L. Classification of 'Granny Smith' apples with different levels of superficial scald severity based on targeted metabolites and discriminant analysis. *J. Appl. Bot. Food Qual.* **2016**, *89*, 49–55.
6. Rudell, D.R.; Mattheis, J.P.; Hertog, M.L. Metabolomic change precedes apple superficial scald symptoms. *J. Agric. Food Chem.* **2009**, *57*, 8459–8466. [CrossRef]
7. Barden, C.L.; Bramlage, W.J. Relationships of antioxidants in apple peel to changes in $\alpha$-farnesene and conjugated trienes during storage, and to superficial scald development after storage. *Postharvest Biol. Technol.* **1994**, *4*, 23–33. [CrossRef]
8. Golding, J.B.; McGlasson, W.B.; Wyllie, S.G. Relationship between production of ethylene and $\alpha$-farnesene in apples, and how it is influenced by the timing of diphenylamine treatment. *Postharvest Biol. Technol.* **2001**, *21*, 225–2533. [CrossRef]
9. Mditshwa, A.; Vries, F.; van der Merwe, K.; Crouch, E.; Opara, U.L. Antioxidant content and phytochemical properties of apple 'Granny Smith' at different harvest times. *S. Afr. J. Plant Soil* **2015**, *32*, 221–226. [CrossRef]
10. Zanella, A.; Stürz, S. Replacing DPA postharvest treatment by strategical application of novel storage technologies controls scald in 1/10th of EU's apples producing area. *Acta Hortic.* **2013**, *1012*, 419–426. [CrossRef]
11. Liu, Y.B. Ultralow oxygen treatment for postharvest control of western flower thrips, *Frankliniella occidentalis* (Thysanoptera: Thripidae), on iceberg lettuce. II. Effects of pre-treatment storage on lettuce quality. *Postharvest Biol. Technol.* **2008**, *49*, 135–139. [CrossRef]
12. Juhņeviča-Radenkova, K.; Radenkovs, V. Influence of 1-Methylcyclopropene and ULO conditions on sensory characteristics of apple fruit grown in Latvia. *J. Hortic. Res.* **2016**, *24*, 37–46. [CrossRef]
13. Wang, Z.; Dilley, D.R. Initial low oxygen stress controls superficial scald of apples. *Postharvest Biol. Technol.* **2000**, *18*, 201–213. [CrossRef]
14. Ramokonyane, T.M. Effects of Dynamic Controlled Atmosphere and Initial Low Oxygen Stress on Superficial Scald of 'Granny Smith' Apples and 'Packham's Triumph' Pears. Master's Thesis, Stellenbosch University, Stellenbosch, South Africa, March 2016. Available online: https://scholar.sun.ac.za/handle/10019.1/98327 (accessed on 20 January 2021).
15. Ghahramani, F.; Scott, K.J. Oxygen stress of 'Granny Smith' apples in relation to superficial scald, ethanol and alpha-farnesene, and conjugated trienes. *Aust. J. Agric. Res.* **1998**, *49*, 207–210. [CrossRef]
16. Chervin, C.; Brouard, L.; Frémondière, G.; Westercamp, P.; Thieffry, N.; Larrigaudiere, C. Superficial scald versus ethanol vapours: A dose response. Superficial scald versus ethanol vapours: A dose response. *Acta Hortic.* **2003**, *600*, 117–121. [CrossRef]
17. Wang, Z.; Dilley, D.R. Initial low oxygen stress (ILOS) controls scald of apples without using postharvest chemical treatments. *Acta Hortic.* **2001**, *553*, 261–266. [CrossRef]
18. Thewes, F.R.; Both, V.; Brackmann, A.; Weber, A.; de Oliveira Anese, R. Dynamic controlled atmosphere and ultralow oxygen storage on 'Gala' mutants quality maintenance. *Food Chem.* **2015**, *188*, 62–70. [CrossRef]

19. Erkan, M.; Pekmezc, M.; Gübbük, H.; Karafiah, I. Effects of controlled atmosphere storage on scald development and postharvest physiology of 'Granny Smith' Apples. *Turk. J. Agric.* **2004**, *28*, 43–48.
20. Tran, D.T.; Verlinden, B.E.; Hertog, M.; Nicolaï, B.M. Monitoring of extremely low oxygen control atmosphere storage of 'Greenstar' apples using chlorophyll fluorescence. *Sci. Hortic.* **2015**, *184*, 18–22. [CrossRef]
21. Prange, R.K.; Wright, A.H.; DeLong, J.M.; Zanella, A. A review on the successful adoption of dynamic controlled-atmosphere (DCA) storage as a replacement for diphenylamine (DPA), the chemical used for control of superficial scald in apples and pears. *Acta Hortic.* **2015**, *1071*, 389–396. [CrossRef]
22. Weber, A.; Brackmann, A.; Both, V.; Pavanello, E.P.; de Oliveira Anese, R.; Thewes, F.R.; Anese, R.D.O.; Rodrigo, F. Respiratory quotient: Innovative method for monitoring 'Royal Gala' apple storage in a dynamic controlled atmosphere. *Sci. Agric.* **2015**, *72*, 28–33. [CrossRef]
23. Feng, F.; Li, M.; Ma, F.; Cheng, L. Effects of location within the tree canopy on carbohydrates, organic acids, amino acids and phenolic compounds in the fruit peel and flesh from three apple (Malus × domestica) cultivars. *Hortic. Res.* **2014**, *1*, 1–7. [CrossRef]
24. Bessemans, N.; Verboven, P.; Verlinden, B.E.; Nicolaï, B.M. A novel type of dynamic controlled atmosphere storage based on the respiratory quotient (RQ-DCA). *Postharvest Biol. Technol.* **2016**, *115*, 91–102. [CrossRef]
25. Wright, A.H.; DeLong, J.M.; Gunawardena, A.H.; Prange, R.K. The interrelationship between the lower oxygen limit, chlorophyll fluorescence and the xanthophyll cycle in plants. *Photosynth. Res.* **2011**, *107*, 223–235. [CrossRef] [PubMed]
26. Veltman, R.H.; Verschoor, J.A.; van Dugteren, J.H.R. Dynamic control system (DCS) for apples (Malus domestica Borkh. cv 'Elstar'): Optimal quality through storage based on product response. *Postharvest Biol. Technol.* **2003**, *27*, 79–86. [CrossRef]
27. Watkins, C.B.; Bramlage, W.J.; Cregoe, B.A. Superficial scald of 'Granny Smith' apples is expressed as a typical chilling injury. *J. Am. Soc. Hortic. Sci.* **1995**, *120*, 88–94. [CrossRef]
28. Chen, L.; Opara, U.L. Texture measurement approaches in fresh and processed foods—A review. *Food Res. Int.* **2013**, *51*, 823–835. [CrossRef]
29. Hussein, Z.; Fawole, O.A.; Opara, U.O. Effects of bruising and storage duration on physiological response and quality attributes of pomegranate fruit. *Sci. Hortic.* **2020**, *267*, 1–7. [CrossRef]
30. Al-Said, F.A.; Opara, L.U.; Al-Yahyai, R.A. Physico-chemical and textural quality attributes of pomegranate cultivars (*Punica granatum* L.) grown in the Sultanate of Oman. *J. Food Eng.* **2009**, *90*, 129–134. [CrossRef]
31. Mayuoni-Kirshinbaum, L.; Daus, A.; Porat, R. Changes in sensory quality and aroma volatile composition during prolonged storage of 'Wonderful' pomegranate fruit. *Int. J. Food Sci. Technol.* **2013**, *48*, 1569–1578. [CrossRef]
32. Poirier, B.C.; Mattheis, J.P.; Rudell, D.R. Extending 'Granny Smith' apple superficial scald control following long-term ultra-low oxygen controlled atmosphere storage. *Postharvest Biol. Technol.* **2020**, *161*, 111062. [CrossRef]
33. Lavilla, T.; Puy, J.; López, M.L.; Recasens, I.; Vendrell, M. Relationships between volatile production, fruit quality, and sensory evaluation in Granny Smith apples stored in different controlled-atmosphere treatments by means of multivariate analysis. *J. Agric. Food Chem.* **1999**, *47*, 3791–3803. [CrossRef] [PubMed]
34. McGhie, T.K.; Hunt, M.; Barnett, L.E. Cultivar and growing region determine the antioxidant polyphenolic concentration and composition of apples grown in New Zealand. *J. Agric. Food Chem.* **2005**, *53*, 3065–3070. [CrossRef]
35. Zanella, A. Control of apple superficial scald and ripening—A comparison between 1-methylcyclopropene and diphenylamine postharvest treatments, initial low oxygen stress and ultra-low oxygen storage. *Postharvest Biol. Technol.* **2003**, *27*, 69–78. [CrossRef]
36. Ju, Z.; Curry, E.A. Stripped corn oil emulsion alters ripening, reduces superficial scald, and reduces core flush in 'Granny Smith' apples and decay in 'd' Anjou' pears. *Postharvest Biol. Technol.* **2000**, *20*, 185–193. [CrossRef]
37. Johnson, D.S.; Colgan, R.J. Low ethylene controlled atmosphere induces adverse effects on the quality of 'Cox's Orange Pippin' apples treated with aminoethoxyvinylglycine during fruit development. *Postharvest Biol. Technol.* **2003**, *27*, 59–68. [CrossRef]
38. Fan, X.; Mattheis, J.P.; Blankenship, S. Development of apple superficial scald, soft scald, core flush, and greasiness is reduced by MCP. *J. Agric. Food Chem.* **1999**, *47*, 3063–3068. [CrossRef]
39. De Ell, J.R.; Khanizadeh, S.; Saad, F.; Ferree, D.C. Factors affecting apple fruit firmness—A review. *J. Am. Pomol. Soc.* **2001**, *55*, 8–27.
40. Rebeaud, S.G.; Gasser, F. Fruit quality as affected by 1-MCP treatment and DCA storage—A comparison of the two methods. *Eur. J. Hortic. Sci.* **2015**, *80*, 18–24. [CrossRef]
41. Zanella, A.; Rossi, O. Post-harvest retention of apple fruit firmness by 1-methylcyclopropene (1-MCP) treatment or dynamic CA storage with chlorophyll fluorescence (DCA-CF). *Eur. J. Hortic. Sci.* **2015**, *80*, 11–17. [CrossRef]
42. Graell, J.; Larrigaudiere, C.; Vendrell, M. Effect of low-oxygen atmospheres on quality and superficial scald of 'Top red' apples. *Food Sci. Technol. Int.* **1997**, *3*, 203–211. [CrossRef]
43. Magazin, N.; Keserović, Z.; Milić, B.; Dorić, M. Fruits quality of granny smith apples treated with 1- methylcyclopropene or diphenylamine and stored under ulo conditions. *Acta Hortic.* **2013**, *981*, 619–624. [CrossRef]
44. Ding, C.K.; Chachin, K.; Hamauzu, Y.; Ueda, Y.; Imahori, Y. Effects of storage temperatures on physiology and quality of loquat fruit. *Postharvest Biol. Technol.* **1998**, *14*, 309–315. [CrossRef]
45. Melgarejo, P.; Salazar, D.M.; Artes, F. Organic acids and sugars composition of harvested pomegranate fruits. *Eur. Food Res. Technol.* **2000**, *211*, 185–190. [CrossRef]

46. Both, V.; Thewes, F.R.; Brackmann, A.; de Freitas Ferreira, D.; Pavanello, E.P.; Wagner, R. Effect of low oxygen conditioning and ultralow oxygen storage on the volatile profile, ethylene production and respiration rate of 'Royal Gala' apples. *Sci. Hortic.* **2016**, *209*, 156–164. [CrossRef]

47. Lafer, G. Effect of different CA storage conditions on storability and fruit quality of organically grown 'Uta' pears. *Acta Hortic.* **2011**, *909*, 757–760. [CrossRef]

48. Matich, A.J.; Banks, N.H.; Rowan, D.D. Modification of α-farnesene levels in cool-stored 'Granny Smith' apples by ventilation. *Postharvest Biol. Technol.* **1998**, *14*, 159–170. [CrossRef]

49. Rupasinghe, H.P.V.; Paliyath, G.; Murr, D.P. Biosynthesis of alpha-farnesene and its relation to superficial scald development in 'Delicious' apples. *J. Am. Soc. Hort. Sci.* **1998**, *123*, 882–886. [CrossRef]

50. Sabban-Amin, R.; Feygenberg, O.; Belausov, E.; Pesis, E. Low oxygen and 1-MCP pretreatments delay superficial scald development by reducing reactive oxygen species (ROS) accumulation in stored 'Granny Smith' apples. *Postharvest Biol. Technol.* **2011**, *62*, 295–304. [CrossRef]

51. Anet, E.F.L.J. Superficial scald, a functional disorder of stored apples. XI. Apple antioxidants. *J. Sci. Food Agric.* **1974**, *25*, 299–304. [CrossRef]

52. Huelin, F.E.; Coggiola, I.M. Superficial scald, a functional disorder of stored apples. IV. Effect of variety, maturity, oiled wraps and diphenylamine on the concentration of alpha-farnesene in the fruit. *J. Sci. Food Agric.* **1968**, *19*, 297–301. [CrossRef] [PubMed]

53. Whitaker, B.D.; Villalobos-Acuña, M.; Mitcham, E.J.; Mattheis, J.P. Superficial scald susceptibility and alpha-farnesene metabolism in 'Bartlett' pears grown in California and Washington. *Postharvest Biol. Technol.* **2009**, *53*, 43–50. [CrossRef]

54. Mditshwa, A. The Potential of Dynamic Controlled Atmospheres and Possible Mechanisms in Mitigating Superficial Scald in Apples cv. 'Granny Smith'. Ph.D. Thesis, Stellenbosch University, Stellenbosch, South Africa, December 2015. Available online: https://scholar.sun.ac.za/handle/10019.1/97690 (accessed on 20 January 2021).

55. Anet, E.F.L.J.; Coggiola, I.M. Superficial scald, a functional disorder of stored apples. X. Control of alpha-farnesene autoxidation. *J. Sci. Food Agric.* **1974**, *25*, 293–298. [CrossRef]

56. Defilippi, B.G.; Dandekar, A.M.; Kader, A.A. Impact of suppression of ethylene action or biosynthesis on flavor metabolites in apple (Malus domestica Borkh) Fruits. *J. Agric. Food Chem.* **2004**, *52*, 5694–5701. [CrossRef]

57. Shaham, Z.; Lers, A.; Lurie, S. Effect of heat or 1-methylcyclopropene on antioxidative enzyme activities and antioxidants in apples in relation to superficial scald development. *J. Amer. Soc. Hort. Sci.* **2003**, *128*, 761–766. [CrossRef]

58. Leja, M.; Mareczek, A.; Ben, J. Antioxidant properties of two apple cultivars during long-term storage. *Food Chem.* **2003**, *80*, 303–307. [CrossRef]

59. Tarozzi, A.; Marchesi, A.; Cantelli-forti, G.; Hrelia, P. Cold-storage affects antioxidant properties of apples in Caco-2 cells. *J. Nutr.* **2004**, *134*, 1105–1109. [CrossRef]

60. Ahn, T.; Paliyath, G.; Murr, D.P. Antioxidant enzyme activities in apple varieties and resistance to superficial scald development. *Food Res. Int.* **2007**, *40*, 1012–1019. [CrossRef]

*agriculture*



*Article*

# Effect of Temperature Sensor Numbers and Placement on Aeration Cooling of a Stored Grain Mass Using a 3D Finite Element Model


Benjamin Plumier [1,*] and Dirk Maier [2]

[1] USDA ARS NCAUR, 1815 N University St., Peoria, IL 61604, USA

[2] Department of Agricultural and Biosystems Engineering, Iowa State University, 3325 Elings605 Bissell Rd., Ames, IA 50011, USA; dmaier@iastate.edu

* Correspondence: benjamin.plumier@usda.gov; Tel.: +1-309-645-7739



**Abstract:** Grain stored in silos in the United States of America is generally cooled with an aeration system to limit mold spoilage and insect infestation. Monitoring efficacy of aeration and real-time conditions of stored grain is generally done using temperature cables with fixed-spaced sensor locations that are hung from the roof of the silo. Numerous placement options exist in terms of the number of cables and their positions. However, little investigation has been done into the effects of cable placement on aeration system operation decisions and real-time monitoring of stored grain conditions. For a one-year period, the temperatures predicted by sensors in three recommended temperature cable configurations were evaluated for conditions in Ames, IA, USA. The average temperatures of each of the cable sensor configurations were lower than the average temperatures of the entire silo, with as much as an 11.4 °C difference. When sensor locations were used as inputs for aeration control, all cable sensor configurations predicted similar average temperatures. However, the temperature averages varied by as much as 3.6 °C depending on the temperature cable distribution chosen. Results demonstrated that temperature cables near the center or near the edges of the silos produce results that are not representative of the grain mass, resulting in less efficient aerations. Simulations were also conducted with randomized horizontal "wireless" sensor locations at fixed grain depths. The average temperatures were similar, but an increase in the number of sensors reduced variability between simulated storage years as the number of randomized sensors increased.

**Keywords:** aeration; finite element modeling; stored products; temperature sensors



**Citation:** Plumier, B.; Maier, D. Effect of Temperature Sensor Numbers and Placement on Aeration Cooling of a Stored Grain Mass Using a 3D Finite Element Model. *Agriculture* **2021**, *11*, 231. https://doi.org/10.3390/agriculture11030231

Academic Editors: Isabel Lara Ayala and Les Copel

Received: 28 January 2021
Accepted: 8 March 2021
Published: 10 March 2021

**Publisher's Note:** MDPI stays neutral with regard to jurisdictional claims in published maps and institutional affiliations.

## 1. Introduction

The extent of food-insecure people in the world has decreased during the past decade but is estimated to still be more than 820 million [1,2]. With continued population growth, more food production will be required with lower resource inputs such as labor, fertilizer, water, and land. This is a challenge that cannot be met by focusing exclusively on increasing food production. As a result, reducing post-harvest loss has been recognized as a vital tool for meeting global food and energy needs [1,3]). Even grain that has been handled properly may develop adverse conditions due to insect infestations, moisture condensation, mold spoilage and weather effects. Monitoring grain conditions during aeration and storage is an important strategy to ensure grain quality and food safety.

Monitoring grain conditions is most useful when implemented in conjunction with grain management strategies to maintain the quality of grain and reduce or prevent post-harvest loss. One of the most popular post-harvest loss prevention technologies is grain aeration, which is commonly used in temperate climates such as the upper Midwest of the United States. Aeration reduces insect reproduction [4] by reducing temperatures below the optimum for insect development, i.e., 28 °C to 38 °C [5]. Once grain has been cooled below the lower limit for stored grain insects, i.e., 13 °C to 20 °C [6], the risk of grain loss



*Agriculture* **2021**, *11*, 231. https://doi.org/10.3390/agriculture11030231

due to insect growth has been mitigated. The use of aeration is complicated by a number of factors, such as the suitability of weather conditions to achieve effective cooling, the need to control moisture content, and the desire to reduce fan run hours. For this reason, much research has been focused on developing effective algorithms to govern the operating systems of grain aeration fans [7–11]. Many effective strategies, such as those utilized [8], make use of information gathered from the grain mass to make aeration decisions. The effectiveness of these strategies depends on the location and number of sensors used to monitor the grain mass, and how representative those readings are of the entire grain mass. However, little research has been conducted to investigate how grain monitoring decisions, such as the number and placement of temperature sensors, can impact grain aeration efficacy.

The most common monitoring system currently used in stored grain bulks relies on steel cables equipped with thermocouple or thermistor type sensors placed 1.82–2.09 m (6–7 ft) apart. The cables are hung in a bin, silo, tank, or building from the roof supports. While temperature cables remain an important tool, they have several limitations and disadvantages. Grain is a good insulator, so sensors indicate temperature within a limited range and cables result in increased friction forces as the grain mass settles over time or is unloaded. This causes substantial loads on the roofs of grain structures, for example, as high as 4700 N (1056 lbs) in the unloading of a 279 MT silo [12]. Particularly for large storage structures, roofs have to be properly engineered to carry the loading force associated with the number of cables needed to monitor a grain mass. Also, with time thermocouple-based temperature sensors can become inaccurate. Many older systems installed in silos have cables that no longer function reliably. Temperature sensors also do not give an indication of moisture content in the grain mass, an important factor for considering grain quality.

Companies selling temperature cables have specific recommendations for the number and location of cables for different storage structure types and sizes. They range from a minimum of a single cable placed in the center of a grain mass to a large number of cables supposedly covering 100% of a large grain mass. No scientific research has been found that documents the relationship between number and placement of sensors and those temperature readings versus actual temperatures in a stored grain mass, or how those readings would affect aeration strategies that utilize data from the grain mass.

Detecting a hotspot (i.e., high temperature as a result of grain spoilage) in a sensor's range occurs only when the hotspot has grown large enough to begin being detected [13]. Unfortunately, by the time hot spots are large enough to detect, significant damage to the grain mass has already occurred. Grain is a good thermal insulator, but a grain mass contains 35–40% air space, and thus the, sensors measure as much interstitial air conditions as actual grain temperature. Air movement can also be an important factor in determining how early hotspots might be detected. Ileleji et al., 2006 determined that temperature cable sensors 0.3 to 1 m distance from a developing "spoilage hotspot" were unable to detect increasing temperatures [14]. According to Mills (1989), relative humidity and temperature sensors are sensitive to grain conditions at a distance of 30 to 60 cm [15]. This poses a problem because many temperature cable companies advertise cable systems based on a percentage of the grain silo that is "covered" by the sensors with no indication of the length of time or amount of grain it would take for a hotspot a particular distance from a sensor to be detected.

Temperature cables are not commonly placed near the sidewalls of a storage structure, limiting the ability to reliably monitor the perimeter of the grain mass where external solar radiation substantially influences grain temperature and moisture changes which can cause wall caking and grain spoilage. This limits the likelihood that potential grain storage problems will be detected with temperature sensors in a timely manner as grain mass surface and perimeter areas are typically the first areas in the storage structure to be infested by insects.

A novel stored grain monitoring technology that allows for tracking temperature (see Figure 1), relative humidity, and other factors wirelessly is being developed by at least one company (i.e., Amber Agriculture and others). Wireless technology offers a new option in grain storage management by allowing a variable number of sensors to be placed in the grain mass, and with a potentially random vertical and horizontal distribution. The sensors can be removed after grain storage with a simple sieve. A random horizontal distribution would have the advantage of placing sensors in vulnerable areas of the grain mass periphery that conventional cables sensors cannot reach. Wireless sensors will also allow for any number of sensors to be placed in the grain mass without negative consequences on roof structural integrity. However, there is a potential risk if sensors are distributed in ways that give an inaccurate representation of grain mass conditions, or if there is no way of knowing where exactly the wireless sensors end up in the grain mass when added to the grain flow during silo filling. Furthermore, if wireless sensor locations are randomly distributed, how would a stored grain manager know whether this distribution results in more or less representative temperature readings than those of traditional cable-based sensors, or improves aeration performance or how the distribution of these sensors can impact aeration performance?

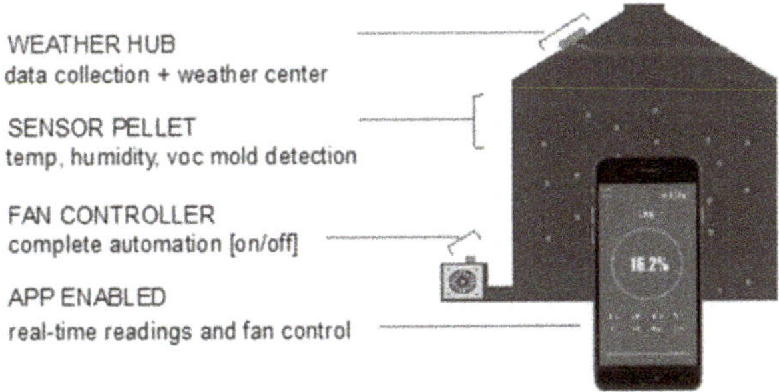


**Figure 1.** Schematic of Amber Agriculture wireless stored grain monitoring system utilizing wireless sensor "pellets" placed randomly within the grain mass reporting temperature and relative humidity values to a weather hub which records and sends data to an app-enabled device. Real-time readings can be used to make stored grain management decisions to manually or automatically control aeration fans.

The 3D MLP (Maier, Lawrence, Plumier) finite element model has been previously used to investigate stored grain ecosystems in a variety of situations [8,16–18]. It has the capacity to analyze the effects that changing sensor distributions have on aeration results by running multiple simulations with sensors distributed in ways consistent with the existing temperature cable technology and the randomized wireless sensor technology. The model also has the ability to simulate the effects of different aeration fan runtime choices made by stored grain managers.

The objectives of this research were to evaluate the effect of temperature sensor numbers and placement on aeration cooling of a stored grain mass utilizing three recommended temperature cable configurations and randomized horizontal placement of "wireless" sensors at fixed depths, and to investigate the impact of these configurations on interpreting grain conditions and associated aeration system operating decisions.

## 2. Materials and Methods

For the purpose of investigating the effect of the number and placement of temperature sensors on aeration cooling of a stored grain mass, the recommended configurations of one commercial supplier of temperature cable monitoring systems, i.e., Tri-States Grain Conditioning Inc. (Spirit Lake, IA, USA) were utilized. The three recommended temperature cable configurations were for a 1346 MT (53,000 bushel) silo holding peaked maize (see Figure 2). In the first configuration, one cable is hung in the center of the silo and claims to monitor temperatures reliably in 201 MT (7950 bushels) of maize in the core, equivalent to 15% of the total grain mass. In the second configuration, three cables are hung in the silo, evenly spaced radially and 1/3 of the distance between the center and the wall of the silo. This placement claims to monitor temperatures reliably in 673 MT (26,500 bushels) of maize within a larger radius of the core, equivalent to 50% of the total grain mass. In the third configuration, seven cables are hung in the silo, with six cables evenly spaced radially and 2/3 of the distance between the center and the wall of the silo plus one cable in the center. This placement claims to monitor temperatures reliably in 1346 MT (53,000 bushels) of maize from core to wall of the silo, equivalent to 100% of the grain mass. These sensor distributions will be referred to as low, medium, and high. Each cable had six evenly spaced sensors 2.09 m apart (7 ft) from bottom to top. The silo specifications (14.6 m diameter, 10.1 m eave height, 14.6 m peak height) were used to create a mesh of elements for 3D simulation of a level stored grain mass using the Abaqus software with 2940 nodes. One year of hourly weather data (2014) (i.e., solar radiation, temperature, relative humidity, and wind speed) was acquired from the Iowa State Mesonet system (https://mesonet.agron.iastate.edu, accessed on 9 September 2019), representing the aerated grain storage period in Ames, IA, USA.

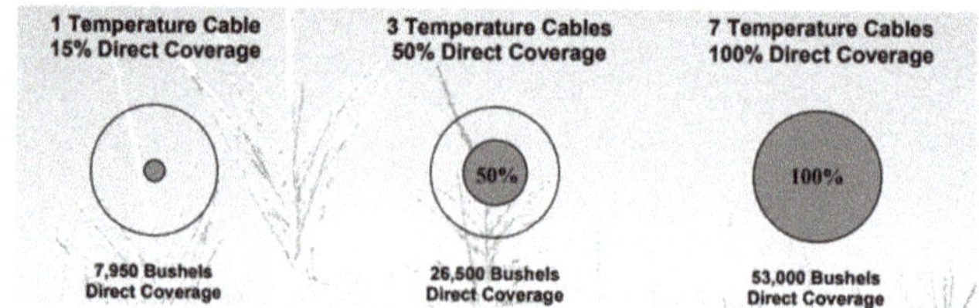


**Figure 2.** Three recommended temperature cable configurations and claimed grain mass coverage for a 14.6 m (48 ft) diameter, 10.1 m (33 ft) eave height, and 14.6 m (48 ft) peak height grain silo rated to hold 1346 MT (53,000 bu) of peaked maize.

In the first investigation, the effectiveness of each of the sensor cable configurations in representing the actual average conditions of the grain mass was analyzed during one year of aerated storage. The aeration strategy ran the fan whenever the center point of the grain mass was warmer than ambient temperature conditions. The fan was turned off whenever the temperature of the sensor in the middle of the grain mass was lower than the ambient air temperature. These aeration strategies are based on successful strategies discussed in Plumier (2018) [9]. While the aeration simulations were identical in this trial, the results shown represent the values observed by monitoring the grain at positions reflective of the sensor locations in each of the three monitoring strategies described in Figure 2, along with one labeled total that represents the temperature values reflected by averaging all 2940 nodes of the numerical solution.

For the second investigation, four aeration simulations were conducted where the aeration control strategy was dependent on the values observed by the sensor locations indicated by the three cable configurations depicted in Figure 2. The aeration fan was turned on whenever the average temperature reported by all sensors for a particular cable configuration was greater than the ambient temperature, and that average grain temperature was above 0 °C. The fan was turned off whenever the average of the sensors was lower than the ambient air temperature. Results were analyzed for each of the three configurations as well as for the total that represents the average of the numerical solution.

The third analysis investigated aeration fans controlled by sensors placed randomly at horizontal locations, that is, in line with the potential new wireless sensor technology. To accomplish this, sensor locations in the code were placed at depths and in numbers that correspond to the sensor locations of the three cable configurations. The horizontal positioning of each of the nodes, however, was assigned using a random number generator. The same aeration strategy was used as in the previous analysis. Ten replicates of the aeration simulations were conducted with numbers corresponding to the low, medium, and high sensor distributions. The averages, standard deviations, and differences between the highest and lowest temperature values were reported.

In order to further investigate the difference between placing a cable in the center or the periphery, the single cable configuration was investigated by also placing it one-third and two-thirds of the distance to the silo wall and along the silo wall in the eastern direction.

### 3. Results

The three cable configurations purportedly represent "direct coverage" for 15%, 50%, and 100% of the grain mass with one, three, and seven cables, respectively. However, it is not known how far from a sensor a point in the grain mass can be located and still represent the same grain temperature as the sensor location. It is also unclear how these percentages were calculated given temperature sensor locations 2.09 m apart along with a cable and cable placement configurations. Based on the radius of a 1.05 m sphere around each sensor without overlap between sensor volumes, the three cable configurations would represent only 1.5%, 4.6%, and 10.8% of the grain mass, respectively. In order to achieve the claimed values, it is necessary to assume a radius of greater length (i.e., 2.23 m) than the distance between the sensors (i.e., 2.09 m) and overcount overlapped areas between sensor volumes. These assumptions would yield 15%, 45%, and 105% of the grain mass represented by the three respective cable configurations. However, this is clearly unrealistic, as there are locations in the grain mass that are not covered and some areas that are double or triple counted. From the perspective of a stored grain manager, this would not make any sense. As a matter of fact, it may give a false sense of security in terms of how quickly a temperature increase due to mold spoilage, for example, would be detected. Accounting for only the vertical overlap within each column, the representative grain mass volumes would reduce to less than 9.6%, 28.7%, and 66.9%, respectively, but the implication stated above remains. Thus, the subsequent simulation results were not presented and discussed in terms of these percentage values.

Figures 3 and 4 show the average temperatures over time predicted for the sensors of the three cable configurations for the one-year aerated storage period. It is important to note that the predicted values are for the same aeration simulation (i.e., fan controlled based on the temperature of the center cable mid-point sensor), and that the observed differences are due to the difference in the number and placement of the sensors within the grain mass. For the low sensor distribution (i.e., one temperature cable in the center), the overall average temperature for the one-year period was $-5.3\,°C$ (8.56 $°C$ SD). For the medium sensor distribution, the one-year average temperature was $-0.6\,°C$ (7.84 $°C$ SD), and for the high sensor distribution, it was $-1.3\,°C$ (8.01 $°C$ SD). The average temperature for the numerical solution that takes into account all 2940 nodes was 6.1 $°C$ (5.34 $°C$ SD).

As is evident in the figures, the medium and high sensor distributions had similar results for most of the simulated period. The low sensor distribution shows a similar pattern but reported average temperature is consistently cooler throughout the year. The pattern for the average temperature of the numerical solution ("total") shows much larger daily variability due to the grain temperature in the periphery as influenced by the daily weather pattern. As a result, the numerical solution showed consistently warmer temperatures than the three cable configurations ranging from 5–10 $°C$ for the medium and high sensor distributions and 7.5–12.5 $°C$ for the low sensor distribution during the January through September storage period.

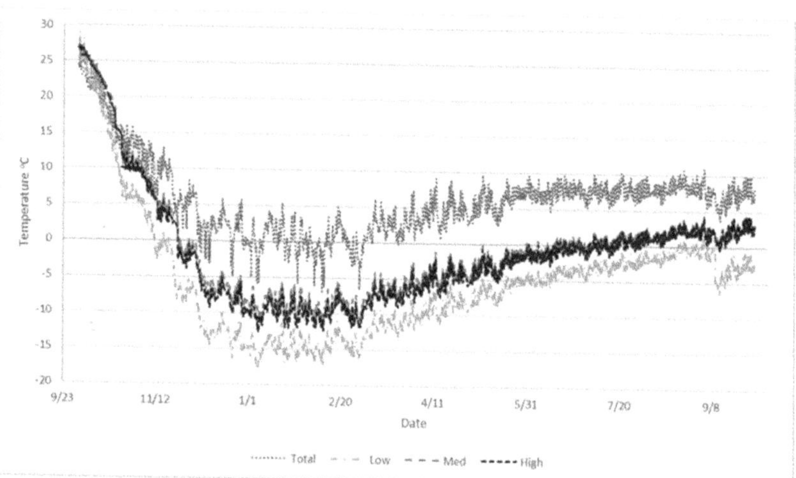

**Figure 3.** The average values of the temperature sensors for each of the three cable configurations defined in Figure 2 as low, medium, and high, versus the average temperature calculated for the numerical solution (total) for a one-year controlled aeration simulation of 1346 MT of maize beginning 1 October, 2014. The fan was turned on whenever the ambient air temperature was cooler than the mid-point sensor in the core of the grain mass.

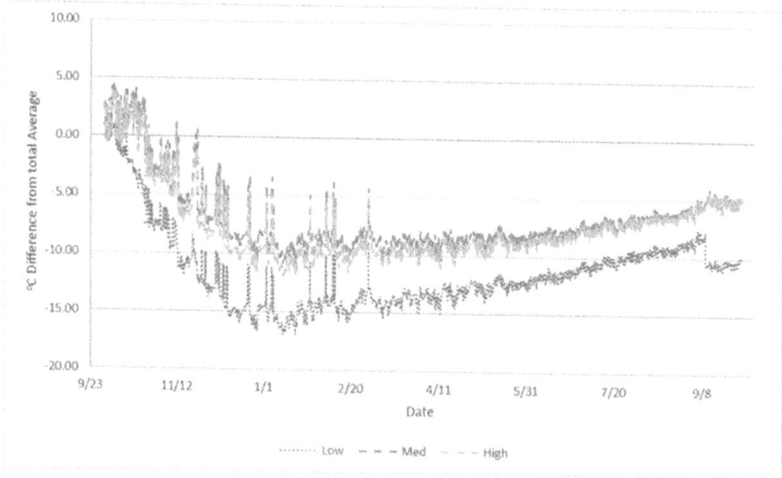

**Figure 4.** The average difference between reported and actual temperatures for each of the three cable configurations defined in Figure 2 as low, medium, and high, for a one-year controlled aeration simulation of 1346 MT of maize beginning 1 October, 2014. The fan was turned off whenever the mid-point sensor in the core of the grain mass was lower than the ambient air temperature.

In order to further compare the predicted temperature patterns and mitigate periphery and center effects, the high sensor distribution results were recalculated by eliminating the center cable sensors (Table 1). This increased the average temperature by 0.64 °C and resulted in close to the same average temperature as for the medium sensor distribution. During the winter months (December–February), the medium sensor distribution, with sensors 1/3 of the way from the center, had comparatively warmer values than those 2/3 of the way towards the wall. During spring (March–May), the sensors closer to the wall began to warm earlier, and the trend reversed, with the high sensor distribution less center cable configuration reporting warmer values than the medium sensor distribution. The break-even point occurred in mid-May. When the temperatures of the medium and modified high sensor densities were combined, the average temperature of $-0.67\,°C$ was essentially the same as the high sensor distribution less center cable configuration.

**Table 1.** The average temperatures and standard deviations reported for each of the three cable configurations defined in Figure 2, as well as the high sensor distribution without the center cable, and that distribution plus the medium, versus the average temperature calculated for the numerical solution (total) for a one-year controlled aeration simulation of 1346 MT of maize beginning 1 October, 2014. The fan was turned on whenever ambient air was cooler than the sensor in the center of the grain mass and the average was above $0\,°C$.

|  | Total | Low | Med | High | High-Center Cable | High-Center + Medium |
|---|---|---|---|---|---|---|
| Average (°C) | 6.10 | −5.31 | −0.64 | −1.32 | −0.68 | −0.67 |
| Std Dev (°C) | 5.34 | 8.50 | 7.84 | 8.01 | 7.93 | 7.90 |
| Number of Cables |  | 1 | 3 | 7 | 6 | 9 |
| Number of Sensors |  | 6 | 18 | 42 | 36 | 54 |

In order to understand the size of the periphery effect, data reported from only one temperature cable near the silo wall was considered. A cable 0.52 m from the wall (6.78 m from the center) reported an overall average temperature of 17.6 °C, a cable at 0.82 m from the wall reported 21.8 °C, and a cable at 1.0 m reported 22.0 °C, and a cable at 1.15 m reported 19.5 °C. This result indicates that after the first meter, temperatures begin to drop and the periphery effect is minimized. The higher temperatures in the periphery are somewhat offset by temperatures near the wall during the cold winter period. However, it appears that highly elevated average temperatures persist for the first 1 m into the grain mass from the sidewall. This result is significant, as more than 25% of the volume in this silo is within 1 m of the wall.

The above simulation results did not take into account how a stored grain manager may make decisions based on all temperature data available to turn on and off aeration fans. In Figures 5 and 6, results are shown for aeration fans turned on whenever ambient air is cooler than the average of the sensors for a given cable configuration, and the average of those sensors was above 0 °C. Figure 5 shows average temperatures during the one-year aerated storage period as the stored grain manager would see them reported by the sensors of the three cable configurations and used by the controller to turn on and off aeration fans. As can be seen, there was little difference in what the stored grain manager would see in terms of average temperatures. The average temperatures over the one-year aerated storage period were 3.6 °C, 3.7 °C, and 4.1 °C for the low, medium, and high sensor distributions, respectively. Most of the observed differences occurred during the initial fall cooling period (October through mid-November). Once the grain mass reached around 0 °C, a stored grain manager would observe little difference between the average grain temperatures. However, the sensors used here do not reveal the whole story, and the stored grain manager should not conclude that number and placement of temperature sensors do not matter when deciding to turn on or off aeration fans to maintain grain quality.

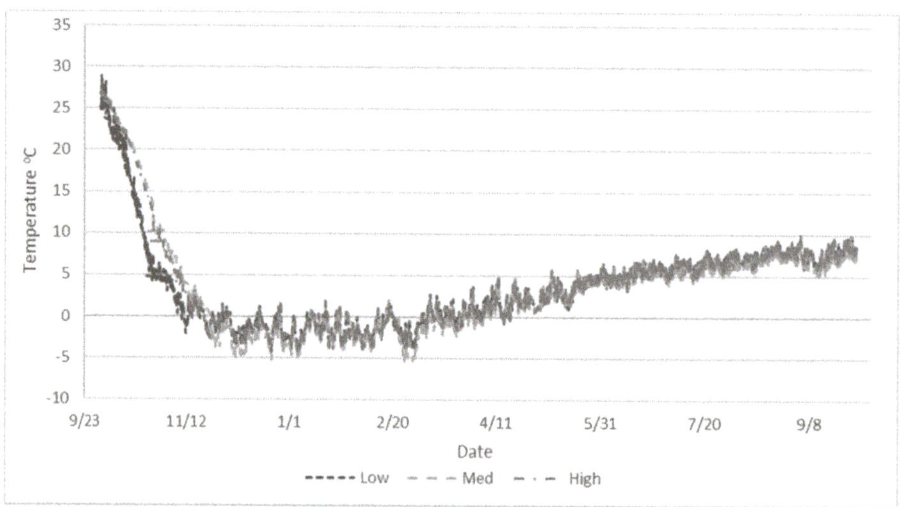

**Figure 5.** The average values of the temperature sensors for each of the three cable configurations defined in Figure 2. as low, medium, and high, for a one-year controlled aeration simulation of 1346 MT of maize beginning 1 October, 2014. The fan was turned on whenever ambient air was cooler than the average of the sensors and the average was above 0 °C.

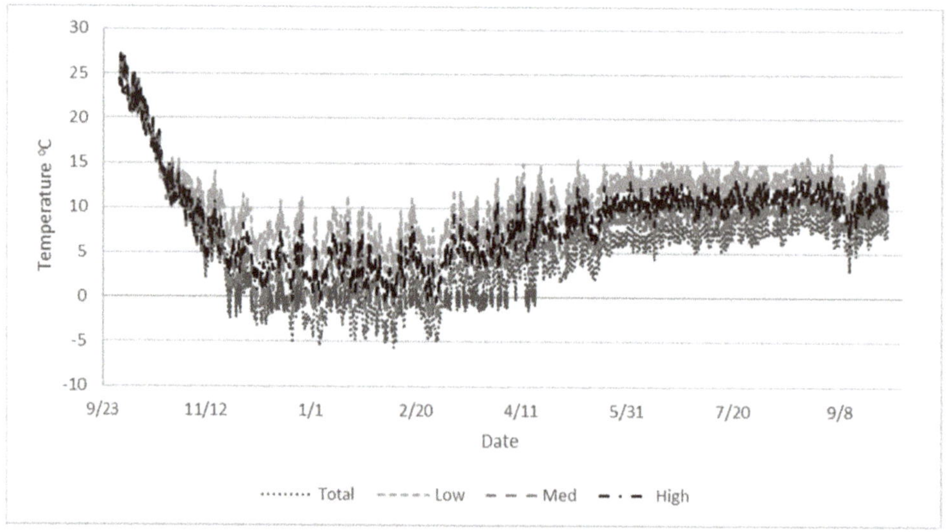

**Figure 6.** The average temperature of the entire grain mass (all nodes) for each of the three cable configurations defined in Figure 2 as low, medium, and high, plus the numerical solution (total) for a one-year controlled aeration simulation of 1346 MT of maize beginning 1 October, 2014. The fan was turned on whenever ambient air was cooler than the average of the sensors (or nodes) and the average was above 0 °C.

Figure 6 shows average temperatures calculated for the entire grain mass based on the aeration simulation results of the three cable configurations. As can be seen, the results show much warmer temperatures overall when compared to the results showing only sensor values, because they include the highly variable periphery values. The average temperatures over the one-year aerated storage period were 10.8 °C, 7.2 °C, 8.5 °C, 10.7 °C, and 8.5 °C for the low, medium, high, and modified high and high plus medium sensor distributions, respectively, and 4.9 °C for the numerical solution.

The trends for each cable configuration and the numerical solution for all nodes over the one-year storage period were consistent, as demonstrated in Table 2. The numerical solution resulted in the coolest grain, with an overall average temperature 2.3 °C lower than the medium sensor distribution, which resulted in the coolest grain among the cable configurations tested. The medium sensor distribution resulted in slightly lower temperatures (by 1.25 °C) than the high and modified sensor distributions despite employing fewer sensors (18 vs. 36, 42 and 54). With the center cable removed, the modified high sensor distribution resulted in warmer grain on average than the medium case (by 3.45 °C), and even warmer than the original high sensor distribution (by 2.2 °C). The medium plus modified high sensor distribution resulted in grain with an average grain temperature equal to the high sensor distribution over the one-year storage and with the same variability. In comparison, the medium sensor distribution with fewer cables (and sensors) resulted in a larger variability, with the highest standard deviation of any of the sensor cable simulations except for the numerical solution. The low sensor distribution resulted in the highest average temperature and least variability, which was primarily caused by the core of the grain mass cooling relatively quickly in October and not being further affected by the rewarming of the periphery grain during non-fan operating periods.

**Table 2.** The average temperatures and standard deviations, aeration fan run hours, and percent of fan run time through the end of February, calculated for the entire grain mass for each of the three cable configurations defined in Figure 2, as well as the high sensor distribution without the center cable, and that distribution plus the medium, versus the average temperature calculated for the numerical solution (Total) for a one-year controlled aeration simulation of 1346 MT of maize beginning 1 October, 2014. The fan was turned on whenever ambient air was cooler than the average temperature reported by the sensors and the average was above 0 °C.

| | Total | Low | Med | High | High-Center Column | High-Center + Medium |
|---|---|---|---|---|---|---|
| Average (°C) | 4.91 | 10.79 | 7.20 | 8.45 | 10.65 | 8.45 |
| Std Dev (°C) | 5.81 | 4.04 | 5.13 | 4.74 | 4.08 | 4.74 |
| Number of Cables | | 1 | 3 | 7 | 6 | 9 |
| Number of Sensors | | 6 | 18 | 42 | 36 | 54 |
| Fan Run Hours | 1636 | 741 | 1058 | 1067 | 984 | 1254 |
| Run Time (%) | 45.1 | 20.5 | 29.2 | 29.4 | 27.2 | 34.6 |

In order to compare the results for a conventional temperature cable system to those of a new system using wireless technology, the same aeration strategy was used with 10 replicates for randomizing horizontal sensor locations. The results are shown in Table 3. A similar trend was seen for the overall average temperatures. The low number of sensors resulted in slightly warmer grain, and the medium number of sensors in slightly cooler grain. The average temperature values were within 0.11 °C for the three sensor distributions likely due to the averaging effect of ten replicates each. The standard deviations and the largest differences (between the maximum and minimum average temperatures) observed between the ten replicates show a clear improvement in the reliability of results as more sensors are included to make aeration decisions. The standard deviation decreased from 0.61 °C for the low number of sensors (one randomly located sensor at each sensor depth) to 0.40 °C for the medium number of sensors (three sensors randomly located at each sensor depth) to 0.33 °C for the high number of sensors (seven sensors randomly located at each sensor depth), i.e., almost by half. Similarly, the average difference between the maximum and minimum average temperature values reduced almost by half. However, fewer cables (and sensors) that result in acceptable stored grain conditions would reduce fixed and variable costs for stored gain managers, and would thus favor randomly placed wireless over fixed-placed cable sensors. The fan-run hours were similar for the three sensor distributions, and less variable with more sensors. They were consistent with those for the medium and high distribution of fixed-placed sensors and resulted in 30% fan run time during the 5-month cool-down and winter holding period.

**Table 3.** Results of ten simulations each of randomized horizontal placement of wireless sensors at fixed grain depths for numbers of sensors corresponding to each of the three cable configurations defined in Figure 2. Average temperature, average standard deviation (SD.), the average difference between the maximum and minimum values, average fan run hours, and average standard deviation of fan-run hours are the respective averages across each set of ten simulations. The last volume lists the corresponding average temperature for each configuration predicted by the temperature cables. The results reflect a one-year controlled aeration simulation of 1346 MT of maize beginning 1 October, 2014. The fan was turned on whenever ambient air was cooler than the average of the sensors and the average was above 0 °C.

| Sensor Numbers | Average Temperature (°C) | Average SD (°C) | Average Max-Min (°C) | Average Fan Run Hours | Average Run Hours SD. (h) | Average Temperature in Fixed Cable Sensor Simulations (°C) |
|---|---|---|---|---|---|---|
| Low (6) | 8.02 | 0.61 | 1.99 | 1075 | 82.3 | 10.78 |
| Medium (18) | 7.91 | 0.40 | 1.41 | 1105 | 53.2 | 7.20 |
| High (42) | 8.01 | 0.33 | 1.07 | 1078 | 43.3 | 8.45 |

In the final analysis, the effectiveness of aeration cooling was evaluated by turning aeration fans on and off based on the average temperature of six sensors on a single cable placed at four different locations in the grain mass. The single temperature cable located at the wall resulted in the warmest grain mass as the average temperature of all 2940 nodes remained above 15 °C until 21 March (Figure 7). The sensors in the periphery reported cool conditions during the winter resulting in no need for additional aeration despite the grain bulk remaining warmer.

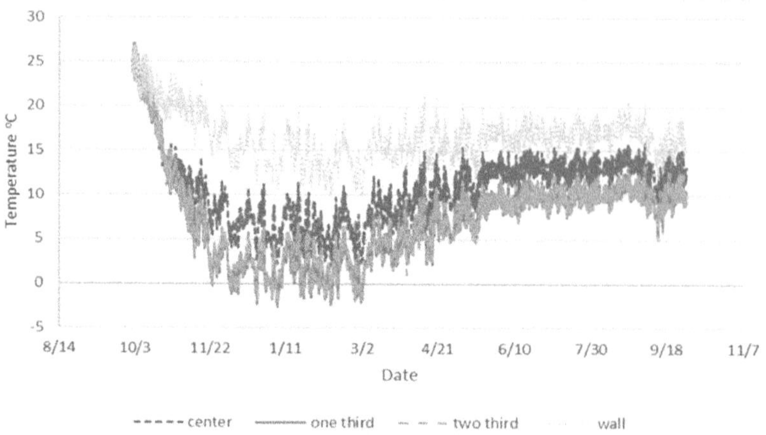


**Figure 7.** The average temperature of the entire grain mass (all nodes) for one cable configuration defined in Figure 2 as low and placed in the center of the grain mass versus one third and two-thirds distance from the wall, and at the wall, for a one-year controlled aeration simulation of 1346 MT of maize beginning 1 October, 2014. The fan was turned on whenever ambient air was cooler than the average of the sensors, and the average was above 0 °C.

When placed in the center, the single temperature cable was the second-worst option, reducing the average grain temperature of the grain mass below 15 °C as fast as the one-third and two-thirds location placements, but maintaining a higher average temperature for the remainder of the storage period. The average temperature of the grain mass when the single temperature cable was placed one-third of the distance to the silo wall was about the same (within 0.08 °C) as the one placed two-thirds of the way, and the patterns essentially overlapped during the aeration cool-down and storage periods.

The average grain temperatures for the center and periphery simulations were higher (by 3–9 °C) than the previous scenarios, while the 1/3 and 2/3 scenarios were similar. Results were consistent in that neither the center nor the periphery of the grain mass was ideal for placing a single cable as neither location represents the grain volume at-large. Temperature sensors placed about one-third of the way between the regions that are most (periphery) and least (core) affected by weather conditions appear to be the most representative for monitoring grain conditions with a single cable (or set of wireless sensors) for the silo size evaluated. It will result in rapid and maximum cooling of the grain mass during the all-important fall aeration period. The fan-run hours for the center, one-third, two-thirds, and periphery simulations were 741, 1019, 1044, and 889 h, respectively, and consistent with previous results. However, these results need to be further investigated for different silo sizes.

## 4. Discussion

These results are of importance to stored grain managers because they demonstrate that they essentially have no control in terms of temperature management over the periphery layer of a grain mass. Additionally, temperature values they rely on to make informed aeration control and inventory management decisions heavily depend on the sensor numbers and placement in the rest of the grain mass. Depending on the temperature cable configuration, the average grain temperature reported may be off by several degrees (e.g., 6–12 °C in these examples) from the overall temperature average in the grain mass. A key reason for this discrepancy is the fact that grain temperatures in the periphery are lowest during winter and highest during summer but are generally not captured because of the lack of cables placed near the silo wall. The silo wall (and thus a 1 m layer of grain closest to the wall) experiences the greatest temperature fluctuations during a one-year storage period. This result highlights one of the problems of temperature cables which are fixed in place and generally not close to the wall.

This result also demonstrates that it is important to select the sensors used to decide when fans are turned on and off. In this case, only the center sensor was used to make that control decision. The core of the grain mass is not influenced by the weather effects on the periphery, and once cooled during the all-important fall harvest period, the core remains cooler throughout the remainder of the storage period. This could mislead a stored grain manager into thinking that the grain mass has cooled to a sufficiently low temperature during the fall cool down phase and is as cool as it could get when the rest of the grain mass is still at a higher temperature and not yet sufficiently cooled to mitigate rewarming of grain during the spring and summer storage phase. As a matter of fact, a 4.7 °C difference in the average temperature was observed between the low and medium sensor distributions even though the actual conditions were identical across all cases. Activating the aeration fans and operating them to achieve the maximum cooling effect is critical for maintaining stored grain quality and avoiding mold spoilage and insect infestation during the storage period. Thus, relying on the center cable alone is not advisable.

Adding more temperature cables (and sensors) did not provide more useful information than was expected to make informed stored grain management decisions except when temperature readings from the cable in the core of the grain mass was eliminated. The high sensor distribution less the center the cable required six cables and 36 sensors while the medium-plus modified high sensor distribution required nine cables and 54 sensors. For this silo size these additional temperature readings did not provide more useful information for controlling the aeration fan and achieving a lower average temperature than the medium sensor distribution, which utilized only three cables and 18 sensors.

An additional consideration is the number of hours that aeration fans are operated by the automatic controller, which affects cost as a result of electricity consumption. The low sensor distribution had the fewest runtime hours operating for about 20% of the time during the 5-month cool-down and winter holding period (i.e., 741 h out of 3624 h) as shown in Table 2. In comparison, the medium, high, and modified high sensor distributions operated for about 30% of the time, which would be 50% more costly. The modified high plus medium distribution operated the fan for slightly more than a third of the time, and the average temperature of all nodes called for nearly 50% run time which would be 2.5 times costlier than the low sensor distribution. For this U.S. Midwestern Maize Belt location, the recommended fan operating practice consists of three cycles of 150 h during the October through December cool-down period, which results in a fan run time of 20.4% (i.e., 450 h out of 2208 h) [19]. Interestingly, only the low sensor distribution matched the recommended practice relatively over time. However, by allowing fans to operate whenever ambient air was cooler than the average temperature reported by the sensors and the average was above 0 °C, fans operated 65–264% more hours (and costlier) than would supposedly be needed based on recommended practice. These findings need to be investigated further in order to refine current aeration decision strategies, especially with regard to the progress of aeration fronts through the grain mass from bottom to top and minimizing fan operating hours and associated electricity costs, which are not considered in this analysis.

In terms of temperatures, these results in Table 3 do not seem to agree closely with those predicted for the three cable-based fixed sensor configurations. For each comparison, the temperature cable results were more than one standard deviation outside the average temperature results for randomized sensor placement. For the low number of sensors case, temperature cable results were more than 4.5 standard deviations (and 2.76 °C) above the average temperature for randomized placement. The medium number of sensors case showed the temperature cables were 1.75 standard deviations (and 0.71 °C) lower than for randomized placement. For the high number of sensors case, temperature cable results were 1.49 standard deviations (and 0.44 °C) above those for randomized placement. The medium distribution case seems to further confirm that excluding sensors placed in the center when deciding whether to turn on or off an aeration fan based on average grain temperature gives the stored grain manager more reliable information to make an aeration control decision than when they are included.

Randomized placement of wireless sensors increases the likelihood that sensors are not placed in the center and that sensors are placed throughout the bulk of the grain mass. This mitigates the core effect by accounting for warmer conditions in the bulk and more variable conditions in the periphery of the grain mass. Randomized placement would result during the filling of a silo assuming wireless sensors are added to the grain stream. Once they hit the grain surface, they will slide or roll at the angle of repose before coming to rest at a location outside the core of the grain mass and sufficient distance away from the silo wall. One disadvantage, however, for not having sensors in the core of the grain mass is that generally stored grain managers do not core the grain mass and remove peaked grain during the harvest season to maximize available storage capacity in a silo. Cooling peaked grain occurs at a much slower rate than the rest of the grain mass due to non-uniform airflow rates. Airflow rates have shown to be almost three times lower through the core of peaked grain than the airflow through the periphery [20].

## 5. Conclusions

The MLP 3D finite element model was used to evaluate the effect of number and placement of cable-based and wireless temperature sensors on stored grain aeration decisions and quality management for a 1346 MT (53,000 bushel) silo located in Ames, IA, USA. The key results are:

1. For a one-year aerated stored grain period, three typical temperature cable configurations predicted average temperatures that varied by 4.7 °C from each other and were as much as 11.4 °C lower than the average temperature predicted by the numerical solution.

2. When the three temperature cable configurations were used to control aeration fans, the average reported temperatures were similar throughout the year of aerated stor-

age, but the predicted average temperatures differed by as much as 2.3 °C to 5.9 °C compared to the average grain mass temperature predicted by the numerical solution.

3. Effect of randomized horizontal sensor placement with the number of sensors of the three cable configurations at fixed depths showed that average temperature values differed slightly, but the standard deviation between overall average temperatures of ten simulations was reduced from 0.61 °C to 0.33 °C when increasing the number of sensors placed in the grain mass from 6 to 42.

4. A medium number of wireless sensors (18 in the silo evaluated) randomly placed in layers approximately 2.09 m apart resulted in a timely cool-down (average temperature below 5 °C) during the important fall aeration period with a reasonable fan run time (30%), and in acceptable grain conditions during the remainder of the storage period (average below 12.5 °C). Reduced fixed (wireless sensors) and variable (fan operating) costs would thus favor randomly placed wireless over fixed-placed cable temperature sensors.

5. Placement of a single temperature cable (or column of wireless sensors) indicated the optimum location for aeration control and grain quality monitoring was about one to two-thirds of the distance between the core and the periphery of the grain mass in the silo size evaluated.

**Author Contributions:** Writing—original draft preparation, B.P.; Writing—review and editing, D.M. All authors have read and agreed to the published version of the manuscript.

**Funding:** This research received no external funding.

**Institutional Review Board Statement:** Not applicable.

**Informed Consent Statement:** Not applicable.

**Data Availability Statement:** Not applicable.

**Conflicts of Interest:** The authors declare no conflict of interest.

**Disclaimer:** Mention of trade names, proprietary products, or specific equipment does not constitute a guarantee or warranty to the exclusion of other products that may be suitable. It does not imply recommendation or endorsement by the U.S. Department of Agriculture (USDA) or Iowa State University (ISU). USDA and ISU each is an equal opportunity provider and employer.

## References

1. Food and Agriculture Organization. *Stemming Post-Harvest Waste Crucial to African Food Security*; United Nations Food and Agriculture Organization (FAO): Rome, Italy, 2014.
2. Food and Agriculture Organization. *The State of Food Insecurity and Nutrition in the World*; United Nations Food and Agriculture Organization (FAO): Rome, Italy, 2019.
3. World Bank; NRI; FAO. *Missing Food: The Case of Postharvest Grain Losses in Sub-Saharan Africa (No. 60371-AFR)*; FAO: Rome, Italy, 2011.
4. Armitage, D.M.; Stables, L.M. Effects of aeration on established insect infestations in bins of wheat. *Prot. Ecol.* **1984**, *6*, 63–73.
5. Burges, H.D.; Burrell, N.J. Cooling bulk grain in the British climate to control storage insects and to improve keeping quality. *J. Sci. Food Agric.* **1964**, *15*, 32–50. [CrossRef]
6. Fields, P. Alternatives to chemical control of stored-product insects in temperate regions. In Proceedings of the 9th International Working Conference on Stored-Product Protection, Sao Paulo, Brazil, 15–18 October 2006; pp. 653–662.
7. Akdogan, H.; Casada, M.E. Climatic humidity effects on controlled summer aeration in the hard red winter wheat belt. *Trans. ASABE* **2006**, *49*, 1077–1087. [CrossRef]
8. Lawrence, J.; Maier, D.E. Development and validation of a model to predict air temperatures and humidities in the headspace of partially filled stored grain silos. *Trans. ASABE* **2011**, *54*, 1809–1817. [CrossRef]
9. Plumier, B.M. 3D Ecosystem Modeling of Aeration and Fumigation in Australian Grain Silos to Improve Efficacy Against Insects. Ph.D. Thesis, Kansas State University, Manhattan, KS, USA, 2018. Unpublished.
10. Sinicio, R.; Muir, W.E. Aeration strategies for preventing spoilage of wheat stored in tropical and subtropical climates. *Trans. Am. Soc. Agric. Biol. Eng.* **1998**, *14*, 517–527. [CrossRef]
11. Wilson, S.G.; Desmarchelier, J.M. Aeration according to seed wet-bulb temperature. *J. Stored Prod. Res.* **1994**, *30*, 45–60. [CrossRef]
12. Casada, M.E.; Thompson, S.A.; Armstrong, P.R.; McNeill, S.G.; Maghirang, R.G.; Montross, M.D.; Turner, A.P. Forces on monitoring cables during grain bin filling and emptying. *Trans. ASABE* **2019**, *35*, 409–415.

13. Muir, W.E. *Grain Preservation Biosystems. Spoilage and Heating of Stored Grain Agriculture Products: Prevention, Detection and Control*; University of Manitoba: Winnipeg, MB, Canada, 2000.

14. Ileleji, K.E.; Maier, D.E.; Bhat, C.; Woloshuk, C.P. Detection of a developing hot spot in stored corn with a $CO_2$ sensor. *Appl. Eng. Agric.* **2006**, *22*, 275–289. [CrossRef]

15. Mills, J.D. *Spoilage and Heating of Stored Agricultural Products: Prevention, Detection and Control*; Canadian Government Publishing Centre Supply and Services Canada: Ottawa, ON, Canada, 1989.

16. Plumier, B.M.; Schramm, M.; Maier, D.E. Developing and verifying a fumigant loss model for bulk stored grain to predict phosphine concentrations by taking into account fumigant leakage and sorption. *J. Stored Prod. Res.* **2018**, *77*, 197–204. [CrossRef]

17. Plumier, B.M.; Maier, D.E. Sensitivity analysis of a fumigant movement and loss model for bulk stored grain to predict effects of environmental conditions and operational variables on fumigation efficacy. *J. Stored Prod. Res.* **2018**, *78*, 18–26. [CrossRef]

18. Plumier, B.M.; Schramm, M.; Ren, Y.; Maier, D.E. Modeling post-fumigation desorption of phosphine in bulk stored grain. *J. Stored Prod. Res.* **2020**, *85*, 101548. [CrossRef]

19. Maier, D.E.; McNeil, S.; Hellevang, K.; Ambrose, K.; Ileleji, K.; Jones, C.; Purshwitz, M. *Grain Drying, Handling, and Storage Handbook, MWPS-13*, 3rd ed.; Iowa State University/Midwest Plan Service: Ames, IA, USA, 2017.

20. Bartosik, R.E.; Maier, D.E. Effect of airflow distribution on the performance of NA/LT in-bin drying of corn. *Trans. ASABE* **2006**, *49*, 1095–1104. [CrossRef]

*Article*

# On-Farm Assessment of Maize Storage and Conservation Technologies in the Central and Northern Republic of Benin

Evelyne Sissinto Gbenou [1,*], Ygué Patrice Adégbola [2], Pélagie Manhoussi Hessavi [3], Segla Roch Cedrique Zossou [3] and Gauthier Biaou [4]

[1] Departement of Economics, Socio-Anthropology and Communication for Rural Development, University of Abomey-Calavi, Abomey-Calavi 01 BP 526, Benin
[2] Agricultural Policy Analysis Program, National Institute of Agricultural Research of Benin, Cotonou 01 BP 884, Benin; patrice.adegbola@yahoo.fr
[3] International Center of Research and Training in Social Science, Porto-Novo 02 BP 238, Benin; hess.pelagie@gmail.com (P.M.H.); rochybuggs@yahoo.fr (S.R.C.Z.)
[4] Rectorate, National University of Agriculture, Porto-Novo 01 BP 55, Benin; gbiaou@yahoo.fr
* Correspondence: evesinto@yahoo.fr; Tel.: +229-97-05-27-74 or +229-95-96-28-57

**Abstract:** The loss rates and financial profitability of maize storage and conservation technologies were assessed in the central and northern regions of the Republic of Benin. The experimentations were conducted specifically in the villages of Boukoumbé and Savalou and were randomly sampled. A total of four storage technologies were offered to 137 producers: the polypropylene bag, the Purdue Improved Cowpea Storage (PICS) bag, the metal silo, and the improved and closed earthen attic (with or without stock processing). The method by Pantenius was used to determine the loss rates, and the economic method of Gittinger was used to evaluate the profitability of technologies. The results showed that the technologies that recorded fewer losses in the two communes during storage were the PICS bag with grain treatment by chemical conservation measures in Savalou (9.42 ± 4.64%) and Boukoumbé (2.69 ± 0.77%), the PICS bag without grain treatment in Savalou (11.71 ± 2.78%), the metal silo with grain treatment in Boukoumbé (4.92 ± 1.36%) and the polypropylene bag with grain treatment in Savalou (10.56 ± 2.80%) and Boukoumbé (4.02 ± 1.23%). Therefore, the financial analysis results indicated that the most profitable storage technologies were the PICS bag with treatment in the center of Benin and the polypropylene bag without treatment in northern Benin.

**Keywords:** maize; storage systems; financial profitability

**Citation:** Sissinto Gbenou, E.; Adégbola, Y.P.; Hessavi, M.P.; Zossou, S.R.C.; Biaou, G. On-Farm Assessment of Maize Storage and Conservation Technologies in the Central and Northern Republic of Benin. *Agriculture* **2021**, *11*, 32. https://doi.org/10.3390/agriculture11010032

Received: 6 November 2020
Accepted: 16 December 2020
Published: 5 January 2021

**Publisher's Note:** MDPI stays neutral with regard to jurisdictional claims in published maps and institutional affiliations.



## 1. Introduction

Maize cultivation occupies almost 70% of the total area devoted to cereals in Benin. The area sown for maize production was 1,003,715 ha in 2016 across the country, representing the production of 1,300,000 tons of maize MAEP [1]. Crop covers 53 municipalities out of the 77 municipalities in Benin and is present in the seven Agricultural Development Hubs (ADHs). In Benin, maize is the staple food and the only cereal that generates exportable surpluses to neighboring countries, namely, Niger and Nigeria Sohinto et Aïna [2].

Efforts have been made over the past 10 years to increase the production of maize by improving productivity and subsidizing seeds through several projects, including the West Africa Agricultural Productivity Program (WAAPP), as well as technical and financial partner projects. Efforts to increase the production of maize have also included the provision of fertilizers and herbicides. Adégbola [3] analyzed the impact of the adoption of improved varieties of maize and concluded that this practice increased the yield by 9.77 kg/ha. Thus, the adoption of improved varieties of maize improved the income from production by 2427 FCFA per hectare. In the distribution of maize production costs, labor is the highest variable cost item in all maize production systems in Benin. The supply of foodstuffs, such as maize, to the population, presents a temporal and spatial gap between production and consumption. This gap is filled by the storage of maize.

143

During this storage stage, maize stocks suffer losses that reduce the potential quantity of the food and lead to a reduction in agricultural income. To mitigate the magnitude of the losses and to allow producers to have a relatively large marketable surplus, improved maize storage and conservation technologies were introduced from 1996 to 2004 Fandohan, Maboudou [4,5]. Despite advances in research in the implementation of innovations that can significantly reduce these loss rates, it has been found that the adoption of these new techniques has not been effective among many producers who prefer to continue their endogenous practices Maboudou [6]. Storage losses lead to losses of seeds, fertilizers, pesticides and labor during the production of stored maize, and significant postharvest losses will reduce farm income.

Adégbola [7] showed that the use of improved storage technologies provided a significantly higher income than previous or endogenous local technologies. These technologies allowed adoptive producers to acquire more material than nonadopters. Additionally, the adoption of these improved technologies contributed to the improvement of production factors (land, capital, labor). Previous research has shown that the adoption of improved maize storage technologies had a positive impact on income and hence on the acquisition of material goods and investments in human capital and production. Based on the analysis of Hinnou and Aloukoutou [8], it appeared that the financial means, the difficulty of construction, the risks of intoxication (due to the smell of sofagrain after several months of storage), the positive impact of accessibility to building materials of granaries and the positive impact of the mastery of construction techniques were the main issues affecting the adoption of these improved technologies. Hinnou and Aloukoutou [8] concluded that in the North, it was necessary to promote storage in bags with non-winnowing after ginning and the use of the repellent leaves in clay granaries. In contrast, in the South, improved granaries made of plant materials with repellent leaves, especially neem, and bags of grain maize with "Phostoxin" were the best possible alternatives. Jones et al. [9] evaluated the storage of maize grains with the use of sofagrain and in Purdue Improved Cowpea Storage (PICS) bags without chemical conservation measures and determined that the weight losses were only 0.5% without the use of chemicals in PICS bags. They concluded that the PICS bag had a good chance of adoption in Ghana, Tanzania, Kenya, Malawi and Mozambique, where tests have been carried out.

Over the past ten years, with the LISA projects of the NGO Louvain Coopération and the Postharvest Project of HELVETAS Swiss Intercooperation, storage technologies have been developed and introduced in rural areas in central and northern Benin. The following storage technologies have been tested in Benin and compared to the existing technology used by producers: the polypropylene storage bag ZeroFly®, the metal silo and the PICS bag. The polypropylene bag is one of the most used materials by producers in Benin for storing maize. This technology can be classified as "peasant know-how". The situation is the same for earthen granaries, which have been improved over time. Maize stored in polypropylene bags with or without treatment undergoes postharvest losses ranging between 3.66% and 13.21% Sissinto-Gbénou [10]. Recently, several studies have highlighted the effectiveness of hermetic storage technologies for reducing losses of maize stored in Benin and in its subregions [11–15]. In addition, the socioeconomic aspects of storage technologies were partially addressed by Adégbola [7] and Adéoti et al. [16]. Therefore, the objective of this research was to evaluate the storage losses and profitability of improved maize storage technologies in central and northern Benin. Financial profitability is an important criterion for producers as economic agents and will allow them to decide whether to adopt these technologies. The method of Pantenius [17] was used for loss rates, and the economic method of Gittinger [18] was used to assess the profitability of different maize storage and conservation technologies.

## 2. Materials and Methods

### 2.1. Experimentation Sites

The experiment was carried out in the municipalities of Savalou and Boukoumbé in the central and northern regions of Benin, respectively. The municipality of Savalou is located in the savannah-humid agroecological zone (800–1000 mm), and that of Boukoumbé is located in northern Benin in the humid agroecological zone (>1000 mm). Boukoumbé is between 10° and 10°40′ N latitude and 0°75′ and 1°30′ E longitude with an annual average rainfall of 1067 mm and average temperature of approximately 27 °C, with variations from 20 to 38 °C Tchegnon [14]. In Savalou, the average annual rainfall is 1150 mm, with annual variations between 864 and 1637.3 mm, and temperatures are high throughout the year, with minimums between 23 and 24 °C and maximums between 35 and 36 °C [19]. Four and ten villages were selected in Savalou and Boukoumbé, respectively, based on the production level and the importance of maize in the production system. The experiments were carried out over three years. In the first year, 23 and 27 farmers were selected in Savalou and Boukoumbé, respectively, based on their willingness to host and ability to manage the experimental storage trials. To perform analyses, we used data from farmers who participated in the trials for two successive years, either the first and second years or the second and third years. This gave rise to a total of 12 producers in central Benin (Savalou) and 13 in northern Benin (Boukoumbé).

### 2.2. Storage Technologies

Four types of storage structures were used to store maize grains during the experiments: (1) Polypropylene bags with a capacity of 100 kg; (2) Purdue Improved Crop Storage (PICS) bags with a capacity of 100 kg; (3) Improved clay granaries of different capacities, the largest of which was 1000 kg; (4) Metal silos with a capacity varying from 250 to 1000 kg. Generally, in both study areas, farmers had been using improved clay granaries and polypropylene bags for grain storage Sissinto-Gbénou [10]. However, as the polypropylene bag was a widely used storage structure, it was considered as the control structure in the experiment. The other storage structures offered alternatives to producers. The polypropylene bag and the improved clay granary were storage technologies that already exist in rural areas. The PICS bag was developed by Purdue University and has been more commonly used to store cowpea than maize. Both the large metal silo and PICS bag were introduced in Benin for product storage at a community level. In this study, we used silos for agricultural product storage at an individual producer level.

In addition, maize was treated using Actellic® Super powder (2% pirimiphos-methyl) to protect the stocks of maize against pest attacks. The insecticide was applied at a rate of 50 g Actellic® Super powder per 100 kg maize grains. This insecticide is recommended by the Plant Protection and Phytosanitary Control Service to farmers in the Republic of Benin. It is an easily biodegradable insecticide that is only slightly persistent. Actellic® Super is effective in combating *Sitophilus zeamais*, but is ineffective against borers (Bostrichidae), including the grain borer (*Prostephanus truncatus*). It is sold under the trade name of Actellic® Super.

### 2.3. Vegetal Material

The grains of the improved maize variety DMR were used for the experiments in both areas. They were provided by the farmers involved in the experimental trials. The grains had been sorted and dried before storage. The grain moisture levels were recorded at the beginning of the trials.

### 2.4. Experimental Setup and Treatments Implemented

Height treatments were designed for the experiments. Each treatment was a combination of one storage structure, with and without the application of Actellic® Super powder. The experiment was arranged in a completely randomized block design, with a 2 * 4 factorial experiment for the first year of experiments and a 2 * 5 factorial experiment for the last two years, including the metal silo in the trials. In addition, the principle of divided plots (split-plot) was used. The experience plan was unbalanced because the treatments did not have the same number of repetitions. Two categories of experimental farmers were defined depending on the application of Actellic® Super powder to protect the stocks of maize against pest attacks. The treatments were as follows:

- Treatment 1: Untreated maize grains stored in a polypropylene bag;
- Treatment 2: Untreated maize grains stored in a PICS bag;
- Treatment 3: Untreated maize grains stored in an improved clay granary;
- Treatment 4: Untreated maize grains stored in a metal silo;
- Treatment 5: Maize grains treated with Actellic® Super in a polypropylene bag;
- Treatment 6: Maize grains treated with Actellic® Super and stored in a PICS bag;
- Treatment 7: Maize grains treated with Actellic® Super and stored in an improved clay granary;
- Treatment 8: Maize grains treated with Actellic® Super and stored in a metal silo.

### 2.5. Sampling and Data Collection

During the three experiments years, technical and economic data were collected each year at monthly intervals for seven months, from the establishment of the experiments to the maize destocking. Data collection covered the period from February or March to August or September each year. This period coincided with the seven or eight months of maize storage each year. Each month, four samples of 750 g per treatment (one per treatment) were taken from each experimental producer for damage and loss evaluation of the storage treatments in the laboratory. A one-meter handle ladle was used to take samples by mixing the grains in the stock enclosure before the sample was taken. Each sample was then saved in an envelope, which was packed in a labeled plastic bag and sealed. The samples were then transported to the laboratory of the Agricultural and Food Technology Program (PTAA) of the National Institute of Agricultural Research of Benin (INRAB) in Porto-Novo for further analysis. In the laboratory, the samples were stored in a refrigerator at a temperature of 4 °C for the two weeks required for the loss rate assessment.

Data on maize storage costs (from harvest to storage) were collected from each trial participant. They included the cost of construction or purchase price of the storage structures, the cost of labor related to maize storage operations (including dispatching and shelling costs), and the cost of the chemical for storing the maize in the structures. Selling prices per kilogram of maize were collected per month for the duration of the experiments at the level of each participant. The interest rates charged by the microfinance institutions (10% and 12%, in Savalou and Boukoumbé, respectively) and the lifespan of the storage structures (1 year for the polypropylene bag, 2 years for the PICS bag, 20 years for the metal silo and 15 years for the improved granary) were considered for the calculation of storage costs. A structured questionnaire was designed to collect these data. The results of the evaluation of the loss rates were used to calculate the amount of financial loss for each storage treatment.

### 2.6. Data Analysis
#### 2.6.1. Loss Rate

Several approaches for determining quantitative losses have been developed. The counting and weighing methods were used to assess the rates of maize loss during storage. This method is the most used in studies and produces better results.

The formula used by Pantenius [17] is as follows:

$$\%Losses = \frac{(E * B) - (C * D)}{(E * A)} * 100 \tag{1}$$

where $A$ is the total number of grains, $B$ is the number of damaged grains, $C$ is the number of healthy grains, $D$ is the weight of damaged grains and $E$ is the weight of healthy grains.

### 2.6.2. Storage Costs

The cost of the maize storage encompassed both fixed and variable costs. Thus, the total cost of maize storage was obtained by summing up the fixed and variable costs. The fixed costs included the costs of the storage structures and small storage equipment, such as basins and baskets. Following Arouna et al. [20] the average monthly cost was calculated as:

$$E(j) = \frac{C - R}{n} + [(C - R) * f + R] * (q - 1) + C * r \tag{2}$$

where $j$ is the type of storage structure or storage small equipment, $E(j)$ is the monthly cost of the storage structure or small storage equipment $j$, $C$ is the storage structure construction cost or purchase price, $R$ is the residual value of the storage structure or small storage equipment, $n$ is its useful lifespan, $(q - 1)$ is the interest rate, $f$ is the capital asset factor, and $r$ is the repair or maintenance cost factor (coefficient). The capital asset factor is estimated using the formula as follows:

$$Capital\,asset\,factor = (Interest\,rate / 100) * maize\,selling\,price \tag{3}$$

In the study area, the small equipment was mostly not repaired and used until it was thrown away. Thus $R = 0$ and $r = 0$ and Equation (2) is written as follows:

$$E(j) = \frac{C}{n} + C * f * (q - 1) \tag{4}$$

The variable costs of maize storage encompassed the costs associated with storage losses and other variable costs. The storage loss in monetary value, also called financial loss, represented the quantified sum of the quantitative losses occurring during storage with a given treatment. Thus, to calculate the quantitative loss, the quantity stored was multiplied by the loss rate. The financial loss was obtained by multiplying the quantitative loss by the average monthly maize selling price. The other variable costs comprised the labor costs associated with maize storage operations—including the dispatching and shelling costs of maize before storage in storage structures, the cost of loading and unloading the maize from the storage structure, the cost of the application of the conservation measure Actellic® Super and interest on the capital asset of the maize stock.

### 2.6.3. Benefit–Cost Ratio (BCR)

The benefit–cost ratio (BCR) was estimated for each treatment.

It was obtained by dividing the flow of the present value of the benefits (benefits) by that of the present value of the cost following the formula of Gittinger [18].

$$\frac{B}{C} = \sum_{t=1}^{n} \frac{B_t}{(1 + i)^{t-1}} \bigg/ \sum_{t=1}^{n} \frac{C_t}{(1 + i)^{t-1}} \tag{5}$$

where $B_t$ is the monthly benefits, $C_t$ is the monthly costs, $n$ is the storage duration (in months), $i$ is the interest rate (expected) and $t$ is a given month.

### 2.6.4. Break-Even Quantity

The break-even quantity was estimated for each treatment, and thus, the break-even quantity of the turnover and the break-even ratio were determined.

The break-even quantity in turnover (BREQ) was calculated using the following formula:

$$BREQ = \left( Fixed\,costs * \frac{Revenue}{Gross\,margin} \right) * 100 \tag{6}$$

The percentage of capacity used, also called the break-even ratio, indicates the percentage of production for which the gross margin covers the fixed costs. The risk increases as the capacity percentage increases; a low percentage (maximum = 1) gives a level of security against unpredictable operating difficulties. Therefore, when the value of this capacity tends towards 1, there is a risk in which the producer may no longer be able to pay for the equipment used. This percentage is computed by the following formula:

$$\%capacityused = \left(\frac{BREQ}{Revenue}\right) * 100 \tag{7}$$

The break-even quantity of maize to be stored in each storage and preservation technology to make the investment profitable is given by the following formula:

$$Break - evenquantity = \%capacityused * technicalcapacity \tag{8}$$

### 2.6.5. Statistical Analysis

The Student–Newman–Keuls (SNK) ANOVA statistical test was used to test the difference between the paired means between the different storage and conservation technologies using SPSS Statistics software. These paired multiple comparison tests captured the difference between paired means and generated a matrix that informed the means of groups of significantly different storage and conservation technologies [21].

### 2.6.6. Sensitivity Analysis

The sensitivity analysis consisted of varying the fixed costs and the selling prices of maize to assess the effect of the change in certain parameters (fixed costs and selling prices of maize) on the benefit–cost ratio and the threshold quantity for each of the technologies. This assessment allowed the impact of the different treatments implemented on the economic profitability of maize storage to be observed when these parameters changed for any reason.

### 3. Results

#### 3.1. Average Loss Rate Recorded for the Different Treatments Implemented

The PICS bag with grain treatment and the PICS bag without grain treatment recorded fewer losses in the two communes during storage (Tables 1 and 2).

**Table 1.** Evolution of the average loss rates of the different treatments during storage in Savalou.

| Treatments | Month 1 | Month 2 | Month 3 | Month 4 | Month 5 | Month 6 |
|---|---|---|---|---|---|---|
| Polypropylene bag − ctrol | 3.17 ± 1.62% ab | 5.17 ± 1.76% b | 7.35 ± 1.88% | 9.66 ± 1.42% | 10.77 ± 5.18% | 12.88 ± 11.20% |
| Polypropylene bag + ctrol | 1.64 ± 0.43% ab | 5.17 ± 1.76% b | 5.62 ± 4.52% | 9.32 ± 2.28% | 9.62 ± 5.25% | 10.56 ± 2.80% |
| PICS bag − ctrol | — | — | 4.85 ± 1.36% | 9.00 ± 3.71% | 10.08 ± 4.61% | 11.71 ± 2.78% |
| PICS bag + ctrol | — | — | 3.77 ± 1.49% | 7.24 ± 1.28% | 9.17 ± 1.82% | 9.42 ± 4.64% |
| Metal silo − ctrol | 2.83 ± 1.67% b | 5.60 ± 1.81% b | 11.94 ± 5.64% | 13.07 ± 4.83% | 13.78 ± 6.79% | 14.61 ± 9.13% |
| Metal silo + ctrol | 1.62 ± 0.65% a | 2.56 ± 0.83% a | 3.03 ± 0.99% | 10.22 ± 3.18% | 11.79 ± 2.56% | 14.07 ± 6.73% |
| Improved clay granary − ctrol | 4.15 ± 2.41% ab | 10.00 ± 5.63% b | 10.19 ± 4.03% | 13.84 ± 3.77% | 21.69 ± 6.23% | 22.64 ± 7.21% |
| Improved clay granary + ctrol | 0.70 ± 0.41% ab | 5.65 ± 1.72% b | 8.99 ± 3.18% | 12.19 ± 0.63% | 16.66 ± 2.10% | 20.90 ± 0.20% |
| Fisher test | 4.62 *** | 2.17 ** | 1.53 ns | 1.28 ns | 0.65 ns | 0.814 ns |

$p >$ F probabilities are indicated by symbols: ns = no significant differences; ** significant differences at $p < 0.05$; *** significant differences at $p < 0.01$. For each column, values with the same letter indicate no significant differences at 5%; Source: Experimentation data, 2015, 2016 and 2017 (− ctrol means: without chemical conservation measure; + ctrol means: with chemical conservation measure).

**Table 2.** Evolution of the average loss rates of the different treatments during storage in Boukoumbé.

| Treatments | Month 1 | Month 2 | Month 3 | Month 4 | Month 5 | Month 6 |
|---|---|---|---|---|---|---|
| Polypropylene bag − ctrol | 0.53 ± 0.16% ab | 1.84 ± 0.90% | 4.79 ± 1.97% | 7.98 ± 1.87% ab | 9.45 ± 1.66% ab | 9.64 ± 2.73% ab |
| Polypropylene bag + ctrol | 0.45 ± 0.24% ab | 0.79 ± 0.27% | 1.27 ± 0.53% | 3.57 ± 0.93% a | 3.28 ± 0.86% a | 4.02 ± 1.23% a |
| PICS bag − ctrol | — | — | 3.42 ± 0.90% | 5.30 ± 1.65% ab | 7.59 ± 2.32% ab | 7.71 ± 1.74% ab |
| PICS bag + ctrol | — | — | 0.85 ± 0.33% | 2.09 ± 0.68% a | 2.52 ± 0.67% a | 2.69 ± 0.77% a |
| Metal silo − ctrol | 1.07 ± 0.40% b | 1.70 ± 0.51% | 3.88 ± 1.27% | 7.77 ± 2.02% b | 8.06 ± 3.49% b | 11.26 ± 2.93% b |
| Metal silo + ctrol | 0.62 ± 0.29% b | 1.69 ± 0.72% | 3.55 ± 1.22% | 4.08 ± 1.97% ab | 4.19 ± 1.49% a | 4.92 ± 1.36% a |
| Improved clay granary − ctrol | 1.08 ± 0.15% b | 2.16 ± 0.90% | 4.89 ± 1.24% | 10.04 ± 1.87% ab | 10.18 ± 2.07% ab | 12.57 ± 3.68% ab |
| Improved clay granary + ctrol | 0.57 ± 0.31% ab | 1.30 ± 0.41% | 2.22 ± 0.71% | 4.05 ± 1.55% ab | 4.09 ± 1.12% ab | 5.00 ± 1.38% ab |
| Fisher test | 3.81 *** | 3.06 ns | 2.44 ns | 3.34 *** | 4.131 *** | 3.093 *** |

$p >$ F probabilities are indicated by symbols: ns = no significant differences; ** significant differences at $p < 0.05$; *** significant differences at $p < 0.01$. For each column, values with the same letter indicate no significant differences at 5%; Source: Experimentation data, 2015, 2016 and 2017 (− ctrol means: without chemical conservation measure; + ctrol means: with chemical conservation measure).

The statistical differences observed showed that a significant increase in the loss of dry matter was observed at the level of each treatment throughout the six months of storage in Boukoumbé ($p < 0.0001$). In Savalou, significant variations were observed during the first two months. However, in the last month of storage, in the commune of Savalou, the PICS bag with a chemical conservation measure ($9.42 \pm 4.64\%$), the polypropylene bag with a chemical conservation measure ($10.56 \pm 2.80\%$) and the PICS bag without a chemical conservation measure ($11.71 \pm 2.78\%$) were the three treatments that recorded fewer losses, while the improved clay granary ($22.64 \pm 7.21\%$) recorded the highest loss rate (Table 2). In Boukoumbé, the PICS bag with a chemical conservation measure ($2.69 \pm 0.77\%$), the polypropylene bag with a chemical conservation measure ($4.02 \pm 1.23\%$) and the metal silo with a chemical conservation measure ($4.92 \pm 1.36\%$) were the treatments that recorded significantly fewer losses in order of priority, while the improved clay granary without a chemical conservation measure ($12.57 \pm 3.68\%$) was the treatment that recorded the most maize loss during storage ($F = 3.093$, $p < 0.0001$) (Table 2).

### 3.2. Pests in Experimental Trials

Regarding the pest infestation, the treatments including the chemical conservation measure recorded the lowest insect populations. Application of the chemical conservation measure decreased the infestation by insects significantly. As a result, the application of the chemical conservation measure improved the efficacy of control of the insect population. The treatments most affected by these pests were, in decreasing order, the improved clay granary without a chemical conservation measure, the polypropylene bag without a chemical conservation measure and the metal silo without a chemical conservation measure. The PICS bag and the ZeroFly® Bag without chemical conservation measure recorded a lower insect pest population. Therefore, these two technologies are likely to better protect stocks against insects.

*Sitophilus zeamaïs* was the primary pest counted, and a few *Dinoderus* spp. were observed during the identification carried out in the laboratory for the first year. In Savalou, there was a heavy infestation of grain stocks by beetles. In Boukoumbé, the number of beetles counted for all treatments remained relatively lower compared to the count carried out in Savalou. For two years, *Sitophilus zeamaïs* (primary) and *Tribolium castaneum* (secondary) were the two main pests (insects) counted during the identification carried out in the laboratory. In Savalou, we noted a high infestation of grain stocks by these two beetles, the most abundant of which was *Sitophilus zeamaïs*. In Boukoumbé, the same insects were counted, and the most abundant was *Tribolium castaneum*. However, the number of insects counted at all the treatments remained relatively lower in Boukoumbé than in Savalou.

### 3.3. Financial Analysis of the Treatments Implemented

The results from the evaluation of the loss rates of the treatments were used to calculate the amount of financial loss at the level of each treatment. Details on monthly costs of maize storage in each storage structure involved in trials are presented in the supplementary materials, Tables S1–S4.

### 3.3.1. Storage Costs

The results showed that the fixed costs for the metal silo with (25.88 FCFA/kg) and without (31.61 CFAF /kg) chemical conservation measures, and the improved granary with (15.67 CFAF/kg) and without (26.1 CFAF/kg) conservation measure were statistically superior to that of the polypropylene bag and PICS bag ($p < 0.001$), whatever the municipality (Tables 3 and 4). The same trends were observed with regard to total costs, which involved both fixed and variable costs. In fact, the storage structures with a long lifespan (metal silo and improved earthen granary) had total maize storage costs of more than CFAF 100/kg while storage structures (polypropylene bag and PICS bag) had total storage costs of less than CFAF 100/kg. This was explained by the fixed costs and costs related to financial losses, which were statistically higher for the structures with a long lifespan, especially without lifespan (F = 3.24, $p < 0.05$). This referred to the case of the metal silo without a chemical conservation measure, whose fixed costs and costs related to financial losses were higher than those of other treatments. The market price of maize does not vary according to the storage structure or whether the producer has used a chemical conservation measure. Thus, the discounted income, which corresponded to the selling price of one kilogram of maize, did not vary according to treatment, and the discounted income was CFAF 498.93 per kilogram of maize stored over a period of 6 months in Savalou.

The total costs of storing maize were all less than CFAF 100 in the commune of Boukoumbé (Table 4). The statistical difference showed that the lowest cost of storage of maize was recorded at the level of the polypropylene bag without a chemical conservation measure (CFAF 54.55) (F = 1.46, $p < 0.05$). The metal silo with chemical conservation measure, on the other hand, recorded the highest total cost of storing and conserving maize (CFAF 90.87). The improved clay granary without a chemical conservation measure recorded the highest financial loss (CFAF 9.24), while the PICS bag with a chemical conservation measure recorded the lowest financial loss (CFAF 2.07) (F = 1.46; $p < 0.05$) (Table 4). The discounted income was CFAF 389.30 in this commune for one kilogram of maize stored for six (06) months.

### 3.3.2. Benefit–Cost Ratio

The calculated benefit–cost ratios were all greater than 1, and, therefore, when the producer invested 1 CFAF for the storage and conservation of one kilogram of maize, he or she obtained an income of more than 1 CFAF /kg. In Savalou, the storage of maize in the PICS bag (10.87), polypropylene bag (7.89) and improved granary (7.53) without chemical conservation measure, by order of priority, had the highest benefit–cost ratio compared to other structures (F = 2.01; $p < 0.05$) (Table 3). In other words, the storage of one kilogram of maize in the PICS bag, polypropylene bag and improved granary without chemical conservation measures generated an income of 10.87, 7.89 and 7.53 CFAF, respectively for an investment of 1 CFAF /kg.

The polypropylene bag had an average benefit–cost ratio that was statistically lower than that of the PICS bag without chemical conservation measure. This means that producers of maize in Savalou can use all the other maize storage structures and chemical conservation measures except the untreated PICS bag for maize storage in place of the polypropylene bag with chemical conservation measures. In the same vein, the lowest value of the income generated was observed, especially at the level of the metal silo, with (4.58 CFAF /kg) and without (4.00 CFAF /kg) chemical conservation measures. Therefore, storing maize in the improved granary without a chemical conservation measure gave the lowest benefit–cost ratio. In Boukoumbé, the storage of maize in the polypropylene bag and PICS bag without a chemical conservation measure generated an income of 7.42 and 7.30 CFAF /kg, respectively, for an investment of 1 FCFA. This was statistically higher than the value generated by the other storage structures (improved clay granary, metal silo with or without grain treatment) (F = 14.01; $p < 0.001$) (Table 4). Unlike Savalou, the income generated with the storage of maize in the polypropylene bag was relatively higher than the PICS bag without a chemical conservation measure in Boukoumbé. Consequently, producers in Boukoumbé should use the PICS bag with a chemical conservation measure in place of the polypropylene bag with a chemical conservation measure or the improved clay granary with or without a chemical conservation measure in place of the polypropylene bag with a chemical conservation measure.

**Table 3.** Benefit–cost ratio and break-even quantity of the different treatments implemented in Savalou.

| | | Savalou | | | | | | |
| | | Discounted Storage Costs (CFAF/kg) | | | | | Financial Profitability Parameters | |
| Treatments | Discoun-ted Total Revenue * (CFAF/Kg) | Fix Costs | Variables Costs | | | Total Costs | BCR | Break-Even Quantity (kg) |
| | | | Storage Loss Cost | Other Variable Costs | Total variable Costs | | | |
|---|---|---|---|---|---|---|---|---|
| Polyp bag − ctrol | 498.93 | 1.21 a | 23.01 b | 63.77 | 86.78 b | 87.99 a | 7.89 b | 31.24 a |
| Polyp bag + ctrol | 498.93 | 1.21 a | 17.38 ab | 71.16 | 88.54 ab | 89.75 ab | 5.76 ab | 30.97 a |
| PICS bag − ctrol | 498.93 | 3.94 a | 10.25 a | 63.77 | 74.02 a | 77.96 a | 10.87 b | 97.94 b |
| PICS bag + ctrol | 498.93 | 3.94 a | 8.68 a | 71.16 | 79.84 ab | 83.78 a | 6.18 ab | 98.52 b |
| Metal silo − ctrol | 498.93 | 31.85 b | 31.61 ab | 63.77 | 95.38 ab | 127.23 b | 4.00 a | 848.86 d |
| Metal silo + ctrol | 498.93 | 31.85 b | 25.88 b | 71.16 | 97.04 b | 128.89 b | 4.58 a | 821.35 d |
| Imp. clay granary − ctrol | 498.93 | 15.08 b | 26.1 ab | 63.77 | 89.87 ab | 104.95 ab | 7.53 b | 351.02 c |
| Imp. clay granary+ ctrol | 498.93 | 15.08 b | 15.67 ab | 71.16 | 86.83 b | 101.91 ab | 5.32 ab | 369.22 c |
| Fisher test | | 1.69 *** | 3.24 ** | 8.33 ns | 0.29 ** | 1.84 ** | 2.01 ** | 16.05 *** |

* = Average selling prices of one kg of maize is 498.93 CFAF in Savalou; $p > F$ probabilities are indicated by symbols: ns = no significant differences; ** = significant differences at $p < 0.05$; *** = significant differences at $p < 0.01$. For each column, values with the same letter indicate no significant differences at 5%. Source: Experimentation data, 2015, 2016 and 2017 (− ctrol = without chemical conservation measure; + ctrol = with chemical conservation measure).

**Table 4.** Benefit–cost ratio and break-even quantity of the different treatments implemented in Boukoumbé.

| | | Boukoumbé | | | | | | |
| | | Discounted Storage Costs (CFAF/kg) | | | | | Financial Profitability Parameters | |
| Treatments | Discoun-ted Total Revenue * (CFAF/Kg) | Fix Costs | Variables Costs | | | Total Costs | BCR | Break-(kg) |
| | | | Storage Loss Cost | Other Variable Costs | Total Variable Costs | | | |
|---|---|---|---|---|---|---|---|---|
| Polyp bag − ctrol | 389.30 | 1.21 a | 6.01 bc | 47.33 | 53.34 ab | 54.55 a | 7.42 c | 37.46 a |
| Polyp bag + ctrol | 389.30 | 1.21 a | 3.52 a | 54.47 | 57.99 ab | 59.2 ab | 6.12 bc | 38.27 a |
| PICS bag − ctrol | 389.30 | 4.30 a | 3.58 ab | 47.33 | 50.91 a | 55.21 a | 7.30 c | 132.32 b |
| PICS bag + ctrol | 389.30 | 4.30 a | 2.07 a | 54.47 | 56.54 ab | 60.84 ab | 6.00 bc | 134.91 b |
| Metal silo − ctrol | 389.30 | 31.54 b | 7.85 bc | 47.33 | 55.18 ab | 86.72 c | 4.71 a | 932.62 d |
| Metal silo + ctrol | 389.30 | 31.54 b | 4.86 ab | 54.47 | 59.33 b | 90.87 c | 4.24 a | 946.72 d |
| Imp. Clay granary − ctrol | 389.30 | 11.63 b | 9.24 c | 47.33 | 56.57 ab | 68.2 bc | 6.08 c | 346.30 c |
| Imp. Clay granary+ ctrol | 389.30 | 11.63 b | 7.61 bc | 54.47 | 62.08 b | 73.71 bc | 5.54 ab | 317.41 c |
| Fisher test | | 5.31 *** | 0.40 *** | 4.31 ns | 0.63 ** | 1.46 ** | 14.01 *** | 1.03 * |

* = Average selling prices of one kg of maize is 389.30 CFAF in Boukoumbé; $p > F$ probabilities are indicated by symbols: ns = no significant differences; ** = significant differences at $p < 0.05$; *** = significant differences at $p < 0.01$. For each column, values with the same letter indicate no significant differences at 5%. Source: Experimentation data, 2015, 2016 and 2017 (− ctrol = without chemical conservation measure treatment; + ctrol = with chemical conservation measure).

### 3.3.3. Break-Even Quantity

The break-even quantity is the minimum quantity that the storage technology can contain to allow the producer to return on an investment for the purchase of the storage structure. In general, in the two municipalities, the results indicated that a large quantity of approximately one ton must be stored in the metal silo to make the investment profitable. In fact, considering its initial investment, the metal silo presented a large break-even quantity compared to all the other technologies (F = 16.05 $p < 0.001$ in Savalou; F = 1.03 $p < 0.001$ in Boukoumbé) (Tables 3 and 4). The break-even quantity was variable depending on the treatment. In Savalou, the metal silo without grain treatment presented a break-even quantity value of 848.86 kg compared to 821.35 kg for the metal silo with grain treatment (Table 3). The polypropylene bag without grain treatment had a threshold capacity of 31.24 kg compared to 30.97 kg for the polypropylene bag with grain treatment (Table 3). The PICS bag without grain treatment presented a break-even quantity of 97.94 kg compared to 98.52 kg for the PICS bag with grain treatment. The closed ground granary without grain treatment presented a break-even quantity of 351.02 kg, and the improved clay granary with grain treatment presented a break-even quantity of 369.22 kg.

In Boukoumbé, the PICS bag presented a break-even quantity exceeding 100 kg. The PICS bag without grain treatment yielded 132.32 kg, while the metal silo with grain treatment yielded 946.72 kg (Table 4). The break-even quantity for the metal silo was approximately one ton, while the improved clay granary presented a break-even quantity of less than 500 kg. The metal silo without grain treatment yielded 932.62 kg, compared to 946.72 kg for the metal silo with grain treatment. The improved clay granary without grain treatment presented a break-even quantity of 346.30 kg, and the improved clay granary without treatment presented a break-even quantity of 317.41 kg.

### 3.3.4. Analysis of the Sensitivity of the Benefit–Cost Ratio and of the Break-Even Quantity

Tables 5 and 6 present the results of the sensitivity analysis of the benefit–cost ratio and the break-even quantity of the treatments implemented in Savalou and Boukoumbé. On the one hand, we increased the fixed costs by 10% and decreased the selling price of maize by 30% and 40%. On the other hand, we increased the fixed costs by 10% and the selling price of maize by 30% and 40%. In the two (02) municipalities, the results showed that the break-even quantity was the most sensitive indicator to variations in the fixed costs of storing maize and the selling price of maize.

Rising fixed costs of 10% and a decrease in maize sales price implies a decrease in the benefit–cost ratio value, whatever the type of structure. The polypropylene bag and PICS bag without grain treatment had the highest benefit–cost ratio compared to other structures, with an increased fixed cost of approximately 10%, and a decrease in maize sales price by 30% and 40% (F = 0.95; $p < 0.05$). The same trends were observed following an increase in the maize sales price by 30% and 40%; however, the improved clay granary also had the highest benefit–cost ratio. In Savalou, the reduction of the selling price of maize by 40% (compared to the current selling price of maize) caused the polypropylene bag without grain treatment to no longer be profitable for maize storage for a break-even quantity of $-1322.34$ kg, given its statistically low value compared to that of other structures ($p < 0.001$) (Table 5). The result was the same for the metal silo without treatment, which became unprofitable following 30% and 40% reductions in the selling price of maize, with the break-even quantity changing from 166.06 to $-22.40$ kg. On the contrary, an increase in the selling price of maize by 30% or 40% caused the metal silo to be profitable for maize storage for break-even quantity.

In Boukoumbé, the metal silo with and without grain treatment constitutes the main storage and preservation technology, and the break-even quantity of this technology is very sensitive to variations in the selling price of maize. However, the storage of maize remained profitable, given its statistically high value compared to that of other structures ($p < 0.001$). Indeed, the break-even quantity increased from 150.63 to 198.40 kg for the storage of maize in the metal silo without grain treatment, with reductions in the selling price of maize of 30% and 40%, respectively (Table 6). For the metal silo with grain treatment, the break-even quantity increased from 147.60 to 181.46 kg and then to 235.60 kg (Table 6). On the contrary, a decrease in the break-even quantity was observed with regard to an increase in the selling price of maize. The same trends were observed in Saval.

**Table 5.** Sensitivity analysis of the benefit–cost ratio and the break-even quantity of the different treatments implemented in Savalou.

| Treatments | 10% Increase in Fixed Costs | 10% Increase in Fixed Costs and 30% Decrease in Maize Sales Price | | | 10% Increase in Fixed Costs and 40% Decrease in Maize Sales Price | | | 10% Increase in Fixed Costs and 30% Increase in Maize Sales Price | | | 10% Increase in Fixed Costs and 40% Increase in Maize Sales Price | | |
|---|---|---|---|---|---|---|---|---|---|---|---|---|---|
| | | 30% Decrease in Maize Sales Price | BCR | Break-Even Quantity (kg) | 40% Decrease in Maize Sales Price | BCR | Break-Even Quantity (kg) | 30% Increase in Maize Sales Price | BCR | Break-Even Quantity (kg) | 40% Increase in Maize Sales Price | BCR | Break-Even Quantity (kg) |
| Polyp bag − ctrol | 0.12 b | 363.38 | 5.71 b | 5.54 a | 311.47 | 4.89 c | −1322.34 ab | 674.86 | 10.60 b | 84.28 a | 726.77 | 11.42 b | 78.25 a |
| Polyp bag + ctrol | 0.12 b | 339.85 | 4.24 ab | 5.02 a | 291.30 | 3.63 ab | 6.39 b | 631.16 | 7.88 ab | 94.69 a | 679.71 | 8.49 ab | 87.90 a |
| PICS bag − ctrol | 0.39 a | 363.38 | 5.83 b | 16.55 b | 311.47 | 5.00 c | 28.79bc | 674.86 | 10.84 b | 104.73 b | 726.77 | 11.68 b | 99.73 b |
| PICS bag + ctrol | 0.39 a | 339.85 | 4.32 ab | 15.75 b | 291.30 | 3.70 ab | 19.84bc | 631.16 | 8.04 ab | 107.26 b | 679.71 | 8.65 ab | 96.50 b |
| Metal silo − ctrol | 3.18 d | 382.63 | 4.32 ab | 166.06 d | 327.97 | 3.70 ab | −22.40 ab | 710.61 | 6.90 a | 285.26 c | 765.27 | 7.43 a | 279.16 c |
| Metal silo + ctrol | 3.18 d | 354.15 | 3.66 a | 138.88 d | 303.55 | 3.14 ab | 185.87 d | 657.71 | 5.22 a | 298.66 c | 708.30 | 5.62 a | 291.59 c |
| Imp. Clay granary − ctrol | 1.51 c | 363.38 | 3.71 ab | 55.22 c | 311.47 | 3.18 ab | 69.11 c | 674.86 | 8.03 ab | 177.09 d | 726.77 | 8.64 ab | 160.86 d |
| Imp. Clay granary+ ctrol | 1.51 c | 345.27 | 2.81 a | 58.33 c | 295.95 | 2.40 a | 72.4 c | 641.23 | 6.81 a | 184.67 d | 690.56 | 7.33 a | 167.88 d |
| Fisher test | 1.30 | | 0.95 ** | 6.61 *** | | 0.47 * | 5.44 ** | | 7.64 ** | 2.96 *** | | 4.71 *** | 9.33 *** |

$p > F$ probabilities are indicated by symbols: ns = no significant differences; ** = significant differences at $p < 0.05$; *** = significant differences at $p < 0.01$. For each column, values with the same letter indicate no significant differences at 5%. Source: Experimentation data from 2015, 2016 and 2017 (− ctrol = without chemical conservation measure; + ctrol = with chemical conservation measure).

**Table 6.** Sensitivity analysis of the benefit–cost ratio and the break-even quantity of the different treatments implemented in Boukoumbé.

| Treatments | 10% Increase in Fixed Costs | 10% Increase in Fixed Costs and 30% Decrease in Maize Sales Price | | | 10% Increase in Fixed Costs and 40% Decrease in Maize Sales Price | | | 10% Increase in Fixed Costs and 30% Increase in Maize Sales Price | | | 10% Increase in Fixed Costs and 40% Increase in Maize Sales Price | | |
|---|---|---|---|---|---|---|---|---|---|---|---|---|---|
| | | 30% Decrease in Maize Sales Price | BCR | Break-Even Quantity (kg) | 40% Decrease in Maize Sales Price | BCR | Break-Even Quantity (kg) | 30% Increase in Maize Sales Price | BCR | Break-Even Quantity (kg) | 40% Increase in Maize Sales Price | BCR | Break-Even Quantity (kg) |
| Polyp bag − ctrol | 0.12 a | 278.13 | 4.93 b | 5.90 a | 238.40 | 4.22 b | 7.49 a | 516.55 | 9.16 b | 87.56 a | 556.28 | 9.87 b | 81.29 a |
| Polyp bag + ctrol | 0.12 a | 267.41 | 4.18 ab | 5.98 a | 229.21 | 3.58 ab | 7.36 a | 496.63 | 7.77 ab | 92.53 a | 534.83 | 8.37 ab | 85.90 a |
| PICS bag − ctrol | 0.43 b | 282.95 | 4.95 b | 20.72 b | 242.53 | 4.24 b | 26.09 b | 525.49 | 9.19 b | 106.06 b | 565.91 | 9.90 b | 101.89 b |
| PICS bag + ctrol | 0.43 b | 267.41 | 4.06 ab | 21.02 b | 229.21 | 3.48 ab | 25.84 b | 496.63 | 7.54 ab | 111.91 b | 534.83 | 8.12 ab | 106.25 b |
| Metal silo − ctrol | 3.15 d | 296.64 | 4.31 ab | 150.63 d | 254.26 | 3.70 b | 198.40 d | 550.91 | 5.96 a | 282.16 c | 593.29 | 6.42 a | 261.84 c |
| Metal silo + ctrol | 3.15 d | 296.74 | 3.87 a | 147.60 d | 254.35 | 3.32 a | 181.46 d | 551.10 | 5.62 a | 288.75 c | 593.49 | 6.06 a | 272.40 c |
| Imp. Clay granary − ctrol | 1.16 c | 281.67 | 3.21 a | 52.13 c | 241.43 | 2.75 a | 62.76 c | 523.11 | 8.02 b | 187.79 d | 563.35 | 8.63 b | 181.50 d |
| Imp. Clay granary+ ctrol | 1.16 c | 297.33 | 3.02 a | 48.76 c | 254.85 | 2.59 a | 59.42 c | 552.19 | 7.20 ab | 184.63 d | 594.67 | 7.76 ab | 198.57 d |
| Fisher test | 0.63 | | 0.41 *** | 1.30 *** | | 0.64 ** | 7.14 *** | | 1.31 *** | 6.95 *** | | 5.35 *** | 2.54 *** |

$p$ > F probabilities are indicated by symbols: ns = no significant differences; ** = significant differences at $p < 0.05$; *** = significant differences at $p < 0.01$. For each column, values with the same letter indicate no significant differences at 5%. Source: Experimentation data from 2015, 2016 and 2017 (− ctrol = without chemical conservation measure; + ctrol = with chemical conservation measure).

## 4. Discussion

Regardless of municipality, the PICS bag recorded the lowest loss rate. The PICS bag with treatment recorded total losses of 9.42% (±4.64%), and the PICS bag without a chemical control recorded total losses of 11.71% (±2.78%) in Savalou. The PICS bag with grain treatment recorded total losses of 2.69% (±0.77%), and the PICS bag without chemical control recorded total losses of 7.71% (±1.74%) in Boukoumbé. Our findings agreed with results obtained by Poudel et al. [22], who evaluated the efficiency of maize storage and conservation structures in Central and Northern Benin and concluded that in Savalou, as in Boukoumbé, the PICS bag without grain treatment with Actellic® Super was more effective than the untreated polypropylene bag for reducing loss rates.

*Sitophilus zeamaïs* and *Tribolium castaneum* were the two main pests counted in stocks, and the better pest control technology was the PICS bag. Pests were more common in Savalou than in Boukoumbé. This may be because Savalou is a humid area, and Boukoumbé is a dry area. However, the assessment of pest levels compared to temperature or moisture was not studied.

The benefit–cost ratio of the PICS bag without a chemical conservation measure was the highest (7.89) in the Savalou region. Ndegwa et al. [23] found a similar result and observed that airtight bags (the PICS bag and the SuperGrain Bag) without preservatives exhibited the highest benefit–cost ratio of 1.6, with a loss rate of 3.9% over four months of storage in Kenya.

The benefit–cost ratio of the metal silo in Savalou was 5.86 without the chemical conservation measure and 4.99 with the chemical control. This ratio for the metal silo was 6.20 without the chemical protectant and 5.28 with the chemical protectant measure in Boukoumbé. Compared to the improved clay granary, the metal silo displayed the highest ratios regardless of region. The metal silo had a ratio of 4.99, compared to the improved clay granary without and with conservation measure, which presented benefit–cost ratios of 4.85 and 3.63 in Savalou, respectively, and 4.35 and 4.18, respectively, in Boukoumbé, and was more profitable. Our findings agreed with results obtained by Nduku et al. [24], who performed a comparative analysis of the metal silo with storage technologies in a traditional improved granary in Kenya. The highest ratios obtained for the metal silo and the improved traditional granary were 2.5 and 1.6, respectively. Similarly, the metal silo was profitable at a break-even quantity of 1000 kg, regardless of the level of variation of maize sales price at Boukoumbé. It was in this context that De Groote et al. [13] proposed assessing the cost per kilogram of grain stored for different capacities of metal silos to determine the type of metal silo that would be financially and economically profitable.

The untreated polypropylene bag had the highest benefit–cost ratio of 6.91 in the northern region. We observed that agroclimatic conditions influenced the efficiency and the profitability of storage technologies; however, our research did not address these aspects.

For both Central and Northern Benin, the improved clay granary with chemical conservation measure was more expensive than the reference technology (here, the polypropylene bag with the chemical control). Adégbola et al. [3] found a similar result and indicated that the improved clay granary was more expensive than the reference system (traditional granary with treatment using local products) in southern Benin.

## 5. Conclusions

To identify strategies that reduced postharvest losses of maize and improved maize storage, conservation technologies were introduced in central and northern Benin. The loss rates and financial profitability of these different storage and conservation technologies were assessed. The analysis showed that, in central and northern Benin, the PICS bag and the polypropylene bag recorded less storage losses and were more profitable than the improved and closed earth granary and the metal silo. Specifically, the PICS bag without treatment was more profitable in Savalou, and the polypropylene bag with treatment was more profitable in Boukoumbé. The PICS bag and the polypropylene bag had a low initial investment cost compared to the improved clay granary and the metal silo, which had a high initial investment cost. The metal silo was also found to be more efficient and profitable than the improved clay granary, but the initial investment for the metal silo was high, and it was profitable at a break-even quantity of 1000 kg. To facilitate the dissemination and adoption of the metal silo, especially by small producers, a reduction in import taxes on galvanized sheet metal, which is the raw material for manufacturing metal silos, may help reduce the cost. In addition, appropriations adapted to the storage of grains in these technologies will have to accompany their diffusion. The PICS bag and the polypropylene bag are plastic bags, the recycling of which takes time after use. Governments should take responsibility for reducing the import taxes on galvanized sheet metal to facilitate the distribution and use of the metal silo; however, storage in bags facilitates transportation but does not protect the environment.

**Supplementary Materials:** The following are available online at https://www.mdpi.com/article/10.3390/agriculture11010032/s1, Table S1: Monthly flow costs of storing one kilogram of maize in the polypropylene bag, Table S2: Monthly flow costs of storing one kilogram of maize in the PICS bags, Table S3: Monthly flow costs of storing one kilogram of maize in the Metal silo, Table S4: Monthly flow costs of storing one kilogram of maize in the Improved clay granary.

**Author Contributions:** Conceptualization, E.S.G., Y.P.A. and G.B.; methodology, E.S.G., Y.P.A., P.M.H., S.R.C.Z. and G.B.; software, E.S.G., P.M.H. and S.R.C.Z.; validation, E.S.G., Y.P.A., S.R.C.Z. and G.B.; formal analysis, E.S.G., P.M.H. and S.R.C.Z.; investigation, E.S.G., P.M.H. and S.R.C.Z.; resources, E.S.G.; data curation, E.S.G., Y.P.A., P.M.H. and S.R.C.Z.; writing—original draft preparation, E.S.G., P.M.H., S.R.C.Z. and G.B.; writing—review and editing, E.S.G., Y.P.A., P.M.H., S.R.C.Z. and G.B.; visualization, E.S.G., Y.P.A., P.M.H., S.R.C.Z. and G.B.; supervision, E.S.G., Y.P.A. and G.B.; project administration, E.S.G., Y.P.A. and G.B.; funding acquisition, E.S.G. All authors have read and agreed to the published version of the manuscript.

**Funding:** This research was funded by the Swiss Cooperation and HELVETAS Swiss Intercooperation.

**Data Availability Statement:** Data sharing not applicable. No new data were created or analyzed in this study. Data sharing is not applicable to this article.

**Acknowledgments:** We acknowledge the Swiss Cooperation and HELVETAS Swiss Intercooperation for their funding and scientific contribution to the implementation of the experimental device. We also thank the National Agricultural Research Institute of Benin (INRAB), the Food and Agricultural Technology Program (PTAA) and the experimental producers from the communes of Savalou and Boukoumbé that mobilized in the context of this research. We also thank Centre International de Recherches et de Formations en Sciences Sociales (CIRFoSS) for data collection, processing and contribution to data analysis.

**Conflicts of Interest:** The authors declare no conflict of interest.

## References

1. MAEP. Statistiques agricoles de 2006 à 2015. 2016. Available online: https://insae.bj/images/docs/insae-statistiques/enquetes-recensements/Recensement-General-des-Entreprises/Rapport-Agriculture-RGE2.pdf (accessed on 29 September 2020).
2. Sohinto, D.; Aina, M.S. Analyse économique et financière de cinq chaînes de valeurs ajoutées (CVA) de la filière maïs au Bénin. *Rapp. d'étude* **2010**, 76.
3. Adégbola, Y.P.; Arouna, A.; Ahoyo, N. Analyse des facteurs affectant l'adoption des greniers améliorés pour le stockage du maïs au Sud-Bénin. *In Bulletin de la Recherche Agronomique du Bénin (BRAB) Numéro Spécial* **2011**, *8*, 43–50.
4. Fandohan, P. Introduction du grenier fermé en terre au Sud Bénin pour le stockage du maïs. *Rapp. Tech. Rech. INRAB-PTAA* **2000**, 29.
5. Maboudou, A.G.; Adégbola, P.Y.; Coulibaly, O.; Hell, K.; Amouzou, E. Factors affecting the use of improved clay store for maize storage in the central and northern Benin. In *New Directions for a Diverse Planet, Proceedings of the 4th International Crop Science Congress, Brisbane, Australia, 26 September–1 October 2004*; Fischer, T., Ed.; Crop Science Society of America: Madison, WI, USA, 2004.
6. Maboudou, A.G. Adoption et diffusion de technologies améliorées de stockage du maïs en milieu paysan dans le centre et le nord du Bénin. In *Mémoire Pour l'obtention du Diplome D'etudes Approfondies*; Unisersit2 de Lomé, Faculté des Lettres et Sciences Humaines: Lomé, Togo, 2003; 108p.
7. Adégbola, Y.P. Economic Analyses of Maize Storage Innovations in Southern Benin. Ph.D. Thesis, Wageningen University, Wageningen, The Netherlands, 2011; p. 182.

8.    Hinnou, C.L.; Aloukoutou, M.A. Stockage et conservation du maïs au Bénin: Techniques efficaces et stratégies d'adoption. *Rapport d'étude* **2011**, 54.

9.    Jones, M.S.; Alexander, C.E.; Lowenberg-Deboer, J. An Initial Investigation of the Potential for Hermetic Purdue Improved Crop Storage (PICS) Bags to Improve Incomes for Maize Producers in Sub-Saharan Africa. In *Working Paper #11–3*; Department of Agricultural Economics Purdue University: West Lafayette, IN, USA, 2011; p. 44.

10.   Sissinto-Gbénou, E.; Adégbola, Y.P.; Dischl, R.; Fischler, M.; Hessavi, M.P.; Ohouko, S.K.; Vodouhe, S.; Biaou, G. Efficacité des structures de stockage et de conservation du maïs au centre et au Nord du Bénin: Cas des communes de Savalou et Boukoumbé, 2018. *Annales de l'Université de Parakou Series « Sciences Naturelles et Agronomie » Décembre* **2018**, *8*, 151–167.

11.   Abass, A.B.; Fischler, M.; Schneider, K.; Daudi, S.; Gaspar, A.; Rust, J.; Kabula, E.; Ndunguru, G.; Madudu, D.; Msola, D. On-farm comparison of different postharvest storage technologies in a maize farming system of Tanzania Central Corridor. *J. Stored Prod. Res.* 2018, 77, pp. 55–65. Available online: www.elsevier.com/locate/jspr (accessed on 29 September 2020).

12.   Murdock, L.L.; Baributsa, D.; Ousmane, B.; Amadou, L.; Baoua, I.B. PICS bags for post-harvest storage of maize grain in West Africa. *J. Stored Prod. Res.* **2014**, *58*, 20–28. [CrossRef]

13.   De Groote, H.; Kimenju, S.C.; Likhayo, P.; Kanampiu, F.; Tefera, T.; Hellin, J. Effectiveness of hermetic systems in controlling maize storage pests in Kenya. *J. Stored Prod. Res.* **2013**, *53*, 27–36. [CrossRef]

14.   Tchegnon, P. Monographie de la commune de Boukoumbé, Ministère de la Décentralisation, Programme D'appui au Démarrage des Communes au Bénin. 2006. Available online: http://www.ancb-benin.org/pdc-sdacmonographies/monographiescommunales/MonographiedeBoukoumbé.pdf (accessed on 29 September 2020).

15.   Hodges, R.J. Post-harvest Weight Losses of Ce-real Grains in Sub-Saharan Africa. 2012. Available online: http://www.erails.net/FARA/aphlis/aph-lis/weightlosses-reviews (accessed on 29 September 2020).

16.   Adéoti, I.; Adégbola, P.Y.; Sodjinou, E.; Adéoti, R.; Sellamna, N. Processus d'adoption des technologies de stockage et de conservation du maïs au sud du Bénin. *Annales de l'Université de Parakou, Bénin Série « Sciences. Naturelles et Agronomie.* **2018**, 137–148.

17.   Pantenius, C.U. *Etat des pertes dans les systèmes de stockage du maïs au niveau des petits paysans de la région maritime du Togo*; GTZ: Hamburg, Germany, 1988; 83p.

18.   Gittinger, P. *Economic Analysis of Agricultural Projets*, 2nd ed.; EDI series in Economic Development: Baltimore, MD, USA; London, UK, 1985.

19.   Capo-Chichi, Y.J. Monographie de la Commune de Savalou, Ministère de la Décentralisation, Programme D'appui au Démarrage des Communes au Bénin. 2006. Available online: http://www.ancb-benin.org/pdc-sdacmonographies/monographiescommunales/MonographiedeSavalou.pdf (accessed on 29 September 2020).

20.   Arouna, A.; Adégbola, P.Y.; Biaou, G. Analyse des coûts de stockage et de conservation du maïs au Sud-Bénin. *Bulletin de la Recherche Agronomique du Bénin, Numéro Spécial 2: Aspects économiques du stockage et de la conservation du maïs au Sud-Bénin Septembre* **2011**, 13–23.

21.   IBM. SPSS Statistics Documentation, Anova One Factor. Available online: https://www.ibm.com/support/knowledgecenter/fr/SSLVMB_23.0.0/spss/base/idh_onew_post.html (accessed on 29 September 2020).

22.   Poudel, K.L.; Nepal, A.P.; Dhungana, B.; Sugimoto, Y.; Yamamoto, N.; Nishiwaki, A. Capital Budgeting Analysis of Organic Coffee Production in Gulmi District of Nepal. In *Proceedings of the International Association of Agricultural Economists Conference*, Beijing, China, 16–22 August 2009; pp. 1–13.

23.   Ndegwa, M.K.; De-Groote, H.; Gitonga, Z.M.; Bruce, A.Y. Effectiveness and economics of hermetic bags for maize storage: Results of a randomized controlled trial in Kenya. *Crop. Prot.* **2016**, *90*, 17–26. [CrossRef]

24.   Nduku, T.M.; De-Groote, H.; Nzuma, J.M. Comparative Analysis of Maize Storage Structures in Kenya. Presented in the 4th International Conference of the African Association of Agricultural Economists, Hammamet, Tunisia, 22–25 September 2013.

 *agriculture*

*Article*

# Effect of Hot-Air and Freeze-Drying on the Quality Attributes of Dried Pomegranate (*Punica granatum* L.) Arils During Long-Term Cold Storage of Whole Fruit

Adegoke Olusesan Adetoro [1,2], Umezuruike Linus Opara [2] and Olaniyi Amos Fawole [1,2,3,*]

[1] Department of Horticultural Science, Stellenbosch University, Private Bag X1, Stellenbosch 7602, South Africa; 21412944@sun.ac.za

[2] Africa Institute for Postharvest Technology, South African Research Chair in Postharvest Technology, Postharvest Technology Research Laboratory, Faculty of AgriSciences, Stellenbosch University, Private Bag X1, Stellenbosch 7602, South Africa; opara@sun.ac.za

[3] Postharvest Research Laboratory, Department of Botany and Plant Biotechnology, University of Johannesburg, P.O. Box 524, Auckland Park, Johannesburg 2006, South Africa

* Correspondence: olaniyif@uj.ac.za; Tel.: +27-11-559-7237; Fax: +27-11-559-2411

Received: 12 September 2020; Accepted: 20 October 2020; Published: 22 October 2020

**Abstract:** This study investigated the effect of hot-air and freeze-drying on the physicochemical, phytochemical and antioxidant capacity of dried pomegranate arils during long-term cold storage ($7 \pm 0.3\ °C$, with $92 \pm 3\%$ relative humidity) of whole fruit over a single experiment. Extracted arils were processed at monthly intervals during 12 weeks of cold storage of whole fruit. After the 12-week storage period, hot-air and freeze-dried arils showed the least (3.02) and highest (23.6) total colour difference (TCD), respectively. Hot-air dried arils also contained 46% more total soluble solids (TSS) than freeze-dried arils. During the storage of pomegranate fruit, total phenolic content (TPC) steadily increased from 20.9 to 23.9 mg GAE/100 mL and total anthocyanin content (TAC) increased from 6.91 to 8.77 mg C3gE /100 mL. Similarly, an increase in TPC and TAC were observed for hot-air (9.3%; 13%) and freeze-dried arils (5%; 5%), respectively. However, the radical scavenging activity (RSA) reduced by 8.5 and 17.4% for hot-air and freeze-dried arils, respectively, after 12 weeks of cold storage. Overall, the parameters such as colour, TPC and TAC as well as the lower degradation in RSA stability during storage showed distinct differences in quality when using the freeze-drying method, which is, therefore, recommended.

**Keywords:** cold storage; fresh arils; dried methods; total soluble solids; total phenolic content; storage stability

---

## 1. Introduction

Pomegranate fruit (*Punica granatum* L.) is renowned for its bioactive phenolic content, including flavonoids, phenolic acids, tannins, ellagitannins, catechin, rutin and epicatechin [1,2]. These antioxidants have been implicated in the protection against heart, cancer, immune system and other chronic diseases [3,4]. South Africa is leading the pomegranate production and export in the Southern Hemisphere, with an estimated production of 540,000 tonnes/1,200,000 cartons [5]. However, 11% of the total production is processed locally, and 9% is considered as waste due to disorders such as cracks, sunburn, scalds and bruises, which could affect the internal quality of the fruit [5,6]. Fruit similar to pomegranate usually has a small harvest window, whereas processing is carried out over a long period, and this requires the storage of raw materials for the production of niche products. Fifty per cent of fruit that do not meet export requirements are often converted into products such as jellies and juices, which have a short shelf life [7]. Caleb et al. [8] reported a maximum flavour-life of seven days for pomegranate arils. However, drying is a preservation method that reduces the moisture contained in food, thereby extending the shelf-life of the product [9].

Dried pomegranate arils are often referred to as 'anardana' and are used in many traditional medicinal formulations to treat neurological and kidney disorders, as well as stomach and cardiac infections [10]. Due to its acidity profile, these dried arils help to improve digestion and mouth-feel [11]. Indian and Pakistani cuisines use 'anardana' as a condiment, but it can also be used as a substitute for tamarind and mango powder, or in culinary preparations of fruit salad, flavoured yoghurt and ice cream [12]. However, different drying methods, packaging and storage conditions are major factors affecting the inherent characteristics of the final product [13].

In addition to the decline in quality of pomegranate fruit during storage, different processing techniques could also have a negative impact on the quality of the finished product. Previous studies suggest that freeze-drying retains more bioactive compounds during the processing of fruit in comparison to other drying methods. For instance, Asami et al. [14] reported higher retention of phenolic concentration in 'Marion' blackberries during freeze-drying than hot-air drying. Shofian et al. [15] reported that the low temperature used to withdraw water from fruit material in freeze-drying helped to preserve the antioxidant capacity of tropical fruits. However, the freeze-drying process could be expensive and energy-consuming [16]. Among several drying methods available, hot-air drying is cost- and energy-efficient, making it one of the most commonly used methods for drying food materials [17]. However, it has a greater effect on the deformation of final products which is often characterised by dislocation of volatile substances and changes in physical properties [18].

'Wonderful' pomegranate is the most cultivated and consumed globally [19,20]. Currently, there has been a considerable rise in the export of pomegranate fruit grown in South Africa, with an estimated at approximately 70% of total production [21] compared to 56% export in 2013 [22]. 'Wonderful' is desirable because its bioactive compounds are better maintained during a prolonged storage duration compared to other cultivars. According to Arendse [23], the quality attributes of 'Wonderful' were maintained over five months in storage. Furthermore, the highest scavenging capacity exhibited by 'Wonderful' compared to the other eight pomegranate cultivars suggests the commercial potential of the cultivar [24].

The concentration of bioactive compounds in dried fruit products is influenced by numerous factors, including cultivar, harvest maturity, processing method and storage conditions [25]. While there are several studies on the effect of cultivar [26,27], there is a dearth of information on the impact of harvest maturity, extended storage of raw material and processing method [28] on the bioactive compounds of dried fruit, including dried pomegranate aril. In practice, fruit are kept in storage to allow processing at intervals based on demand or processing capacity. Fruit quality attributes degrade over time and will affect the quality of processed products, and hence, important to establish the maximum holding time of raw materials before processing. Therefore, this study aimed to examine the effects of hot-air and freeze-drying on the quality attributes of dried pomegranate arils during prolonged cold storage of whole fruit (raw material).

## 2. Materials and Methods

### 2.1. Fruit Supply and Storage Condition

Pomegranate fruit (cv. Wonderful) was handpicked at commercial harvest period from Blydeverwacht orchard in Wellington, (latitude 33°01′00″ S, longitude 18°58′59″ E) Western Cape Province, South Africa during the 2018/2019 growing season. Fruit were transported in an air-conditioned vehicle to the Postharvest Technology Research Laboratory at Stellenbosch University. Fruit without visible external discolouration or injuries were sorted to include fruit of uniform colour and size. After sorting, fresh fruit were packed inside standard open top cartons with the following dimensions: width 0.3 m, length 0.4 m, height 0.133 m and a total of 22 perforations and stored at 7 ± 0.3 °C, with 92 ± 3% relative humidity (RH). Fruit were sampled at 0, 4, 8 and 12 weeks as described in the experimental flow chart (Figure 1).

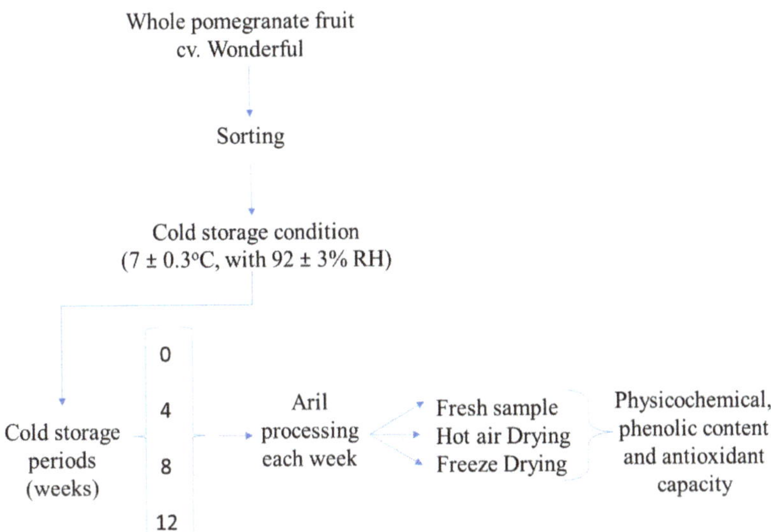


**Figure 1.** Shows a description of the experimental flowchart.

Temperature (°C) and relative humidity (% RH) within the cold rooms were taken every hour throughout storage. This was carried out with the use of a Tiny Tag TV-4500 data loggers (Gemini Data Logger, Sussex, UK) with a functional range of −40 °C to +85 °C and 0% to 100% RH.

### 2.2. Characterisation of Fresh Arils

Fresh pomegranate arils were periodically evaluated before processing for total soluble solids (TSS) by a refractometric method and titratable acidity (TA) by titrating to pH 8.1 with 0.1 N NaOH. Additionally, moisture content was measured using a digital moisture analyser. The Folin-Ciocalteu method was used to quantify the total phenolic content (TPC) and expressed as mean ± SE (milligram) gallic acid equivalent (GAE) per 100 mL of crude juice, while the pH differential method was used to determine the total anthocyanin content (TAC) [1,29], which was expressed as mean ± SE (milligrams) cyanidin-3-glucoside per 100 mL of crude juice. The antioxidant capacity (radical scavenging activity, RSA; ferric ion reducing antioxidant power, FRAP) was also measured in triplicate, according to Fawole and Opara [30] and expressed as Trolox equivalent (mM) per 100 mL of crude juice.

### 2.3. Drying Procedure

#### 2.3.1. Freeze-Drying

Arils were placed in a freeze-drying paper bag and frozen in a static air freezer at −80 °C. Frozen samples were freeze dried in triplicates. The specimen jar containing samples were carefully taken to a laboratory-scale freeze-dryer (VirTis Co., Gardiner, NY, USA) operating at condenser temperature −85 °C and pressure 45 mTorr. Sample weight was measured every third hour until no change in weight was detected, which was after 96 h.

#### 2.3.2. Hot-Air Drying

Arils were dried at 60 °C in a hot-air oven for 11 h to achieve a 10–12% moisture content. Constant air velocity was maintained at 1.0 m s$^{-1}$ for each treatment. To ensure an inner temperature of 60 °C was reached, the hot-air dryer was switched on at least an hour before drying, and the temperature was confirmed using a thermometer, before spreading the arils in glassy Petri dishes and placing them in the drying chamber. Dried arils were packed and sealed in food-grade moisture-resistant plastic bags and stored in glass desiccators containing calcium sulphate (Sigma-Aldrich Pty. Johannesburg, South Africa).

### 2.4. Colour Measurement

By the direct reading using a chromo-meter (Minolta model CR-200, Osaka, Japan), dried aril colour was determined to obtain the colour values: $L^*$ (brightness/darkness), $a^*$ (redness/greenness), $C^*$ (colour intensity) and $h°$ (colour purity). The measurements were recorded at three different times from a transparent petri dish and averaged. The maximum for $L^*$ value is 100 (white), and the minimum is zero (black). Positive $a^*$ value is red and negative $a^*$ is green, while positive $b^*$ value is yellow and negative $b^*$ is blue. The total colour difference (TCD) was calculated [4,31] as:

$$TCD = \left[(\Delta L^*)^2 + (\Delta a^*)^2 + (\Delta b^*)^2\right]^{\frac{1}{2}} \tag{1}$$

where $\Delta L^*$, $\Delta a^*$ and $\Delta b^*$ represents the value before and after drying at each treatment levels and results were expressed as means ± S.E.

### 2.5. Characterisation of Dried Arils

Dried pomegranate arils were ground into powder using liquid nitrogen followed by extraction of 5 g sample in 50 mL of distilled water. For 5 min the mixture was vortexed and sonicated for 15 min in an ultrasonic bath. This was followed by centrifugation at 10,000 rpm for 25 min and recovery of the supernatant for TSS, TA and pH measurements. For phytochemical properties and antioxidant capacity, the same extraction procedure was followed using 50% methanol.

### 2.6. Chemical Properties

Total Soluble Solids and Titratable Acidity Determination

TSS was estimated using a digital hand refractometer (model PT-32; ATAGO, Tokyo, Japan) with the range of 0–32 °Brix, which was blanked with distilled water. For TA, 2 mL of the supernatant was diluted in seventy millilitres of distilled water and titrated against 0.2 N of sodium hydroxide (NaOH) to a pH of 8.2 using a Metrohm 862 Compact titrosampler (Herisau, Switzerland).

### 2.7. Determination of Phytochemical Properties

2.7.1. Total Phenolic Content (TPC)

Folin–Ciocalteu method using a methanolic extract of dried arils was used to determine the TPC [30]. A 0.05 mL of the supernatant was mixed with 0.45 mL of 50% methanol in a test tube followed by adding 0.5 mL Folin–Ciocalteu after 2 min. The mixture was then vortexed and kept in the dark for 10 min before adding 2% $Na_2CO_3$ and further incubated for 40 min in the dark. The absorbance of each sample was read at 520 nm in a UV-visible spectrophotometer (Thermo Scientific technologies, Madison, USA) against a blank containing 50% methanol. Absorbance was compared with a standard curve (Gallic acid, 0–10 mg), and results were expressed as mg gallic acid equivalent per gram pomegranate dry matter (mg GAE/g DM).

2.7.2. Total Anthocyanin Content

By using the pH differential method, total anthocyanin content (TAC) was quantified [29]. In triplicates, 1 mL of extract was separately mixed with 9 mL of pH 1.0 and pH 4.5 buffers. Absorbance was measured at 520 and 700 nm in pH 1.0 and 4.5 buffers, and the result was expressed as cyanidin 3-glucoside using Equations (2) and (3):

$$A = (A_{510} - A_{700})_{\text{pH } 1.0} - (A_{510} - A_{700})_{\text{pH4}} \tag{2}$$

$$\text{Total monomeric anthocyanin}(\text{mg}/\text{mL}) = \frac{A \times MW \times DF}{\varepsilon \times L} \tag{3}$$

where A = Absorbance, $\varepsilon$ = Cyd-3-glucoside molar absorbance (26,900), MW = anthocyanin molecular weight (449.2), DF = dilution factor and L = cell path length (1 cm). Results are expressed as equivalent per gram dry matter (mg $C_3gE$/g DM).

### 2.8. Antioxidant Capacity

#### 2.8.1. Radical-Scavenging Activity (RSA)

The RSA was quantified in triplicate, according to Fawole et al. [30]. Aqueous methanolic extract of dried aril (0.015 mL) was diluted with methanol (0.735 mL) in test tubes, briefly under dim light shade, followed by adding 0.75 mL, 0.1 mM methanolic DPPH solution. For 30 min in the dark and at room temperature, the mixtures were incubated, and the absorbance was measured at 517 nm using a UV-vis spectrophotometer (Thermo Scientific technologies, Madison, USA). Absorbance was compared with the standard curve (Trolox equivalent, 0–2.0 mM). The free-radical activity of dried aril was expressed as Trolox equivalent (mM) equivalent per gram dry matter (mM TE/g DM).

#### 2.8.2. Ferric Ion Reducing Antioxidant Power (FRAP)

The antioxidant power of dried aril was measured using the colourimetric method according to [30,32]. The FRAP working solution was freshly prepared in mixtures of 300 mM acetate buffer (50 mL), 10 mM 2,4,6-tripyridyl-s-triazine (TPTZ) (5 mL) and 20 mM ferric chloride (5 mL) at 37 °C. In triplicates, diluted aqueous methanolic dried aril extracts (0.15 mL) were added to 2.85 mL of the FRAP working solution followed by a 30 min incubation in the dark. By measuring the absorbance at 593 nm, the reduction of the $Fe^{3+}$-TPTZ complex to a coloured $Fe^{2+}$-TPTZ complex at low pH of dried aril extracts was monitored. Trolox (0–10 mM) was used for the calibration curve, and the results were expressed as Trolox (mM) equivalents per gram dry matter (mM TE/g DM).

#### 2.8.3. Stability of RSA and FRAP

Several studies have reported the use of simple first-order reaction kinetics to describe storage and thermal degradation of bioactive compounds from various sources. Li et al. [33] and Moldovan et al. [34] described the degradation kinetics as in Equation (4):

$$\ln[\text{RSA}] = \ln[\text{RSA}_0] - kt \tag{4}$$

where RSA = antioxidant capacity, mM TE/g dried aril at time t; $RSA_0$ = initial RSA, mM TE/g; $k$ = reaction rate constant, weeks$^{-1}$; t = reaction time, weeks. The half-life of antioxidant capacity from the investigated extracts during storage can be calculated using Equation (4):

$$t_{1/2} = -\ln 0.5/k \tag{5}$$

where $t_{1/2}$ = half-life (weeks) and $k$ = reaction rate constant (weeks$^{-1}$).

### 2.9. Statistical Analysis

The measurement made from chemical properties, colour and phytochemical properties were subjected to statistical evaluation. STATISTICA (Statistica 13.0, StatSoft Inc., Tulsa, OK, USA) was used to process the data and expressed as means ± standard error. All analysis was done in triplicates. For fresh aril characterisation, data were subjected one-way analysis of variance (ANOVA) and for dried aril characterisation with different drying methods, data were subjected to two-way ANOVA. Means were separated according to Fisher's LSD test at a level of significance of 95%. The graphs were presented using GraphPad Prism software 4.03 (GraphPad Software, Inc., San Diego, CA,

USA), while the XLSTAT software version 1 April 2012 (Addinsoft, France) was used to estimate Pearson's correlation.

## 3. Results

### 3.1. Effect of Cold Storage on Moisture Content of Pomegranate Aril

The moisture content of fresh pomegranate arils decreased gradually with storage time from 74.7% to 57.4% (Table 1), which affected the weight of the fruit. Pomegranate fruit has been reported to be highly susceptible to weight loss [35], which lead to the visible dehydration observed in Figure 2. The reduced weight observed during storage could be attributed to transpiration through large pores in the fruit peel [4,36]. The reduction in the weight of the whole fruit consequently resulted in a weight reduction of the arils. These findings were corroborated by Fawole and Opara [4], who reported a significant reduction in weight of pomegranate fruit during cold storage.

**Table 1.** Changes in physicochemical attributes of fresh pomegranate arils during 12 weeks of cold storage at 7 ± 0.3 °C, 92 ± 3% RH (wet basis, w.b).

| Storage Period (Weeks) | Moisture Content (%) | TSS (°Brix) | TA (% Citric Acid) | TSS:TA | TCD |
|---|---|---|---|---|---|
| 0 | 74.7 ± 1.25 [a] | 13.7 ± 0.25 [c] | 0.38 ± 0.03 [a] | 36.7 ± 2.01 [c] | - |
| 4 | 71.9 ± 0.92 [a] | 14.4 ± 0.22 [b] | 0.33 ± 0.01 [ab] | 44.2 ± 2.25 [c] | 5.69 ± 1.18 [b] |
| 8 | 67.8 ± 0.73 [b] | 14.8 ± 0.05 [ab] | 0.28 ± 0.01 [bc] | 53.2 ± 2.10 [b] | 4.31 ± 0.77 [b] |
| 12 | 57.4 ± 1.08 [c] | 15.1 ± 0.06 [a] | 0.24 ± 0.01 [c] | 62.5 ± 2.97 [a] | 11.2 ± 1.43 [a] |

TSS, total soluble solids; TA, titratable acidity; TCD, total colour difference. Data presented as means ± SE in each column followed by different letters are significantly different ($p < 0.05$) according to Fisher's LSD.

**Figure 2.** Pomegranate whole fruit (raw material) at harvest (**A**) and during cold storage at 7 ± 0.3 °C, 92 ± 3% RH (w.b) at 4 weeks (**B**), 8 weeks (**C**) and 12 weeks (**D**) storage period. Fresh pomegranate arils show no noticeable differences visible to the naked eye for the period of 12 weeks.

Visual browning of 5% was observed in the arils immediately after peeling the fruit after eight weeks of storage, gradually increasing to 15% at the 12-week storage period. However, differences in the arils over time were unnoticeable in the pictorial representation in Figure 2. A similar study by Konopacka and Plocharski [37] reported increasing tissue browning in apple subjected to long term storage. Conversely, chemical dipping of 'Taify' pomegranate fruit before cold storage showed no browning of the aril tissue [38].

### 3.2. TCD of Fresh and Dried Pomegranate Arils

Storage of pomegranate fruit contributed to the changes in the TCD of fresh arils, and subsequently had a significant effect on the TCD of dried arils. A notable variation was observed in the TCD with increased storage period, with the highest TCD being 11.2 after the 12-week storage period (Table 1). For dried arils processed with hot-air and freeze-dryers, there was a significant ($p < 0.0001$) interaction in TCD (Table 2). Hot-air drying had the least (3.02), while freeze-dried arils had the highest (23.6) TCD after the 12-week storage period (Table 2). A change in TCD is an important attribute of a dried product, expressing the capacity of the human eye to distinguish between various colours attributed to different products [27]. Coklar et al. [39] reported similar findings where hawthorn fruit dried using a freeze dryer had a better colour appearance than fruit dried with oven and microwave dryers. Ali et al. [40] reported that freeze-dried guava fruit preserved its colour the best compared to sunlight and convective oven dryer. The colour change of dried arils could be influenced by the drying method involved and also by the naturally occurring biochemical changes happening during storage of pomegranate fruit.

**Table 2.** Changes in the physicochemical properties of dried pomegranate arils during 12 weeks of cold storage at $7 \pm 0.3$ °C, $92 \pm 3\%$ RH (w.b).

| Drying Method | Storage Period (Weeks) | TCD | TSS (°Brix) | TA (% Citric Acid) | TSS:TA |
|---|---|---|---|---|---|
| Hot-air drying | 0 | - | $22.2 \pm 0.67$ [a] | $3.15 \pm 0.17$ [b] | $7.03 \pm 0.19$ [c] |
| | 4 | $7.15 \pm 0.86$ [b] | $22.7 \pm 0.73$ [a] | $3.23 \pm 0.01$ [a] | $7.00 \pm 0.21$ [c] |
| | 8 | $1.81 \pm 0.71$ [c] | $23.7 \pm 0.44$ [a] | $3.13 \pm 0.00$ [bc] | $7.55 \pm 0.15$ [c] |
| | 12 | $3.02 \pm 1.09$ [bc] | $23.5 \pm 0.58$ [a] | $3.10 \pm 0.02$ [c] | $7.58 \pm 0.22$ [c] |
| Freeze-drying | 0 | - | $17.5 \pm 1.00$ [b] | $1.14 \pm 0.01$ [e] | $15.4 \pm 0.86$ [a] |
| | 4 | $19.6 \pm 2.77$ [a] | $15.0 \pm 0.29$ [c] | $1.20 \pm 0.01$ [d] | $12.5 \pm 0.36$ [b] |
| | 8 | $3.94 \pm 1.32$ [bc] | $14.0 \pm 0.50$ [cd] | $1.24 \pm 0.03$ [d] | $11.3 \pm 0.62$ [b] |
| | 12 | $23.6 \pm 2.55$ [a] | $12.8 \pm 0.33$ [d] | $1.14 \pm 0.01$ [e] | $10.2 \pm 0.36$ [b] |
| Drying method (A) | | 0.0001 | 0.0001 | 0.0001 | 0.0001 |
| Storage period (B) | | 0.0001 | 0.0910 | 0.0001 | 0.0020 |
| A × B | | 0.0001 | 0.0007 | 0.0060 | 0.0002 |

TSS, total soluble solids; TA, titratable acidity; TCD, total colour difference. Data presented as means ± SE in each row followed by different letters are significantly different ($p < 0.05$) according to Fisher's LSD.

### 3.3. Total Soluble Solids (TSS) and Titratable Acidity (TA) of Fresh and Dried Arils

The investigated chemical attributes (TSS and TA) in fresh pomegranate arils were significantly ($p < 0.05$) different from those measured after a period of storage (Table 1). For instance, the TSS of fresh aril increased from 13.7 to 15.1 °Brix after storage (Table 1), while the TA decreased from 0.38 to 0.24 at 12 weeks' storage. In agreement with our study, Arendse et al. [6] reported that pomegranate cultivar Wonderful stored at 5 °C showed an increase in TSS as the storage period progressed. A decrease in TA could be attributed to organic acid break down during the storage period [41]. Fawole and Opara [30] also observed a decrease in TA values for two South African grown cultivars, Bhagwa and Ruby, due to the ongoing metabolism in the fruit during storage.

In dried arils, all chemical attributes showed significant ($p < 0.0001$) interactions with storage period and drying methods (Table 2). Total soluble solids gradually increased with storage period in the hot-air dried arils to almost twice the amount of TSS in freeze-dried arils after storage. The high TSS value could be attributed to drying under high temperature, which resulted in the caramelisation of the product [42].

Throughout the trial, TA was more than double in arils processed with hot-air (3.10–3.15% citric acid) compared to freeze-dried arils (1.14–1.24% citric acid); this could be attributed to the different drying temperatures used (Table 2). Titratable acidity increased after four and eight weeks in hot-air and freeze-dried arils, respectively, before declining with prolonged storage. Ashebir et al. [43] also noted a significant change in the TSS and TA concentrations of dried tomatoes due to variations in the level of drying temperatures.

The TSS:TA ratio is a good indication of flavour and used as one of the quality indexes of pomegranate fruit [44]. Opposite trends of TSS:TA were observed in dried arils after storage—a slight increase from 7.0 to 7.58 in hot-air dried arils and a significant decrease from 15.4 to 10.2 in freeze-dried arils (Table 2). This implies that storage followed by higher temperature drying enhances the caramelisation and Maillard reaction, breaking down the disaccharides into monosaccharides, and seemingly increasing the TSS content in pomegranate. TSS:TA values ranged between in hot-air dried arils and freeze-dried arils. Higher TSS:TA values observed in freeze-dried arils compared to hot-air dried arils reflect a higher percentage of sugar to acid ratio in dried aril.

### 3.4. Total Phenolic Content (TPC) and Total Anthocyanin Content (TAC) of Fresh and Dried Arils

During storage of pomegranate fruit, a steady increase in both TPC (from 20.9 to 23.9 mg GAE/100 mL) and TAC (from 6.91 to 8.77 mg C3gE /100 mL) was observed (Table 3). Arendse et al. [6] reported a similar increase in TPC of pomegranate arils cv. 'Wonderful' stored at 5 °C, 7.5 °C and 10 °C for 5 months. Labbe et al. [45] also reported an increase in the total phenolic content of 'Chilean Chaca' pomegranate cultivar at 5 °C for 12 weeks.

**Table 3.** Changes in the phytochemical properties and antioxidant capacity of fresh pomegranate arils during 12 weeks of cold storage at $7 \pm 0.3$ °C, $92 \pm 3\%$ RH (w.b.).

| Storage Period (Weeks) | TPC mg GAE/100 mL | TAC Cyanidin-3-Glucoside (mg/100 mL) | RSA mM TE/100 mL | FRAP mM TE/100 mL |
|---|---|---|---|---|
| 0 | $20.9 \pm 6.27$ [c] | $6.91 \pm 3.11$ [c] | $12.4 \pm 1.66$ [a] | $2.36 \pm 0.36$ [a] |
| 4 | $22.1 \pm 0.59$ [b] | $7.56 \pm 4.88$ [bc] | $10.4 \pm 1.66$ [b] | $2.27 \pm 0.05$ [a] |
| 8 | $22.9 \pm 0.65$ [ab] | $8.44 \pm 1.62$ [ab] | $8.40 \pm 1.71$ [c] | $2.09 \pm 0.34$ [b] |
| 12 | $23.9 \pm 2.35$ [a] | $8.77 \pm 0.37$ [a] | $4.92 \pm 1.79$ [d] | $2.07 \pm 0.68$ [b] |

RSA, radical scavenging activity; FRAP, ferric reducing antioxidant power; TPC, total phenolic content; TAC, total anthocyanin content; w.b. wet basis; Data presented as means $\pm$ SE in each column followed by different letters are significantly different ($p < 0.05$) according to Fisher's LSD.

Anthocyanin compounds exhibit the main characteristic red colour in pomegranate fruit [35]. Increase in anthocyanin concentration during storage could be related to the increase in biosynthesis and accumulation of anthocyanin, which is induced at lower temperatures in pomegranate fruit [46]. Results from this study agree with those reported by Arendse et al. [6], who attributed an increase in TAC in pomegranate 'Wonderful' to the continued accumulation of anthocyanins at lower temperatures during storage.

After 12 weeks of cold storage, TPC increased, albeit insignificantly, from 105.9 to 116.7 mg GAE/g DM in hot-air dried pomegranate arils, and from 135.6 to 142.7 mg GAE/g DM in freeze-dried arils. Drying methods contributed to the retention of TPC ($p < 0.0001$), as shown in Figure 3. The freeze-drying method retained approximately 18.2% more TPC than hot-air dried arils. This is in support of the study by Shishehgarha et al. [47], who reported that the freeze-drying method is a precision technology utilised to produce high-quality dried products. Additionally, the increased TPC in freeze-dried pomegranate arils could be attributed to mild fruit cell destruction during freezing and ice sublimation, which consequently enhances extraction of biochemical components [14].

The combined effect of drying method ($p < 0.0001$) and storage period ($p < 0.003$) influenced retention of TAC (Figure 4). This figure shows an increase throughout the 12-week storage period of approximately 13 and 5% in TAC of hot-air and freeze-dried pomegranate arils, respectively. A similar trend was observed in fresh arils during cold storage. However, the TAC of freeze-dried arils was higher compared to hot-air dried arils. This is in agreement with other authors who reported higher anthocyanin content in freeze-dried compared to hot-air dried blackberries [13] and blueberries [48].

The vacuum pressure combined with minimal temperature used during the freeze-drying process preserves bioactive compounds from oxidation [13,48].

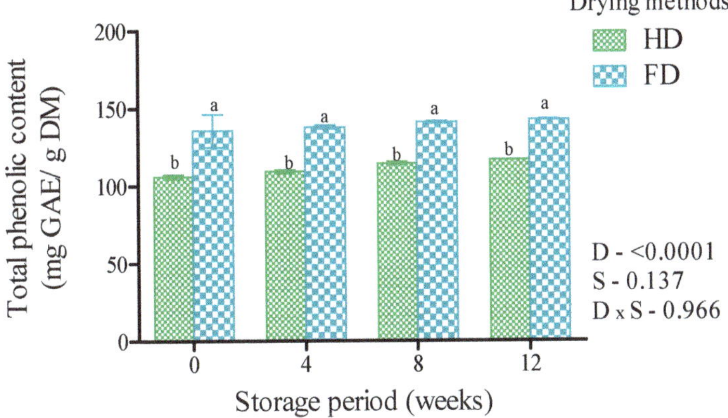

**Figure 3.** Changes in the total phenolic content of pomegranate dried arils during 12 weeks of cold storage at 7 ± 0.3 °C, 92 ± 3% RH (w.b). HD, hot-air drying; FD, freeze-drying. D, drying methods; S, storage period (week). Different letters are significantly different (*p* < 0.05).

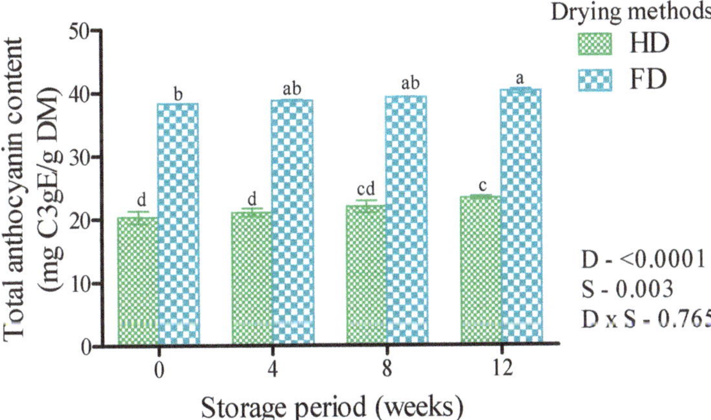

**Figure 4.** Changes in the total anthocyanin content of pomegranate dried arils during 12 weeks of cold storage at 7 ± 0.3 °C, 92 ± 3% RH (w.b). HD, hot-air drying; FD, freeze-drying. D, drying methods; S, storage period (week). Different letters are significantly different (*p* < 0.05).

### 3.5. Antioxidant Capacity of Fresh and Dried Arils

The antioxidant capacity of fresh pomegranate arils decreased significantly with storage time from 12.4 to 4.92 mM TE/ 100 mL RSA and 2.36 to 2.07 mM TE/ 100 mL FRAP (Table 3). In relation to their nutritional benefits, phenolic compounds in fruit contribute to the total antioxidant capacity and its subsequent human health benefits [49]. The observed increase in both TPC and TAC was inversely related to the antioxidant capacity (RSA and FRAP) exhibited by pomegranate fruit during storage at 7 ± 0.3 °C, with 92 ± 3% RH. This suggested that antioxidants often react differently depending on the type of antioxidant assay [50]. Siddhuraju et al. [51] reported that a decrease in reducing power could be attributed to the bioactive compounds—total phenolics, flavonoids, ascorbic acids and other hydrophilic antioxidants—associated with the component of the antioxidants present in the fruit.

For dried pomegranate arils, there were significant interactions on the antioxidant capacity (RSA, $p < 0.023$; FRAP, $p < 0.0001$) (Figure 5). The trend showed a general decrease in RSA and FRAP for both hot-air and freeze-dried arils after storage. The FRAP of freeze-dried arils was unchanged at 4 weeks also followed by gradual decline with the storage period (Figure 5) During the storage period, the RSA (26.9 and 29.5 mM TE/g DM) of hot-air dried arils was close to the values reported previously for hot-air dried pomegranate (22.7 to 30.6 mM TE/g) [52] and higher than papaya (9.72 mM TE/g) [53]. Our study showed similar FRAP (2.49–3.27 mM TE/g DM) in freeze-dried arils to those reported for pomegranate cv. Mollar de Elche (3.4 mM TE/g) [54].

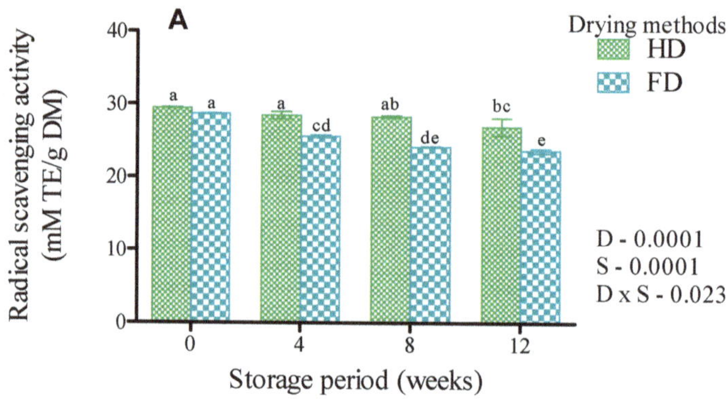


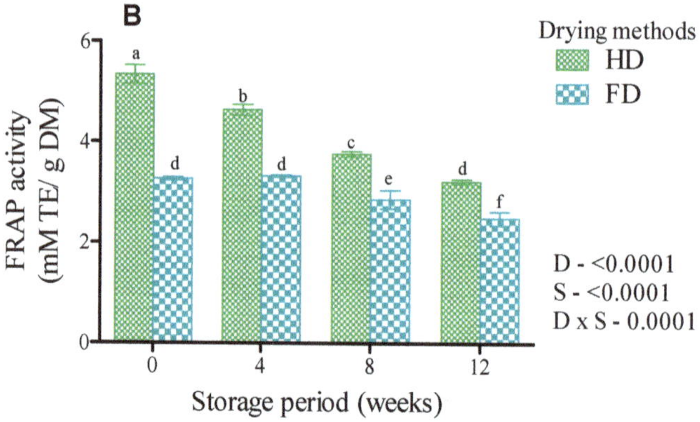


**Figure 5.** Changes in the antioxidant capacity (**A**) RSA and (**B**) FRAP activity of pomegranate dried arils during 12 weeks of cold storage at $7 \pm 0.3$ °C, $92 \pm 3\%$ RH (w.b). HD, hot-air drying; FD, freeze-drying. D, drying methods; S, storage period (week). Different letters are significantly different ($p < 0.05$).

The decrease in the antioxidant capacity of dried arils could be attributed not only to the same observed decrease in the antioxidant capacity in fresh arils, but also to the thermal degradation of heat-sensitive phenolics—since TPC is reported to be the major contributors to antioxidant capacity. Additionally, Moser et al. [55] reported up to 25% reduction in antioxidant capacity in grape powder after 45 days of storage due to the formation of antioxidant polymers, such as low molecular weight procyanidins. This explanation was also supported by Mrad et al. [56]. However, Michalczyk et al. [57] reported that the antioxidant capacity of dried berries was retained during prolonged storage to a remarkably high degree, which is in contrast with the results from this present study.

Furthermore, freeze-drying amounted to approximately 12.1 and 22.9% lower antioxidant capacity (RSA and FRAP) after the storage period compared to the hot-air dried arils, respectively (Figure 5).

Fracassetti et al. observed a similar decline in antioxidant activity while studying the storage of freeze-dried wild blueberry powder [58]. Mphahlele et al. [2] also reported better retention of antioxidants in the oven drying at higher temperatures 60 °C than in freeze-dried pomegranate peel. The authors attributed this to the concentration of compounds contained in the peel, as these are considered scavengers of free radicals produced during oxidation.

### 3.6. Stability of Antioxidant Capacity (RSA and FRAP) of Dried Pomegranate Arils

Understanding the stability or degradation mechanisms of food products is essential to maximise the nutritional and sensory quality of products [55]. The stability of antioxidant capacity (RSA and FRAP) of pomegranate arils after hot-air and freeze-drying were evaluated based on changes in their concentrations (Table 4). This table further shows the kinetic parameters (kinetic rate constants and the half-life values) determined for the thermal degradation of the antioxidant capacity. A lower degradation rate indicates lower kinetic rate constants ($k$) and higher half-life [55].

**Table 4.** Effect of drying methods on the kinetic parameters of antioxidants (RSA and FRAP) degradation in dried pomegranate arils.

| Drying Methods | Antioxidant (mM TE/g) | $k \times 10^{-3}/(\text{Week}^{-1})$ | $t_{1/2}$/Week | $R^2$ |
|---|---|---|---|---|
| Hot-air | RSA | 0.151 | 5.654 | 0.9949 |
| | FRAP | 0.129 | 7.306 | 0.9949 |
| Freeze-drying | RSA | 0.146 | 5.844 | 0.9031 |
| | FRAP | 0.143 | 6.597 | 0.8582 |

$k$, kinetic rate constants; $t_{1/2}$; half-life; $R^2$, coefficients of determination; RSA, radical-scavenging activity; FRAP, ferric ion reducing antioxidant power.

The RSA activity in freeze-dried arils had a lower degradation rate ($k = 0.146$; $t_{1/2} = 5.844$) than hot-air dried arils ($k = 0.151$; $t_{1/2} = 5.654$); however, the FRAP activity in hot-air dried arils had a lower degradation rate ($k = 0.129$; $t_{1/2} = 7.306$) than the freeze-dried arils ($k = 0.143$; $t_{1/2} = 6.597$). Considering the calculated degradation $k$ and $t_{1/2}$ as an indicator of the amount of antioxidant loss, with a half-life ($t_{1/2}$/week), the stability of RSA in the hot-air dried arils was approximately 3.3% lower than freeze-dried arils. However, the stability in FRAP activity in freeze-dried arils was 9.7% lower than the hot-air dried arils (Table 4). Several researchers have reported a decrease in the bioactive compounds in fruit after drying [19,59,60]. Zhou et al. [61] reported high degradation in antioxidant capacity (DPPH, FRAP and ABTS) of red pepper.

Similarly, Garau et al. [62] also found that air-drying decreased the antioxidant capacity in orange fruit matrix. This is consistent with the results of this study. Moreover, the values of coefficients of determination ($R^2$) ranging from 0.85–0.99 were obtained for all linear regressions, indicating that the degradation process of these bioactive compounds for both hot-air and freeze-drying methods followed first-order reaction kinetics.

### 3.7. Correlations among Quality Attributes for Dried Arils at 12 Weeks of Cold Storage

Significant relationships that exist among attributes measured for dried arils are presented in Table 5.

Pearson's correlation tests indicated a strong positive relationship between TPC and TAC ($r = 0.998$) (Table 5). Additionally, there were strong negative correlations between TPC and RSA ($r = -0.894$) as well as TPC and FRAP values ($r = -0.998$); TAC and RSA ($r = -0.910$) as well as TAC and FRAP ($r = -1.000$). However, a positive correlation was found between RSA and FRAP ($r = 0.919$) (Table 5). A similar result was reported by Cano-Lamadrid et al. [54] between antioxidant ABTS and FRAP in osmotically dehydrated pomegranate arils cv. Mollar de Elche. Strong correlations were found between TSS and phytochemical properties (TPC and TAC), but none of the relationships seems to be applicable in practice. For instance, a strong correlation ($r = 0.937$) was found between TSS and TAC

(Table 5). In practice, no relevant prediction of dried aril flavour could be made using total anthocyanin content since soluble solids measurement technique applies only to the sweetness (°Brix) of aril tissues. However, a moderately negative correlation was observed between TSS and TA ($r = -0.555$) (Table 5). This relationship clearly showed that the increase in total soluble solids of dried arils could also contribute to a decrease in TA. Other relationships found a moderately negative correlation between TA and TSS:TA (Table 5).

**Table 5.** Pearson's correlation coefficients among the investigated parameters of dried pomegranate arils during the 12-week storage period.

| Variables | TCD | TSS | TA | TSS:TA | TPC | TAC | FRAP | RSA |
|---|---|---|---|---|---|---|---|---|
| TCD | 1 | | | | | | | |
| TSS | 0.052 * | 1 | | | | | | |
| TA | 0.687 ** | −0.555 * | 1 | | | | | |
| TSS:TA | −0.237 ns | 0.941 ** | −0.804 ** | 1 | | | | |
| TPC | 0.067 ns | 0.946 ** | −0.649 ** | 0.944 ** | 1 | | | |
| TAC | 0.122 ns | 0.937 ** | −0.612 ** | 0.922 ** | 0.998 ** | 1 | | |
| FRAP | −0.132 ns | −0.944 ** | 0.599 * | −0.922 ** | −0.998 ** | −1.000 ** | 1 | |
| RSA | 0.430 ns | −0.905 ** | 0.244 ns | −0.749 ** | −0.894 ** | −0.910 ** | 0.919 ** | 1 |

95% confidence interval. TPC, total phenolic content; TAC, total anthocyanin content; RSA, radical scavenging activity; FRAP, ferric reducing antioxidant power; TCD, total colour difference; TSS, total soluble solids; TA, titratable acidity. ns; non-significant, $* = p < 0.05$ and $** = p < 0.001$ (two-tailed).

In consideration of the noted benefits of consuming fruit with high phytochemical properties, it is therefore not surprising that antioxidant capacity (RSA and FRAP) showed a strong positive correlation with TPC. Therefore, the overall quality of dried arils investigated showed that only the interactions among the bioactive components seem promising and practicable.

## 4. Conclusions

In practice, pomegranate fruit is stored for batch processing. During this time, quality attributes degrade, and hence, the quality of dried pomegranate arils. This study has established the effect of long-term storage of whole fruit on the quality of the final products using hot-air and freeze-drying methods. Prolonged cold storage of raw material considerably affected the total soluble solids and titratable acidity of hot-air and freeze-dried pomegranate arils. The TSS of fresh arils increased while TA decreased with storage period. Freeze-dried aril had a significantly higher total colour difference (TCD) than hot-air dried arils after the 12-week storage period. Hot-air dried arils presented the highest TSS and TA compared to freeze-drying after the storage period. A steady increase in the total phenolic content (TPC) and total anthocyanin content (TAC) of both fresh and dried arils was also observed. Cold storage negatively affected the antioxidant activity (RSA and FRAP) in both fresh and dried arils. At the end of the storage period, freeze-drying presented higher stability of antioxidant capacity (RSA) than hot-air drying.

In contrast, hot-air drying showed higher stability of antioxidant capacity (FRAP) with the highest half lifetime, suggesting that the preservation of antioxidant capacity in dried arils is dependent on the type of assay and choice of drying method. Due to the significantly broad total colour difference (TCD) in fresh fruit after the 12-week storage period, as well as the decline in antioxidant capacity in both the raw material and dried arils in the same period, this study suggests processing fresh pomegranate fruit between harvest and eight-week storage duration. Additionally, based on the importance of colour in marketability and consumer preference, the freeze-drying method is recommended.

**Author Contributions:** The following authors contributed to the work: conceptualisation, O.A.F. and U.L.O.; methodology, A.O.A. and O.A.F.; software, A.O.A.; formal analysis, A.O.A.; investigation, A.O.A.; resources, O.A.F. and U.L.O.; writing—original draft preparation, A.O.A.; writing—review and editing, O.A.F. and U.L.O.; supervision, O.A.F. and U.L.O.; project administration, O.A.F.; funding acquisition, O.A.F. and U.L.O. All authors have read and agreed to the published version of the manuscript.

**Funding:** This work is based on the research supported by the National Research Foundation (NRF) of South Africa (IUD: 64813 and (IUD: 105722). The financial support of the World Academy of Science and National Research Foundation of South Africa through the award of doctoral scholarship to Mr Adetoro is gratefully acknowledged (IUD: 105483). The opinions, findings and conclusions or recommendations expressed are those of the author(s) alone, and the NRF accepts no liability whatsoever in this regard.

**Conflicts of Interest:** The authors declare no conflict of interest.

## References

1. Fawole, O.A.; Opara, U.L.; Theron, K.I. Chemical and phytochemical properties and antioxidant activities of three pomegranate cultivars grown in South Africa. *Food Bioproc. Technol.* **2012**, *5*, 2934–2940. [CrossRef]

2. Mphahlele, R.R.; Caleb, O.J.; Fawole, O.A.; Opara, U.L. Effects of different maturity stages and growing locations on changes in chemical, biochemical and aroma volatile composition of 'Wonderful' pomegranate juice. *J. Sci. Food Agric.* **2016**, *96*, 1002–1009. [CrossRef] [PubMed]

3. Opara, U.L.; Al-Ani, M.R.; Al-Shuaibi, Y.S. Physico-chemical properties, vitamin C content and antimicrobial properties of pomegranate fruit (*Punica granatum* L.). *Food Bioproc. Technol.* **2009**, *2*, 315–321. [CrossRef]

4. Fawole, O.A.; Opara, U.L. Effects of storage temperature and duration on physiological responses of pomegranate fruit. *Ind. Crop. Prod.* **2013**, *47*, 300–309. [CrossRef]

5. Pomegranate Association of South Africa (POMASA). Pomegranate Industry Overview. 2018. Available online: https://www.sapomegranate.co.za/statistics-and-information/pomegranate-industry-overview/ (accessed on 26 October 2019).

6. Arendse, E.; Fawole, O.A.; Opara, U.L. Influence of storage temperature and duration on postharvest physico-chemical and mechanical properties of pomegranate fruit and arils. *J. Food Sci.* **2014**, *12*, 389–398. [CrossRef]

7. Belay, Z.A.; Caleb, O.J.; Opara, U.L. Impacts of low and super-atmospheric oxygen concentrations on quality attributes, phytonutrient content and volatile compounds of minimally processed pomegranate arils (cv. Wonderful). *Postharvest Biol. Technol.* **2017**, *124*, 119–127. [CrossRef]

8. Caleb, O.J.; Opara, U.L.; Mahajan, P.V.; Manley, M.; Mokwena, L.; Tredoux, A.G.J. Effect of modified atmosphere packaging and storage temperature on volatile composition and postharvest life of pomegranate arils (cv. 'Acco' and 'Herskawitz'). *Postharvest Biol. Technol.* **2013**, *79*, 54–61. [CrossRef]

9. Kingsly, A.R.P.; Singh, D.B. Drying kinetics of pomegranate arils. *J. Food Eng.* **2007**, *79*, 741–744. [CrossRef]

10. Jalikop, S.H.; Tiwari, R.B.; Kumar, S. Amlidana: A new pomegranate hybrid. *Indian J. Hortic.* **2002**, *21*, 22–23.

11. Singh, D.B.; Kinglsley, A.R.P. Effect of convective drying on quality of Anardana. *Indian J. Hortic.* **2008**, *65*, 413–416.

12. Sharma, A.; Thakur, N.S. Effect of different packaging treatments on some chemical constituents of Anardana. *Int. J. Farm Sci.* **2016**, *6*, 64–69.

13. Wu, R.; Frei, B.; Kennedy, J.A.; Zhao, Y. Effects of refrigerated storage and processing technologies on the bioactive compounds and antioxidant capacities of 'Marion' and 'Evergreen' blackberries. *LWT Food Sci. Technol.* **2010**, *43*, 1253–1264. [CrossRef]

14. Asami, D.K.; Hong, Y.; Barrett, D.M.; Mitchell, A.E. Comparison of the total phenolic and ascorbic acid content of freeze-dried and air-dried Marion berry, strawberry, and corn grown using conventional, organic, and sustainable agricultural practices. *J. Agric. Food Chem.* **2003**, *51*, 1237–1241. [CrossRef] [PubMed]

15. Shofian, N.M.; Hamid, A.A.; Osman, A.; Saari, N.; Anwar, F.; Pak Dek, M.S.; Hairuddin, M.R. Effect of freeze-drying on the antioxidant compounds and antioxidant activity of selected tropical fruits. *Int. J. Mol. Sci.* **2011**, *12*, 4678–4692. [CrossRef]

16. Ratti, C. Hot air and freeze-drying of high-value foods: A review. *J. Food Eng.* **2001**, *49*, 311–319. [CrossRef]

17. Vega-Galvez, A.; Scala, K.D.; Rodrıguez, K.; Lemus-Mondaca, R.; Miranda, M.; Lopez, J.; Perez-Won, M. Effect of air- drying temperature on physico-chemical properties, antioxidant capacity, colour and total phenolic content of red pepper (*Capsicum annuum*, L. var. Hungarian). *Food Chem.* **2009**, *117*, 647–653. [CrossRef]

18. Lewicki, P.P.; Jakubczyk, E. Effect of hot air temperature on mechanical properties of dried apples. *J. Food Eng.* **2004**, *64*, 307–314. [CrossRef]

19. Holland, D.; Hatib, K.; Bar-Ya'akov, I. Pomegranate: Botany, Horticulture, Breeding. *Hort. Rev.* **2009**, *35*, 127–191.

20. Fawole, O.A.; Atukuri, J.; Arendse, E.; Opara, U.O. Postharvest physiological responses of pomegranate fruit (cv. Wonderful) to exogenous putrescine treatment and effects on physico-chemical and phytochemical properties. *Food Sci. Hum. Well.* **2020**, *9*, 146–161. [CrossRef]

21. Pomegranate Association of South Africa (POMASA). Statistics and Information. 2019. Available online: https://www.sapomegranate.co.za/focus-areas/statistics-and-information-2019/ (accessed on 18 February 2020).

22. Pomegranate Association of South Africa (POMASA). Pomegranate Industry Overview. 2016. Available online: https://www.sapomegranate.co.za/wp-content/uploads/2017/08/Pomegranate-Industry-Overview-2016_USE-Repaired.pdf (accessed on 26 October 2019).

23. Arendse, E. Determining Optimum Storage Conditions for Pomegranate Fruit (cv. Wonderful). Ph.D. Thesis, Stellenbosch University, Stellenbosch, South Africa, 2014.

24. Fawole, O.A.; Opara, U.L. Physicomechanical, phytochemical, volatile compounds and free radical scavenging properties of eight pomegranate cultivars and classification by principal component and cluster analyses. *Br. Food J.* **2014**, *116*, 544–567. [CrossRef]

25. Rickman, J.C.; Barrett, D.M.; Bruhn, C.M. Review: Nutritional comparison of fresh, frozen and canned fruits and vegetables. Part 1. Vitamins C and B and phenolic compounds. *J. Sci. Food Agric.* **2007**, *87*, 930–944. [CrossRef]

26. Wojdyło, A.; Figiel, A.; Legua, P.; Lech, K.; Carbonell-Barrachina, Á.A.; Hernández, F. Chemical composition, antioxidant capacity, and sensory quality of dried jujube fruits as affected by cultivar and drying method. *Food Chem.* **2016**, *207*, 170–179. [CrossRef]

27. Zhang, M.; Hettiarachchy, N.S.; Horax, R.; Chen, P.; Over, K.F. Effect of maturity stages and drying methods on the retention of selected nutrients and phytochemicals in bitter melon (*Momordica charantia*) leaf. *J. Food Sci.* **2009**, *74*, 441–448. [CrossRef] [PubMed]

28. Beaudry, C.; Raghavan, G.S.V.; Ratti, C.; Rennie, T.J. Effect of four drying methods on the quality of osmotically dehydrated cranberries. *Dry. Technol.* **2004**, *22*, 521–539. [CrossRef]

29. Wrolstad, R.E. Colour and pigment analyses in fruit products. In *Agricultural Experiment Station*; Oregon State University Station Bulletin: Corvallis, OR, USA, 1993; Volume 624.

30. Fawole, O.A.; Opara, U.L. Changes in physical properties, chemical and elemental composition and antioxidant capacity of pomegranate (cv. 'Ruby') fruit at five maturity stages. *Sci. Hortic.* **2013**, *150*, 37–46. [CrossRef]

31. Pathare, P.B.; Opara, U.L.; Al-Said, F.A.J. Colour measurement and analysis in fresh and processed foods: A review. *Food Bioproc. Technol.* **2013**, *6*, 36–60. [CrossRef]

32. Benzie, I.F.F.; Strain, J.J. The ferric reducing ability of plasma (FRAP) as a measure of "antioxidant power". The FRAP assay. *Anal. Biochem.* **1996**, *239*, 70–76. [CrossRef]

33. Li, N.; Taylor, L.S.; Ferruzzi, M.G.; Mauer, L.J. Kinetic study of catechin stability: Effects of pH, concentration and temperature. *J. Agric. Food Chem.* **2012**, *60*, 12531–12539. [CrossRef]

34. Moldovan, B.; David, L.; Popa, A. Effects of storage temperature on the total phenolic content of Cornelian Cherry (*Cornus mas* L.) fruits extracts. *JABFQ* **2016**, *89*, 208–211.

35. Artés, F.; Marín, J.G.; Martínez, J.A. Controlled atmosphere storage of pomegranate. *Z. Lebensm. Unters. Forsch.* **1996**, *203*, 33–37. [CrossRef]

36. Arendse, E.; Fawole, O.A.; Opara, U.L. Effects of postharvest handling and storage on physiological attributes and quality of pomegranate fruit (*Punica granatum* L.): A review. *Int. J. Postharvest Technol. Innov.* **2015**, *5*, 13–31. [CrossRef]

37. Konopacka, D.; Plocharski, W.J. Effect of raw material storage time on the quality of apple chips. *Dry. Technol.* **2001**, *19*, 559–570. [CrossRef]

38. Awad, M.A.; Al-Qurashi, A.D.; Elsayed, M.I. Effect of pre-storage salicylic acid and oxalic acid dipping on chilling injury and quality of 'Taify' pomegranates during cold storage. *J. Food Agric. Environ.* **2013**, *11*, 117–122.

39. Coklar, H.; Akbulut, M.; Kilinc, S.; Yildirim, A.; Alhassan, I. Effect of freeze, oven and microwave pretreated oven drying on color, browning index, phenolic compounds and antioxidant activity of hawthorn (*Crataegus orientalis*) fruit. *Not. Bot. Horti. Agrobo.* **2018**, *46*, 449–456. [CrossRef]

40. Ali, M.A.; Yusof, Y.A.; Chin, N.L.; Ibrahim, M.N. Effect of different drying treatments on colour quality and ascorbic acid concentration of guava fruit. *Int. Food Res. J.* **2016**, *23*, S155–S161.

41. Kader, A.A.; Chordas, A.; Elyatem, S.M. Responses of pomegranates to ethylene treatment and storage temperature. *Calif. Agric.* **1984**, *38*, 4–15.

42. Vanhal, I.; Blond, G. Impact of melting conditions of sucrose on its glass transition temperature. *J. Agric. Food Chem.* **1999**, *47*, 4285–4290. [CrossRef]

43. Ashebir, D.; Jezik, K.; Weingartemann, H.; Gretzmacher, R. Change in colour and other fruit quality characteristics of tomato cultivars after hot-air drying at low final-moisture content. *Int. J. Food Sci. Nutr.* **2009**, *60*, 308–315. [CrossRef]

44. Al-Said, F.A.; Opara, U.L.; Al-Yahyai, R.A. Physico-chemical and textural quality attributes of pomegranate cultivars (*Punica granatum* L.) grown in the Sultanate of Oman. *J. Food Eng.* **2009**, *90*, 129–134. [CrossRef]

45. Labbe, M.; Peria, A.; Saenz, C. Antioxidant capacity and phenolic composition of juices from pomegranates stored in refrigeration. In Proceedings of the International Conference on Food innovation, Valencia, Spain, 25–29 October 2010.

46. Miguel, M.G.; Fontes, C.; Antunes, D.; Neves, A.; Martins, D. Anthocyanin concentration of 'Assaria' pomegranate fruits during different cold storage conditions. *J. Biomed. Biotechnol.* **2004**, *5*, 338–342. [CrossRef]

47. Shishehgarha, F.; Makhlouf, J.; Ratti, C. Freeze-drying characteristics of strawberries. *Dry. Technol.* **2002**, *20*, 131–145. [CrossRef]

48. Mejia-Meza, E.I.; Yanez, J.A.; Davies, N.M.; Rasco, B.; Younce, F.; Remsberg, C.M.; Clary, C. Improving nutritional value of dried blueberries (*Vaccinium corymbosum* L.) combining microwave-vacuum, hot-air drying and freeze drying technologies. *Int. J. Food Eng.* **2008**, *4*. [CrossRef]

49. Tzulker, R.; Glazer, I.; Bar-Ilan, I.; Holland, D.; Aviram, M.; Amir, R. Antioxidant activity, polyphenol content, and related compounds in different fruit juices and homogenates prepared from 29 different pomegranate accessions. *J. Sci. Food Agric.* **2007**, *55*, 9559–9570. [CrossRef]

50. Çam, M.; Hisil, Y.; Durmaz, G. Classification of eight pomegranate juices based on antioxidant capacity measured by four methods. *J. Food Chem.* **2009**, *112*, 721–726. [CrossRef]

51. Siddhuraju, P.; Mohan, P.S.; Becker, K. Studies on the antioxidant activity of Indian laburnum (*Cassia fistula* L.): A preliminary assessment of crude extracts from stem bark, leaves, flowers and fruit pulp. *Food Chem.* **2002**, *79*, 61–67. [CrossRef]

52. Golukcu, M. The effects of drying methods, packaging atmosphere and storage time on dried pomegranate aril quality. *J. Agric. Sci.* **2014**, *21*, 207–219.

53. Chong, C.H.; Law, C.L.; Figiel, A.; Wojdyło, A.; Oziembłowski, M. Colour, phenolic content and antioxidant capacity of some fruits dehydrated by a combination of different methods. *J. Food Chem.* **2013**, *141*, 3889–3896. [CrossRef] [PubMed]

54. Cano-Lamadrid, M.; Lech, K.; Michalska, A.; Wasilewska, M.; Figiel, A.; Wojdyło, A.; Carbonell-Barrachina, Á.A. Influence of osmotic dehydration pre-treatment and combined drying method on physico-chemical and sensory properties of pomegranate arils, cultivar Mollar de Elche. *J. Food Chem.* **2017**, *232*, 306–315. [CrossRef]

55. Moser, P.; Telis, V.R.N.; de Andrade Neves, N.; García-Romero, E.; Gómez-Alonso, S.; Hermosín-Gutiérrez, I. Storage stability of phenolic compounds in powdered BRS Violeta grape juice microencapsulated with protein and maltodextrin blends. *Food Chem.* **2017**, *214*, 308–318. [CrossRef]

56. Mrad, N.D.; Boudhrioua, N.; Kechaou, N.; Courtois, F.; Bonazzi, C. Influence of air-drying temperature on kinetics, physicochemical properties, total phenolic content and ascorbic acid of pears. *Food Bioprod. Process.* **2012**, *90*, 433–441. [CrossRef]

57. Michalczyk, M.; Macura, R.; Matuszak, I. The effect of air-drying, freeze-drying and storage on the quality and antioxidant activity of some selected berries. *J. Food Process. Pres.* **2009**, *33*, 11–21. [CrossRef]

58. Fracassetti, D.; Del Bo', C.; Simonetti, P.; Gardana, C.; Klimis-Zacas, D.; Ciappellano, S. Effect of time and storage temperature on anthocyanin decay and antioxidant activity in wild blueberry (*Vaccinium angustifolium*) powder. *J. Agric. Food Chem.* **2013**, *61*, 2999–3005. [CrossRef] [PubMed]

59. Di Scala, K.; Crapiste, G. Drying kinetics and quality changes during drying of red pepper. *LWT Food Sci. Technol.* **2008**, *41*, 789–795. [CrossRef]

60. Devic, E.; Guyot, S.; Daudin, J.D.; Bonazzi, C. Kinetics of polyphenol losses during soaking and drying of cider apples. *Food Bioproc. Technol.* **2010**, *3*, 867–877. [CrossRef]

61. Zhou, L.; Cao, Z.; Bi, J.; Yi, J.; Chen, Q.; Wu, X.; Zhou, M. Degradation kinetics of total phenolic compounds, capsaicinoids and antioxidant activity in red pepper during hot air and infrared drying process. *Int. J. Food Sci. Technol.* **2016**, *51*, 842–853. [CrossRef]

62. Garau, M.C.; Simal, S.; Rossello, C.; Femenia, A. Effect of air-drying temperature on physico-chemical properties of dietary fibre and antioxidant capacity of orange (*Citrus aurantium* v. Canoneta) by-products. *Food Chem.* **2007**, *104*, 1014–1024. [CrossRef]


**Publisher's Note:** MDPI stays neutral with regard to jurisdictional claims in published maps and institutional affiliations.

*Review*

# Mango Fruit Processing: Options for Small-Scale Processors in Developing Countries

Willis O. Owino [1,*] and Jane L. Ambuko [2]

[1] Department of Food Science and Technology, Jomo Kenyatta University of Agriculture and Technology, Nairobi P.O. Box 62000-00200, Kenya

[2] Department of Plant Science and Crop Protection, University of Nairobi, Kangemi P.O. Box 29053-00625, Kenya; jane.ambuko@uonbi.ac.ke

* Correspondence: willis@agr.jkuat.ac.ke

**Abstract:** Postharvest losses of mango fruit in a number of developing countries in Africa and Asia have been estimated to be as high as over 50%, especially during the main harvest season. Micro, small, and medium scale food processing enterprises play an important economic role in developing economies in processing of a diversity of healthy food products as a sustainable way to reduce postharvest losses and food waste, extend shelf life of food, boost food security, and contribute to national gross domestic product. Processing of mango fruit into the diverse shelf-stable products makes the seasonal fruit conveniently available to consumers all year round. Over the years, research and food product development have contributed substantially to a number of unique and diverse processed mango products with specific qualities and nutritional attributes that are in demand by a wide array of consumers. These mango products are derived from appropriate food processing and value-addition technologies that transform fresh mango into shelf-stable products with ideal organoleptic, nutritional, and other quality attributes. Some of the common processed products from mango fruit include pulp (puree), juice concentrate, ready-to-drink juice, nectar, wine, jams, jellies, pickles, smoothies, chutney, canned slices, chips, leathers, and powder. Minimum processing of mango fruit as fresh-cut product has also gained importance among health-conscious consumers. Apart from the primary products from mango fruit, mango pulp or powder can be used to enrich or flavor secondary products such as yoghurt, ice cream, beverages, and soft drinks. Byproducts of mango processing, such as the peel and kernel, have been shown to be rich in bioactive compounds including carotenoids, polyphenols, and dietary fibers. These byproducts of mango processing can be used in food fortification and manufacture of animal feeds, thereby gaining greater value from the fruit while reducing wastage. This review focuses on the current trends in processing and value addition of mango applicable to small-scale processors in developing countries.

**Keywords:** postharvest loss; shelf stable; nutrition; bioactive; byproducts

**Citation:** Owino, W.O.; Ambuko, J.L. Mango Fruit Processing: Options for Small-Scale Processors in Developing Countries. *Agriculture* **2021**, *11*, 1105. https://doi.org/10.3390/agriculture 11111105

Academic Editor: Massimiliano Renna

Received: 20 August 2021
Accepted: 21 October 2021
Published: 6 November 2021



## 1. Introduction

Food processing is one of the strategic sectors where developing countries can use their natural base in agriculture to reach the next level of economic development [1]. Food processing in developing countries was, until one or two decades ago, dominated by multinational companies headquartered in advanced economies. However, economic liberalization increased the competitiveness of the structure of the food industry, thereby contributing to more rapid food product and process innovations [2]. In developing countries, population increase, rapid urbanization, rise of the middle class and changing food habits led to a gradual increase in demand for processed, nutritious and healthy food products This has in turn contributed to the rise of micro, small, and medium scale food processing enterprises (MSMSFPE) that process a diversity of healthy and nutritious food products as a sustainable way to reduce postharvest losses and food waste, extend shelf life of food, boost food security, and contribute to national employment and national gross domestic product. The small-scale nature of these food processing enterprises and low level of bureaucracy enables them to rapidly make strategic decisions to respond to demand or change in the local market. These MSMSFPE are however plagued with a number of both upstream and downstream supply chain challenges such as; poor road network especially in rural areas which increase the cost of sourcing of raw materials or distribution of processed products; in a number of developing countries, the food processing sector is still informal thus contributing to inefficiencies in the food value chain that lead to high retail cost of processed products; duplicity and overlaps in laws and regulations governing the food processing sector; confinement of the market of processed food products in urban areas; high cost of food processing equipment; high cost of energy, credit and taxation. However, these enterprises hold potential to economic development in the developing countries if made more competitive, through increased government initiatives, interventions and conducive regulatory and taxation policies, adoption of novel food products and processing innovations, with more stringent quality and safety management systems. Only then will they be able to ward off the challenge of competitive imports from more advanced economies.

One of the most important fruits with a greater potential for food processing in some of the developing countries is the mango. Mango fruit is the second most traded tropical fruit globally and ranks seventh in terms of production [3]. Mango, "also referred to as the 'king of fruits'", is a major fruit of the tropics and subtropics. Although the fruit is mainly consumed in its fresh state, mango can be processed into many nutritious and shelf-stable products. Mango production postharvest losses in developing countries such as Kenya have been estimated to be as high as over 50%, especially during the main harvest season [4]. Other countries such as Rwanda, India, Benin, and Ghana reported mango postharvest losses in the range of 30–80% during harvesting, packing, and distribution in retail and wholesale markets [5]. Processing of mango fruit into diverse shelf-stable products makes the seasonal fruit conveniently available to consumers all year round. Some of the common processed products from mango fruit are derived from the pulp. Apart from the primary products from mango pulp, derivatives of mango pulp can be used to enrich or flavor secondary products such as yoghurt, ice cream, beverages, and soft drinks. Byproducts of mango processing such as the peel and kernel have been shown to be rich in bioactive compounds including carotenoids, polyphenols, and dietary fibers. The byproducts of mango processing can be used in food fortification and manufacture of feeds, thereby gaining greater value from the fruit while reducing wastage. Although mango is amenable to processing into all these products, smallholder farmers and processors in developing countries have not fully exploited this potential. Over the years, research and food product development have contributed substantially to a number of unique and diverse processed mango products with specific qualities and nutritional attributes that are in demand by a wide array of consumers. These mango products are derived from appropriate food processing and value-addition technologies that transform fresh mango into shelf-stable products with ideal organoleptic, nutritional, and other quality attributes. The status of processing technologies and products from mango has been reviewed in the recent past by DeeptiSalvi and Karwe [6], Evans et al. [7], and Siddiq et al. [8]. This review focuses on the current trends in processing and value addition of mango applicable to micro, small and medium food processing enterprises in developing countries.

The mango products of interest described in this mini review include the following as illustrated in Figure 1.

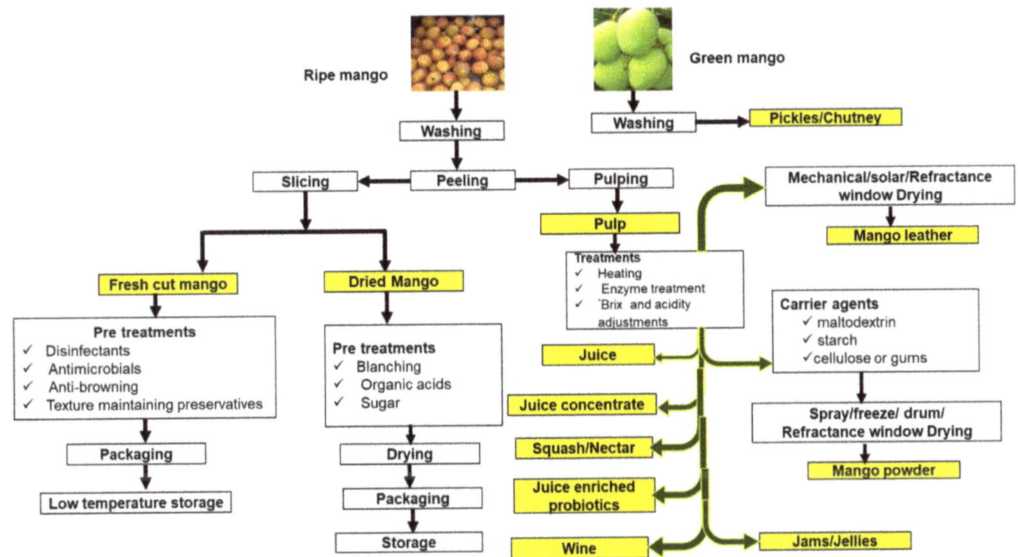


**Figure 1.** Processing of different products derived from mango. (Modified from [8,9]).

## 2. Fresh-Cut Mango (FCM)

FCM is among the minimally processed fruits and vegetables with increased market demand within ready-to-eat fresh fruit products [10,11]. In general, the factors that are fundamental to the quality of FCM include quality of intact mango, mango cultivar, preharvest agronomic practices, harvest maturity, postharvest handling procedures, interval between harvest and processing of the FCM, and the preparation methods, i.e., sharp cutting tools, size and surface area of the slices, washing and removal of surface moisture [12,13]. Nevertheless, peeling and cutting operations involved in processing FCM eliminate the protective pericarp and stimulate the physiological and biochemical activities that predispose the product to dehydration, accelerated tissue softening, and surface browning. Hence, there is a much higher rate of deterioration compared to intact fruit. As a consequence, even with preservation treatments to extend their shelf life, FCM have a consumption window of just a few days. To assure fresh-like quality and extend the shelf life of FCM, currently a combination of treatments and preservation methods are utilized. The dip pretreatments incorporate disinfectants, antimicrobials, antibrowning, and texture-maintaining preservatives [8,14].

The most common disinfectants with antimicrobial activity for FCM are sodium hypochlorite (NaOCl), and calcium hypochlorite (CaCl$_2$O$_2$). The recommended dose of chlorine ranges between 50 and 200 ppm, pH 6.0–7.5, with a contact time of 2–5 min [10,15]. Some other available alternative sanitizers that can be used for FCM and that are available in the market include aqueous chlorine dioxide (<3 mg L$^{-1}$ in water) and hydrogen peroxide (an effective sanitizer especially against *Salmonella* spp., *E. coli* O157:H7, *B. subtilis*, and other foodborne microbes at a dose of <0.3 mg L$^{-1}$, (in vapor form, otherwise it can be phytotoxic). Other sanitizers include calcium solutions (calcium chloride, calcium carbonate and calcium citrate, calcium lactate, calcium phosphate, calcium propionate, and calcium gluconate at a dose of 0.5% to 3% for 1–5 min). In addition, organic acids (0.5–1% ascorbic combined with 1–2% citric acids) are useful alternatives to sulfites in preventing browning and discoloration of cut slices. Acetic acid (vinegar) at a dose of 4% is also an effective antimicrobial [10,16]. The combination of these dipping treatments with edible or polysaccharide-based coating such as chitosan or alginate has also been demonstrated to be useful in extending the shelf life of fresh-cut products [9,17,18].

## 3. Pulp

Depending on the cultivar, mango pulp constitutes about 40–60% of the total fresh fruit weight, and is the main consumable part of the fruit due to the presence of nutritional and functional compounds [19]. The nutritional compounds and bioactive composition of mango are factors of the cultivar, the agroecological condition of the region, and the maturity of the fruit [20–23]. Sucrose, fructose, and glucose (in decreasing order of their concentration) comprise the principal carbohydrates present in mature and ripe mango. The carbohydrates content of pulp averages about 15 g/100 g, total dietary fiber (pectins, hemicellulose, and celluloses) averages 1.6 g/100 g while the protein content is about 0.8 g/100 g. The pulp also contains important micronutrients, vitamins, and bioactive compounds. The vitamin C of mango pulp ranges between 98 mg to 18 g/kg depending on variety and stage of maturity [23]. The nutritional quality of mangoes is to a great extent contributed by carotenoids, particularly β-carotene at about 4.138 mg/100 g [24]. The Tommy Atkins variety has been reported to contain 0.64 mg β-carotene, 0.009 mg α-carotene, 0.01 mg β-cryptoxanthin and lutein, and 0.023 mg zeaxanthin per 100 g [22,25]. This indicates that there is variation of carotenoids naturally among mangoes as a result of climatic effects, variety differences, stage of maturity at harvesting period, and storage. The ripe mango pulp contains all the B complex vitamins except biotin, ranging from 1.5 to 2.5 mg/100 g of fresh fruit pulp [23,26]. Mango pulp is a good source of many micro- and macro-minerals such as calcium, sodium, copper, iron, phosphorus, manganese, magnesium, zinc, boron (0.6–10.6 mg/kg), and selenium. The pulp is also rich in organic acids including citric acid, malic acid, oxalic acid, succinic acid, ascorbic acid, and tartaric acid, and bioactive compounds such as phenolic acids, sterols, and alkaloids [27,28].

The ability of the mango pulp to retain a wide range of nutrients and bioactive compounds is what makes it an ideal base material in the processing and value addition of various products) [8,22]. The pulp is rich in fiber due to the presence of fruit membrane and hence is more advantageous in comparison to juice concentrate in the processing of products. For storage purposes, the pulp is generally standardized to 14–18°Brix and 4–6% acidity by the use of either sugar syrup or citric acid, respectively. The sugar-standardized pulp is then pasteurized at 85 °C, filled when hot into bulk containers and sealed or heated at 100 °C for 20 min, cooled, packaged into bulk containers, and stored at room temperature (~25 °C). Addition of ascorbic acid, sorbic acid, sodium metabisulfite, and sodium benzoate into mango pulp helps in color, flavor, and carotene retention, resulting in a much longer shelf life. Both sodium metabisulfite and sodium benzoate have antimicrobial effects, but metabisulfite is more effective. However, minimal negative effects have been reported on the sensory characteristics of juices prepared from mango pulp preserved by metabisulfite and benzoate.

Mango pulp serves as the base for the processing of a variety of mango products including the following.

### 3.1. Mango Juice

Mango pulp can be mixed with a specific ratio of water to produce mango juice of a final TSS ranging between 12 and 15% of °Brix and 0.4 and 0.5% acidity [8,29,30]. The mango juice can be used as a single strength juice or blended with other fruit juices as juice blends or incorporated in fruit smoothies/shakes.

### 3.2. Mango Juice Concentrate

Mango juice concentrate is processed from mango juice or pulp as the base material. When the concentrate is derived from pulp, the pulp is subjected to polygalacturonase, pectinase, or cellulase enzymes to break down the pectins and cellulose. The juice concentrate has a sugar content of between 28 and 60% of °Brix) [8,30].

### 3.3. Mango Squash

Mango squash is a concentrated drink consisting of 25% juice, 45% TSS and 1.2 to 1.5% acidity with either sulfur dioxide or sodium metabisulfite as a preservative [31].

### 3.4. Cordials

Cordials are simply crystal-clear squashes obtained through filtration of the juice, using either special juice filters or a hygienic muslin cloth or strainer. Cordials have a TSS concentration of 12–14% of °Brix and 3.5% acidity, adjusted by addition of sugar and citric acid, respectively, and preserved by either sodium benzoate or sodium metabisulfite. Mango cordial can be produced on its own or blended with other fruits or vegetables such as pineapple or carrot juice.

### 3.5. Mango Nectar

Mango nectar is similar in composition to squash, except for the presence of a preservative in squash [32–34]. Mango nectar consists of 20–33% pulp content, TSS of 15°Brix and 0.3% acidity as citric acid, other ingredients (sugar, citric acid, vitamin C), and carboxymethylcellulose as a stabilizer.

### 3.6. Mango-Juice-Enriched Probiotic Dairy Drinks

Mango juice in combination with other fruit juices has the potential to be used as a new food matrix alternative to dairy products as a delivery vehicle for probiotics [35]. Mango juice improves the quality characteristics of fermented beverages and the viability of probiotics [36]. Mango pulp can also be used as a thickener or texture modifier or replacement for sugar in mango-flavored probiotic milk drinks [37–40].

### 3.7. Mango Wine

Mango wine is another beverage product derived from mango that can improve the value of mangos and reduce postharvest losses [41]. Due to its high sugar content (total soluble solids content > 16), mango pulp is an appropriate substrate for fruit wine fermentation [42]. The ethanol and aromatic components in mango wine have been shown to be comparable to those of grape. However, mango wine characteristics are affected by a number of factors including fermentation temperature, which affects not only the rate of yeast fermentation and duration but yeast metabolism. This in turn affects the chemical composition and the quality of the wine. The incorporation of sulfur dioxide, which is both an antioxidant and antimicrobial that is critical in inhibiting any spoilage microorganisms in wine production, can affect the volatile compound synthesis during fermentation such as increased acetaldehyde formation in mango wine. Furthermore, the type of yeast strain has an impact on the character and quality of mango wine) [43–45].

## 4. Dried Products

Dried mango products (slices or flakes) are generally prepared from ripe mangoes and dehydrated using a variety of methods including solar, hot-air cabinet, vacuum, spray, or freeze dryers. The dehydrated mango products are intended for either direct market or used in other formulations such as mango leather and powder [46–48]. The production process for dehydrated mango slices, dices, and chips are similar, other than the shape and size of the product. The ripe mango fruits are washed, peeled, pitted, and the pulp is sliced longitudinally into uniform thickness. The slices are then subjected to different specific pretreatments such as blanching, 0.5–1% citric acid, 0.2% ascorbic acid, and 40° Brix sugar to preserve product color and improve product stability. The pretreated slices are then dried at a temperature of 60–65 °C. Citric acid and ascorbic acid pretreatments before drying at 50 °C and 65 °C have the optimal outcome and produce the best physical quality parameters [49]. Different pretreatments prior to drying have significant effects on the moisture content, equilibrium relative humidity (ERH), water activity, and color parameters. Rehydration characteristics are affected by the different pretreatments with the most effective being 0.5% citric acid having the maximum rehydration ratio and coefficient of rehydration [49]. The dried mango slices have better antioxidant properties compared to fresh, probably due to synergistic effects of polyphenols and flavonoids) [50].

### 4.1. Mango Leather

Fruit leathers are dried sheets of fruit pulp which have a soft, rubbery texture and a sweet taste [51]. Leathers can be produced from a variety of fruits, although mango, apricot, banana, and tamarind leathers are amongst the most popular. Mango leather is produced by spreading the pulp evenly in a thin layer on a tray coated with vegetable oil to a depth of 1 cm and drying in mechanical or solar dryers to a final moisture content of 15–20% [51,52]. Solar drying can take a much longer time, leading to discoloration of the pulp. Addition of guar gum, pectin, and ascorbic acid reduces the discoloration of the mango leather. Preservatives such as sodium metabisulfite can be added to extend the shelf life [51]. Incorporation of sucrose, pectin, and maltodextrin reduced the drying rate of mango leather [8].

Mango leather can also be produced by refractance window drying (RWD) which is synonymous with cast-tape drying (CTD) [53,54]. This drying method is characterized by the fruit pulp that is to be dried being spread on a transparent polyester film, commercially known as Mylar (DuPont®). The lower surface of Mylar is kept in contact with hot water which supplies the heat for the product drying. RWD is a drying technique developed for drying of food pulp and purees to retain nutritional quality at relatively low processing temperatures with reasonable capital costs [55]. RWD of mango resulted in much shorter drying times and the mango leather obtained had better quality with higher nutrient retention compared to conventional drying. In addition, scanning electron microscopy showed that RWD resulted in powder particles of irregular shape and smooth surface with uniform thickness. On the other hand, tray and oven drying resulted in powder particles of corrugated, irregular, and crinkled surface with uneven shape and thickness [13].

### 4.2. Mango Powder

Mango powder is used as a flavor enhancer in various foods and beverages such as in ice cream, yoghurt, and the bakery and confectionery industries. Dried mango powder is processed by dehydrating mango pulp to a moisture content of 3% moisture using spray, freeze, vacuum, or drum dryers. However, it has been demonstrated that physiochemical properties of Refractance Window®-dried mango (RW-M) powder are comparable to the freeze-dried counterpart and are better than drum- and spray-dried mango powder [56]. One of the challenges in obtaining physically stable powder from dry fruits is their susceptibility to caking during processing and storage [57]. Caking is characterized by powder agglomeration, consolidation, and adhesion and has a negative impact on shelf life of powder. Caking also results in poor rehydration, lower reduced sensory properties, and short shelf life. To mitigate caking and improve hygroscopic properties of powder, carrier agents such as maltodextrin, starch, cellulose, or gums are used at a concentration of between 1 and 20% dry basis [58].

Mango jams and jellies are semisolid gels, which are made using the same general process [8,59]. Both of these products are made from fruit pulp, with added sugar, pectin, calcium chloride, and citric acid. Jelly is a clear or translucent fruit spread made from sweetened fruit juice and set using naturally occurring pectin. It is made by a process similar to that used for making jam, with the additional step of filtering out the fruit pulp after the initial heating. The incorporation of stem extract of medicinal plant marjoram into mango jam inhibited food spoilage bacteria viz. *Bacillus cereus* and *Bacillus megaterium*, indicating its potential use as a natural preservative in mango jam production [60].

Pickles are made mostly from green mangoes in India and are categorized as salty, oily, or sweet pickle based on the type of preservation used. They can be produced from peeled or unpeeled fruit with or without stones and with different kinds of proportions of spices.

## 5. Utilization of Mango Processing Waste

Mango processing and value addition generate an enormous amount of waste consisting of mainly the peels and the seeds, also known as stones [61–63]. Depending on the variety, 20–60% of the fruit weight comprises the seed while the kernel within the seed accounts for 45–75% of the seed's weight [64]. It has been reported that the mango seed is among the dominant agroindustrial wastes, generating about 123,000 metric tons of wastes annually in the world. Mango peels account for 7–24% of the fruit's weight [65,66]. Hence in general, mango processing generates millions of tons of solid waste approximated at 30−50% of the raw material. Furthermore, the volumes of mango processing waste are on the rise due to growth in the mango fruit production and processing industry [65]. The current standard waste disposal for industrial mango agro wastes and by products comprise of recovery (e.g., co product processing), recycling (e.g., internal upcycling of industrial side-streams into animal feeds or composting into manure), or solid waste disposal (e.g., into land fill or dried and incinerated as a source of energy) [66]. Food processing solid waste disposal has an adverse effect on the environment, such as water pollution, unpleasant odors, asphyxiation, vegetation damage, and greenhouse gas emissions. In addition, waste disposal is costly and adds to the total cost of production [67]. In addressing these challenges, there have been attempts to valorize the waste materials into value-added products.

The nutritional, physiochemical, and bioactive composition of mango seed and peels has been reviewed by Sharma et al., [68] and Mwaurah et al. [64]. There are some potential industrial applications of the value-added products derived from the seed and the peels as illustrated in Figure 2. The mango kernel contains about 15% of edible oil that is comparable to 18−20% oil content in soybeans and cotton seeds [64]. However, the oil from mango kernel comprises low free fatty acid and peroxide value and hence does not require further processing prior to consumption Blending oil from mango kernel and palm oil in the ratio of 80:20 (*w*/*w*) produces an oil with palmitic, oleic, and stearic acids comparable to cocoa butter [64]. Oil from the mango kernel has been considered to be a novel, cheaper, and readily available alternative to cocoa butter due to its phytochemical and physicochemical properties. The seed kernel has also been demonstrated to have antimicrobial activity, probably due to high content in different phenolic compounds, fatty acids, tocopherols, squalene, and sterols [67]. The mango kernel contains anti-nutritional factors and has to be preprocessed by dehulling, washing, soaking, boiling, and drying. The dried kernels can then be ground into flour and used as a functional ingredient in bakery products due to the presence of essential vitamins such as provitamin A and vitamin E and antioxidant activities [69].

Mango peel is a major byproduct of the mango processing industry and it constitutes about 15–20% of the total weight of mango fruit. The peel has been found to be a good source of biologically active substances such as polyphenols, carotenoids, flavonoids, anthocyanins, dietary fiber, vitamin E, vitamin C, and enzymes and hence has a potential use as a functional food [18,70,71].

The peel has been demonstrated to have more polyphenols than the pulp, and has a potential use as a functional food that can be used to supplement various food formulations such as bakery products, ice cream, breakfast cereals, pasta products, beverages, and meat products. It can also be used as a replacement in products such as cream, cheese, and yogurt.

Mango peel has been demonstrated to be a substantive source of odor-active compounds, that could be revalorized and used directly as a flavoring ingredient or even as a natural source out of which volatile compounds could be extracted [72]. Both the extract and the peel byproduct itself would be feasible to be used in food and cosmetic industries to provide or enhance the mango aroma of the product [73–76].

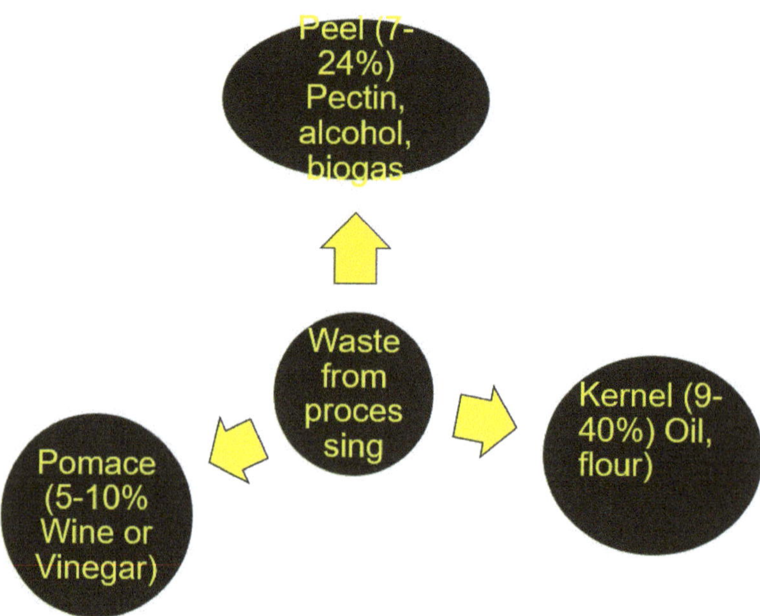



**Figure 2.** Utilization of mango processing waste.

## 6. Increased Value from Processed Mango Fruit

Although marketing of mango as fresh whole fruit is the most common practice among small-scale farmers in developing countries, processing the fruit into nutritious and safe products has greater value as shown in Figure 2 below [77]. In the profit margin calculation described in the Figure 3, the most lucrative processed product from mango fruit is wine with a net profit of USD 5500 per ton of mango fruit. However, processing of mango wine requires a more sophisticated system to produce the quantity and quality required by the market. Besides, market entry for small-scale processors is a challenge because of competition with established market brands. Mango puree, which only requires capacity to pulp and pasteurize, is a common product for many small-scale processors but with the lowest returns. In the cited study, the net profit on pulp from one ton of fruit is USD 700. Drying (dehydration) of mango fruit into products such as chips and leather does not require sophisticated equipment or facilities. According to the cited study, the mango chips and leather can fetch a net profit of USD 1300 and 1600 for mango chips and mango leather, respectively. If drying follows good manufacturing practices that ensure preservation of quality (nutritional and aesthetic) and safety of the products, such products may be the most recommended ones for small-scale farmers/processors in developing countries.

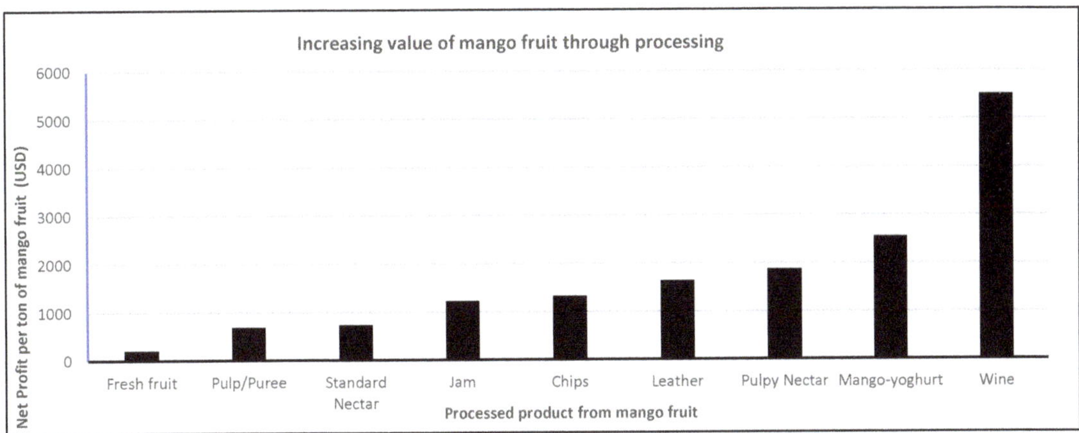


**Figure 3.** Net profits (USD) derived from processing 1 ton of mango fruit into various products [77].

### 7. Conclusions

Mango fruit is a nutritious fruit that is commonly consumed in its fresh state. Processing it into the diverse products described in this mini review has potential to not only contribute to the amelioration of high postharvest losses reported in mango but also to making the fruit available to consumers all year round as nutritious and convenient products. In addition, proper market linkages and demand for the diverse products from mango fruit will ensure better returns for small-scale mango producers who are often exploited by traders who buy the fresh fruits at very low prices.

**Author Contributions:** Conceptualization, writing—original draft preparation, W.O.O. writing—review and editing, W.O.O. and J.L.A.; funding acquisition, J.L.A. All authors have read and agreed to the published version of the manuscript.

**Funding:** The Consortium for Innovation in Post-Harvest Loss and Food Waste Reduction funded by the Rockefeller Foundation (Grant 2018 FOD 004) and the Foundation for Food and Agriculture Research (Grant DFs-18-0000000008).

**Acknowledgments:** Funding support for this publication was provided through the Consortium for Innovation in Post-Harvest Loss and Food Waste Reduction by the Rockefeller Foundation (Grant 2018 FOD 004) and the Foundation for Food and Agriculture Research (Grant DFs-18-0000000008).

**Conflicts of Interest:** The authors declare no conflict of interest.

### References

1. World Bank. *World Development Indicators 2012*; World Bank: Washington, DC, USA, 2012.
2. Brooks, K.; Place, F. Global food systems: Can foresight learn from hindsight. *Glob. Food Secur.* **2019**, *20*, 66–71. [CrossRef]
3. FAO. *Food and Agriculture Organization of the United Nations Statistical Database (FAOSTAT)*; FAO: Rome, Italy, 2018.
4. Maloba, S.; Ambuko, J.; Hutchinson, M.; Owino, W. Off-Season Flower Induction in Mango Fruits Using Ethephon and Potassium Nitrate. *J. Agric. Sci.* **2017**, *9*, 158–167. [CrossRef]
5. Sheahan, M.; Barrett, C.B. Food loss and waste in Sub-Saharan Africa. *Food Policy* **2017**, *70*, 1–12. [CrossRef]
6. DeeptiSalvi, E.A.; Karwe, M. Innovative processing technologies for mango products. In *Handbook of Mango Fruit: Production, Postharvest Science, Processing Technology and Nutrition*; John Wiley & Sons: Hoboken, NJ, USA, 2017; p. 169.
7. Evans, E.A.; Ballen, F.H.; Siddiq, M. Mango production, global trade, consumption trends, and postharvest processing and nutrition. In *Handbook of Mango Fruit*; John Wiley & Sons: Chichester, UK, 2017; pp. 1–16.
8. Siddiq, M.; Sogi, D.S.; Roidoung, S. Mango processing and processed products. In *Handbook of Mango Fruit: Production, Postharvest Science, Processing Technology and Nutrition*; John Wiley & Sons: Hoboken, NJ, USA, 2017; pp. 195–216.
9. Salinas-Roca, B.; Soliva-Fortuny, R.; Welti-Chanes, J.; Martín-Belloso, O. Combined effect of pulsed light, edible coating and malic acid dipping to improve fresh-cut mango safety and quality. *Food Control* **2017**, *66*, 190–197. [CrossRef]

10. De Corato, U. Improving the shelf-life and quality of fresh and minimally-processed fruits and vegetables for a modern food industry: A comprehensive critical review from the traditional technologies into the most promising advancements. *Crit. Rev. Food Sci. Nutr.* **2020**, *60*, 940–975. [CrossRef]

11. Yousuf, B.; Qadri, O.S.; Srivastava, A.K. Recent developments in shelf-life extension of fresh-cut fruits and vegetables by application of different edible coatings: A review. *LWT* **2018**, *89*, 198–209. [CrossRef]

12. Leneveu-Jenvrin, C.; Apicella, A.; Bradley, K.; Meile, J.C.; Chillet, M.; Scarfato, P.; Incarnato, L.; Remize, F. Effects of maturity level, steam treatment or active packaging to maintain the quality of minimally-processed mango (*Mangifera indica* cv. José). *J. Food Process. Preserv.* **2021**, *45*, e15600. [CrossRef]

13. Shende, D.; Kour, M.; Datta, A.K. Evaluation of sensory and physico-chemical properties of Langra variety mango leather. *J. Food Meas. Charact.* **2020**, *14*, 3227–3237. [CrossRef]

14. Yildiz, G.; Aadil, R.M. Comparative analysis of antibrowning agents, hot water and high-intensity ultrasound treatments to maintain the quality of fresh-cut mangoes. *J. Food Sci. Technol.* **2021**, 1–10. [CrossRef]

15. Suriati, L.; Utama, I.S.; Harsojuwono, B.A.; Gunam, I.B.W.; Adnyana, I. Differences in Physicochemical Characters of Fresh-Cut Mango, Mangosteen and Rambutan Due to Calcium Chloride Application. *J. Food Sci. Nutr.* **2021**, *7*, 2.

16. Aldana, D.S.; Aguilar, C.N.; Contreras-Esquivel, J.C.; Souza, M.P.; das Graças Carneiro-da-Cunha, M.; Nevárez-Moorillón, G.V. Use of a Mexican lime (*Citrus aurantifolia* Swingle) edible coating to preserve minimally processed mango (*Mangifera indica* L.). *Hortic. Environ. Biotechnol.* **2021**, *62*, 765–775. [CrossRef]

17. Salinas-Roca, B.; Guerreiro, A.; Welti-Chanes, J.; Antunes, M.D.; Martín-Belloso, O. Improving quality of fresh-cut mango using polysaccharide-based edible coatings. *Int. J. Food Sci. Technol.* **2018**, *53*, 938–945. [CrossRef]

18. Sharma, L.; Saini, C.S.; Sharma, H.K.; Sandhu, K.S. Biocomposite edible coatings based on cross linked-sesame protein and mango puree for the shelf-life stability of fresh-cut mango fruit. *J. Food Process Eng.* **2019**, *42*, e12938. [CrossRef]

19. Zafar, T.A.; Sidhu, J.S. Composition and nutritional properties of mangoes. In *Handbook of Mango Fruit: Production Postharvest Science, Processing Technology and Nutrition*; John Wiley & Sons: Hoboken, NJ, USA, 2017.

20. Akin-Idowu, P.E.; Adebo, U.G.; Egbekunle, K.O.; Olagunju, Y.O.; Aderonmu, O.I.; Aduloju, A.O. Diversity of mango (*Mangifera indica* L.) cultivars based on physicochemical, nutritional, antioxidant, and phytochemical traits in south west Nigeria. *Int. J. Fruit Sci.* **2020**, *20* (Suppl. 2), S352–S376. [CrossRef]

21. Ambuko, J.; Kemunto, N.; Hutchinson, M.; Owino, W. Comparison of the Postharvest Characteristics of Mango Fruits Produced under Contrasting Agro-Ecological Conditions and Harvested at Different Maturity Stages. *J. Agric. Sci.* **2017**, *9*, 181. [CrossRef]

22. Lebaka, V.R.; Wee, Y.J.; Ye, W.; Korivi, M. Nutritional composition and bioactive compounds in three different parts of mango fruit. *Int. J. Environ. Res. Public Health* **2021**, *18*, 741. [CrossRef] [PubMed]

23. Maldonado-Celis, M.E.; Yahia, E.M.; Bedoya, R.; Landázuri, P.; Loango, N.; Aguillón, J.; Restrepo, B.; Ospina, J.C.G. Chemical composition of mango (*Mangifera indica* L.) fruit: Nutritional and phytochemical compounds. *Front. Plant Sci.* **2019**, *10*, 1073. [CrossRef]

24. Mirza, B.; Croley, C.R.; Ahmad, M.; Pumarol, J.; Das, N.; Sethi, G.; Bishayee, A. Mango (*Mangifera indica* L.): A magnificent plant with cancer preventive and anticancer therapeutic potential. *Crit. Rev. Food Sci. Nutr.* **2020**, *61*, 2125–2151. [CrossRef] [PubMed]

25. Olale, K.; Walyambillah, W.; Mohammed, S.A.; Sila, A.; Shepherd, K. FTIR-DRIFTS-based prediction of $\beta$-carotene, $\alpha$-tocopherol and l-ascorbic acid in mango (*Mangifera indica* L.) fruit pulp. *SN Appl. Sci.* **2019**, *1*, 279. [CrossRef]

26. Meena, N.K.; Choudhary, K.; Negi, N.; Meena, V.S.; Gupta, V. Nutritional Composition of Stone Fruits. In *Production Technology of Stone Fruits*; Springer: Singapore, 2021; pp. 227–251.

27. Agatonovic-Kustrin, S.; Kustrin, E.; Morton, D.W. Phenolic acids contribution to antioxidant activities and comparative assessment of phenolic content in mango pulp and peel. *S. Afr. J. Bot.* **2018**, *116*, 158–163. [CrossRef]

28. Quirós-Sauceda, A.E.; Sañudo-Barajas, J.A.; Vélez-de la Rocha, R.; Domínguez-Avila, J.A.; Ayala-Zavala, J.F.; Villegas-Ochoa, M.A.; González-Aguilar, G.A. Effects of ripening on the in vitro antioxidant capacity and bioaccessibility of mango cv. 'Ataulfo' phenolics. *J. Food Sci. Technol.* **2019**, *56*, 2073–2082. [CrossRef] [PubMed]

29. Adedeji, O.E.; Ezekiel, O.O. Chemical composition and physicochemical properties of mango juice extracted using polygalacturonase produced by *Aspergillus awamori* CICC 2040 on pre-treated orange peel. *LWT* **2020**, *132*, 109891. [CrossRef]

30. Sakhale, B.K.; Pawar, V.N.; Gaikwad, S.S. Studies on effect of enzymatic liquefaction on quality characteristics of Kesar mango pulp. *Int. Food Res. J.* **2016**, *23*, 860–865.

31. Muslim, S.; Saleem, A.; Mehmood, Z.; Iqbal, A.; Shah, F.; Khan, Z.U.; Shah, S.; Hamayun, M.; Hussain, A.; Yue, Z.; et al. An environmentally safe and healthy mango squash from natural ingredients. *Fresenius Environ. Bull.* **2021**, *30*, 2410–2415.

32. Huang, B.; Zhao, K.; Zhang, Z.; Liu, F.; Hu, H.; Pan, S. Changes on the rheological properties of pectin-enriched mango nectar by high intensity ultrasound. *LWT* **2018**, *91*, 414–422. [CrossRef]

33. Kumar, R.; Vijayalakshmi, S.; Rajeshwara, R.; Sunny, K.; Nadanasabapathi, S. Effect of storage on thermal, pulsed electric field and combination processed mango nectar. *J. Food Meas. Charact.* **2019**, *13*, 131–143. [CrossRef]

34. Xess, R.; Singh, P.; Patel, D.; Singh, Y. Evaluation of mango (*Mangifera indica* L.) varieties for processing of nectar beverage on organoleptic parameters. *J. Pharmacogn. Phytochem.* **2018**, *7*, 772–774.

35. Acevedo-Martínez, E.; Gutiérrez-Cortés, C.; García-Mahecha, M.; Díaz-Moreno, C. Evaluation of viability of probiotic bacteria in mango (*Mangifera indica* L. Cv. "Tommy Atkins") beverage. *Dyna* **2018**, *85*, 84–92. [CrossRef]

36. de Oliveira, P.M.; BRC, L.J.; Martins, E.M.F.; Martins, M.L.; Vieira, É.N.R.; de Barros, F.A.R.; Cristianini, M.; de Almeida Costa, N.; Ramos, A.M. Mango and carrot mixed juice: A new matrix for the vehicle of probiotic lactobacilli. *J. Food Sci. Technol.* **2020**, *58*, 98–109. [CrossRef] [PubMed]

37. Dhillon, H.S.; Gill, M.S.; Kocher, G.S.; Panwar, H.; Arora, M. Preparation of Lactobacillus acidophilus enriched probiotic mango juice. *J. Environ. Biol.* **2021**, *42*, 371–378.

38. Mayulu 2021, N.; Assa, Y.A.; Kepel, B.J.; Nurkolis, F.; Rompies, R.; Kawengian, S.; Natanael, H. Probiotic drink from fermented mango (*Mangifera indica*) with addition of spinach flour (*Amaranthus*) high in polyphenols and food fibre. *Proc. Nutr. Soc.* **2021**, *80*. [CrossRef]

39. Ryan, J.; Hutchings, S.C.; Fang, Z.; Bandara, N.; Gamlath, S.; Ajlouni, S.; Ranadheera, C.S. Microbial, physico-chemical and sensory characteristics of mango juice-enriched probiotic dairy drinks. *Int. J. Dairy Technol.* **2020**, *73*, 182–190. [CrossRef]

40. Wang, J.; Xie, B.; Sun, Z. Quality parameters and bioactive compound bioaccessibility changes in probiotics fermented mango juice using ultraviolet-assisted ultrasonic pre-treatment during cold storage. *LWT* **2021**, *137*, 110438. [CrossRef]

41. Musyimi, S.M. Production, Optimization and Characterization of Mango Fruit Wine: Towards Value Addition of Mango Produce. Master's Thesis, Jomo Kenyatta University of Agriculture and Technology, Nairobi, Kenya, 21 June 2017.

42. Ogodo, A.C.; Ugbogu, O.C.; Agwaranze, D.I.; Ezeonu, N.G. Production and evaluation of fruit wine from *Mangifera indica* (cv. Peter). *Appl. Microbiol.* **2018**, *4*, 144.

43. Lu, Y.; Chan, L.J.; Li, X.; Liu, S.Q. Effects of sugar concentration on mango wine composition fermented by Saccharomyces cerevisiae MERIT. ferm. *Int. J. Food Sci. Technol.* **2018**, *53*, 199–208. [CrossRef]

44. Patel, V.; Tripathi, A.D.; Adhikari, K.S.; Srivastava, A. Screening of physicochemical and functional attributes of fermented beverage (wine) produced from local mango (*Mangifera indica*) varieties of Uttar Pradesh using novel saccharomyces strain. *J. Food Sci. Technol.* **2021**, *58*, 2206–2215. [CrossRef] [PubMed]

45. Wattanakul, N.; Morakul, S.; Lorjaroenphon, Y.; Jom, K.N. Integrative metabolomics-flavoromics to monitor dynamic changes of 'Nam Dok Mai' mango (*Mangifera indica* Linn) wine during fermentation and storage. *Food Biosci.* **2020**, *35*, 100549. [CrossRef]

46. Dereje, B.; Abera, S. Effect of pre-treatments and drying methods on the quality of dried mango (*Mangifera Indica*, L.) slices. *Cogent Food Agric.* **2020**, *6*, 1747961. [CrossRef]

47. Isaac, N.; Owino, W.; Ambuko, J.; Imathiu, S. Moisture sorption properties of two varieties of dehydrated mango slices as determined by gravimetric method using Guggenheim–Anderson–de Boer model. *J. Food Process. Preserv.* **2021**, *45*, e15041. [CrossRef]

48. Sulistyawati, I.; Verkerk, R.; Fogliano, V.; Dekker, M. Modelling the kinetics of osmotic dehydration of mango: Optimizing process conditions and pre-treatment for health aspects. *J. Food Eng.* **2020**, *280*, 109985. [CrossRef]

49. Nyangena, I.; Owino, W.; Ambuko, J.; Imathiu, S. Effect of selected pre-treatments prior to drying on physical quality attributes of dried mango chips. *J. Food Sci. Technol.* **2019**, *56*, 3854–3863. [CrossRef] [PubMed]

50. Nyangena, I.O.; Owino, W.O.; Imathiu, S.; Ambuko, J. Effect of pre-treatments prior to drying on antioxidant properties of dried mango slices. *Sci. Afr.* **2019**, *6*, e00148.

51. Sarkar, T.; Chakraborty, R. Formulation, physicochemical analysis, sustainable packaging-storage provision, environment friendly drying techniques and energy consumption characteristics of mango leather production: A review. *Asian J. Water Environ. Pollut.* **2018**, *15*, 79–92. [CrossRef]

52. Sarkar, T.; Salauddin, M.; Hazra, S.K.; Chakraborty, R. Effect of cutting-edge drying technology on the physicochemical and bioactive components of mango (*Langra variety*) leather. *J. Agric. Food Res.* **2020**, *2*, 100074. [CrossRef]

53. da Silva Simão, R.; de Moraes, J.O.; de Souza, P.G.; Carciofi, B.A.M.; Laurindo, J.B. Production of mango leathers by cast-tape drying: Product characteristics and sensory evaluation. *LWT* **2019**, *99*, 445–452. [CrossRef]

54. Zotarelli, M.F.; da Silva, V.M.; Durigon, A.; Hubinger, M.D.; Laurindo, J.B. Production of mango powder by spray drying and cast-tape drying. *Powder Technol.* **2017**, *305*, 447–454. [CrossRef]

55. Raghavi, L.M.; Moses, J.A.; Anandharamakrishnan, C. Refractance window drying of foods: A review. *J. Food Eng.* **2018**, *222*, 267–275. [CrossRef]

56. Caparino, O.A.; Nindo, C.I.; Tang, J.; Sablani, S.S.; Chew, B.P.; Mathison, B.D.; Fellman, J.K.; Powers, J.R. Physical and chemical stability of Refractance Window®—dried mango (Philippine 'Carabao' var.) powder during storage. *Dry. Technol.* **2017**, *35*, 25–37. [CrossRef]

57. Fongin, S.; Granados, A.E.A.; Harnkarnsujarit, N.; Hagura, Y.; Kawai, K. Effects of maltodextrin and pulp on the water sorption, glass transition, and caking properties of freeze-dried mango powder. *J. Food Eng.* **2019**, *247*, 95–103. [CrossRef]

58. Tonin, I.P.; Ferrari, C.C.; da Silva, M.G.; de Oliveira, K.L.; Berto, M.I.; da Silva, V.M.; Germer, S.P.M. Performance of different process additives on the properties of mango powder obtained by drum drying. *Drying Technol.* **2018**, *36*, 355–365. [CrossRef]

59. Bekele, M.; Satheesh, N.; Sadik, J.A. Screening of Ethiopian mango cultivars for suitability for preparing jam and determination of pectin, sugar, and acid effects on physico-chemical and sensory properties of mango jam. *Sci. Afr.* **2020**, *7*, e00277. [CrossRef]

60. Bhardwaj, K.; Dubey, W. Exploring potential of hydro-alcoholic extract of stem of marjoram as natural preservative against food spoilage bacteria *Bacillus cereus* and *Bacillus megaterium* in homemade mango jam. *Vegetos* **2021**, 1–11. [CrossRef]

61. Aggarwal, P.; Kaur, A.; Bhise, S. Value-added processing and utilization of mango by-products. In *Handbook of Mango Fruit: Production, Postharvest Science, Processing Technology and Nutrition*; John Wiley & Sons: Hoboken, NJ, USA, 2017; pp. 279–293.

62. Cheok, C.Y.; Mohd Adzahan, N.; Abdul Rahman, R.; Zainal Abedin, N.H.; Hussain, N.; Sulaiman, R.; Chong, G.H. Current trends of tropical fruit waste utilization. *Crit. Rev. Food Sci. Nutr.* **2018**, *58*, 335–361. [CrossRef] [PubMed]

63. Jahurul, M.H.A.; Zaidul, I.S.M.; Ghafoor, K.; Al-Juhaimi, F.Y.; Nyam, K.L.; Norulaini, N.A.N.; Sahena, F.; Omar, A.M. Mango (*Mangifera indica* L.) by-products and their valuable components: A review. *Food Chem.* **2015**, *183*, 173–180. [CrossRef]

64. Mwaurah, P.W.; Kumar, S.; Kumar, N.; Panghal, A.; Attkan, A.K.; Singh, V.K.; Garg, M.K. Physicochemical characteristics, bioactive compounds and industrial applications of mango kernel and its products: A review. *Compr. Rev. Food Sci. Food Saf.* **2020**, *19*, 2421–2446. [CrossRef]

65. Marçal, S.; Pintado, M. Mango peels as food ingredient/additive: Nutritional value, processing, safety and applications. *Trends Food Sci. Technol.* **2021**, *114*, 472–489. [CrossRef]

66. Wall-Medrano, A.; Olivas-Aguirre, F.J.; Ayala-Zavala, J.F.; Domínguez-Avila, J.A.; Gonzalez-Aguilar, G.A.; Herrera-Cazares, L.A.; Gaytan-Martinez, M. Health Benefits of Mango By-products. In *Food Wastes and By-products: Nutraceutical and Health Potential*; Wiley: Hoboken, NJ, USA, 2020; pp. 159–191.

67. Mutua, J.K.; Imathiu, S.; Owino, W.O. Evaluation of the Proximate Composition, Antioxidant Potential and Antimicrobial Activity of Mango Seed Kernel Extracts. *Food Sci. Nutr.* **2017**, *5*, 349–357. [CrossRef] [PubMed]

68. Sharma, S.K.; Bansal, S.; Mangal, M.; Dixit, A.K.; Gupta, R.K.; Mangal, A.K. Utilization of food processing by-products as dietary, functional, and novel fiber: A review. *Crit. Rev. Food Sci. Nutr.* **2016**, *56*, 1647–1661. [CrossRef] [PubMed]

69. Gómez, M.; Martinez, M.M. Fruit and vegetable by-products as novel ingredients to improve the nutritional quality of baked goods. *Crit. Rev. Food Sci. Nutr.* **2018**, *58*, 2119–2135. [CrossRef]

70. Sagar, N.A.; Pareek, S.; Sharma, S.; Yahia, E.M.; Lobo, M.G. Fruit and vegetable waste: Bioactive compounds, their extraction, and possible utilization. *Compr. Rev. Food Sci. Food Saf.* **2018**, *17*, 512–531. [CrossRef]

71. Serna-Cock, L.; García-Gonzales, E.; Torres-León, C. Agro-industrial potential of the mango peel based on its nutritional and functional properties. *Food Rev. Int.* **2016**, *32*, 364–376. [CrossRef]

72. Bonneau, A.; Boulanger, R.; Lebrun, M.; Maraval, I.; Valette, J.; Guichard, É.; Gunata, Z. Impact of fruit texture on the release and perception of aroma compounds during in vivo consumption using fresh and processed mango fruits. *Food Chem.* **2018**, *239*, 806–815. [CrossRef] [PubMed]

73. Li, L.; Ma, X.W.; Zhan, R.L.; Wu, H.X.; Yao, Q.S.; Xu, W.T.; Luo, C.; Zhou, Y.G.; Liang, Q.Z.; Wang, S.B. Profiling of volatile fragrant components in a mini-core collection of mango germplasms from seven countries. *PLoS ONE* **2017**, *12*, e0187487. [CrossRef] [PubMed]

74. Musharraf, S.G.; Uddin, J.; Siddiqui, A.J.; Akram, M.I. Quantification of aroma constituents of mango sap from different Pakistan mango cultivars using gas chromatography triple quadrupole mass spectrometry. *Food Chem.* **2016**, *196*, 1355–1360. [CrossRef] [PubMed]

75. Oliver-Simancas, R.; Muñoz, R.; Díaz-Maroto, M.C.; Pérez-Coello, M.S.; Alañón, M.E. Mango by-products as a natural source of valuable odor-active compounds. *J. Sci. Food Agric.* **2020**, *100*, 4688–4695. [CrossRef] [PubMed]

76. Oliver-Simancas, R.; Díaz-Maroto, M.C.; Pérez-Coello, M.S.; Alañón, M.E. Viability of pre-treatment drying methods on mango peel by-products to preserve flavouring active compounds for its revalorisation. *J. Food Eng.* **2020**, *279*, 109953. [CrossRef]

77. Ambuko, J.; Abong, G.; Gekonge, G.; Maitha, I.; Amwoka, E. Small-scale processing of mango fruits: Putting more money in farmers' pockets while enhancing access to nutritious fruit products. In Proceedings of the 4th All Africa Horticultural Congress, International Horticultural Society, Dakar, Senegal, 29–31 March 2021.

MDPI

St. Alban-Anlage 66

4052 Basel

Switzerland

Tel. +41 61 683 77 34

Fax +41 61 302 89 18

www.mdpi.com

*Agriculture* Editorial Office

E-mail: agriculture@mdpi.com

www.mdpi.com/journal/agriculture